Thomas Hockey
Naked-Eye Astronomy

Thomas Hockey

Naked-Eye Astronomy

Stars, the Sun, the Moon, Planets, and the Deep Sky

DE GRUYTER

Author
Prof. Dr. Thomas Hockey
University of Northern Iowa
Department of Earth and Environmental Sciences
1227 W 27th St
Cedar Falls, Iowa 50614
U.S.A.
hockey@uni.edu

ISBN 978-3-11-144118-4
e-ISBN (PDF) 978-3-11-144124-5
e-ISBN (EPUB) 978-3-11-144158-0

Library of Congress Control Number: 2025939966

Bibliographic information published by the Deutsche Nationalbibliothek
The Deutsche Nationalbibliothek lists this publication in the Deutsche Nationalbibliografie;
detailed bibliographic data are available on the Internet at http://dnb.dnb.de.

© 2026 Walter de Gruyter GmbH, Berlin/Boston, Genthiner Straße 13, 10785 Berlin
Cover image: Nicolas Ivan Gentili/iStock/Getty Images Plus
Typesetting: Integra Software Services Pvt. Ltd.
Printing and binding: CPI books GmbH, Leck

www.degruyterbrill.com
Questions about General Product Safety Regulation:
productsafety@degruyterbrill.com

Preface

Perhaps a good way to start is to define what this textbook is *not* about.

This is not a textbook on introductory astronomy. As you are introduced to sights in your naked-eye sky, the physical nature of what is responsible for that view is described briefly. Though by no means required, you definitely will benefit from having taken a survey course in astronomy prior to using this book.

Moreover, this is not a "how-to" book. Ways of practicing naked-eye astronomy are mentioned. Yet the text's goal is not to turn you into an expert observer.

There are so many things to see in your sky! However, this textbook is limited to those sights that are astronomical in nature. This means that they belong to the universe beyond the Earth.

Meteorological phenomena are those strictly taking place in the Earth's atmosphere. They are excluded. Examples include rainbows, clouds, and "sundogs." This restriction still leaves you with a myriad of upward sightseeing opportunities at your disposal.

The purpose of this textbook is to allow you to "tap into" cultural knowledge, some of it on the verge of being lost from circulation. Your author buys into the idea that any knowledge gleaned from the liberal arts is a source of intellectual power now more than ever – anything that causes you to think in a new way.

Our world is changing so quickly that the content presented in courses taken for vocational purposes becomes obsolete in little time. Indeed, the particular vocation, which you may be preparing for, may itself become obsolete in your lifetime! Yet the liberal arts, e.g., naked-eye astronomy, endure.

This is not to telegraph that naked-eye astronomy is of no practical use. A mere failure of electronics renders most of our current navigational and timekeeping devices useless. That is an unfortunate, "weak link" in contemporary civilization, inasmuch as (more and more frequently) you are required to be able to do both of these tasks.

For such a reason, this textbook goes into some detail on these two closely related subjects. It will not bring you to the level of expertise – there are other books for that – but it will serve as an introduction.

As we peruse naked-eye astronomy, examples of its use appear throughout the pages of this textbook. Once again, though, fairness requires pointing out that this is not an anthropological work. It is not a text on the subject of practices long ago or in lesser-known corners of the world today. There are other textbooks on what we will come to call archaeoastronomy and ethnoastronomy.

One almost stereotypical example may seem to be missing. That is the famous ancient monument in England named "Stonehenge." This is intentional.

Probably half of all archaeoastronomical literature has been written about Stonehenge. In it, an incredible variety of explanations as to edifice's function are presented.

https://doi.org/10.1515/9783111441245-202

Some of these are plausible, just as many are not. Your author has chosen to leave Stonehenge to others so as to have room to share less well-known (and, often, better) examples from elsewhere in the world.

In this book, the sky view assumed defaults to be from a site in the mid-latitudes of the Earth's northern hemisphere. This is where most people happen to live, particularly between the latitudes of 25° and 40° N. Nevertheless, so as to make this book useful around the world, from time to time, its narrative will pause to point out how things might appear differently from elsewhere.

A word on the plentiful illustrations to be found within the pages of this textbook: Photographic images were chosen, if possible, so as to represent a naked-eye view. To make such images is not normally the intent of an "astrophotographer," though. Indeed, it runs counter to one of the purposes of photography in the first place.

Finding real photographs that accurately convey the naked-eye perspective is difficult. Available images instead present detail, color, and apparent size inaccessible to the naked eye. I decided to include them, anyway, inasmuch as they provide context about the object or phenomenon under discussion.

In cases where there were multiple photographs available that equally serve this book's purpose, I tried to select those that you may not have seen before. In other words, I attempted to choose those that have not become cliché due to familiarity or are the first product of a simple Google search.

✶✶✶✶✶

While any errors in this text are mine, their number is diminished through the effort of these colleagues, who were willing to set aside their own projects in order to review a draft of the book.

I appreciate the help provided by each of the following:

Professor Kevin Krisciunas, Texas A&M University
Professor Donald Olson, Texas State University
Professor Jesse Wilcox, University of Northern Iowa
Professor Steve Gullberg, Oklahoma University
Dr. Jennifer Bartlett, Harvard & Smithsonian Center for Astrophysics
Dr. Jarita Holbrook, University of Edinburgh

Christine Tarte produced a series of original gouache-and-ink paintings to illustrate these pages. To my knowledge, it is the only such collection to grace a textbook in astronomy.

Based upon best practices, Jesse Wilcock wrote the self-assessment questions that appear at the end of each chapter (Chapter Review). He is the Association of Science Teacher Education's Outstanding Science Teacher Educator of the Year for 2024.

Sierra Ameen graduated with a Digital Media Administration degree from the University of Northern Iowa. Once the images for this textbook were chosen, she curated them and provided ALT TEXT.

Aaron Spurr created a series of photographs exclusively for this textbook.

Thanks also go to my student assistant, Abbey Mandick.

Ms. Nikki Talebi at the University of Chicago Press kindly granted permission to include selected passages from "Hockey, Thomas. *How We See the Sky*, 2011."

Contents

Appendices

Abbreviations

AAS   American Astronomical Society
CIA   United States Central Intelligence Agency
DNR   Department of Natural Resources
ESA   European Space Agency
ESO   European Southern Observatory
IAU   International Astronomical Union
NASA   United States National Aeronautics and Space Administration
NOAA   United States National Oceanographic and Atmospheric Administration
NSF   United States National Science Foundation
US(A)   United States (of America)
USDA   United States Department of Agriculture
USDI   United States Department of the Interior
USGS   United States Geological Survey
USNO   United States Naval Observatory

https://doi.org/10.1515/9783111441245-204

1 Introduction to your book

Thank you! By taking a course on naked-eye astronomy, you are actively participating in the protection of "endangered" knowledge. Or, at least, erudition in danger of falling from the common into the esoteric.

This is not talk about some sort of "secret knowledge." It is simply knowledge that has dropped out of wide circulation. For instance, vestiges "hide" in introductory chapters of books on navigation and timekeeping. Here it is gathered in one place.

1.1 Chapter learning outcomes

After reading this chapter, you will be able to:
– justify the value of naked-eye astronomy today.
– distinguish between astronomy and astrology.
– look for evidence of naked-eye astronomy practiced long ago.

1.2 Why take a naked-eye astronomy class?

During most of the time that it has existed as an academic discipline, astronomy was a required element of higher education. Before the invention of the telescope in about 1608, the observational part of astronomy was conducted only with the naked eye. The astronomy that you would have heard lectured about and read about in a book was accessible only with the naked eye.

Today, introductory astronomy textbooks include less-and-less material on this aspect of the science, so as to cover more recent discoveries about the Universe gleaned from huge instruments and both computational and theoretical astrophysics. They must do this in order to offer a realistic view of what astronomy is in the twenty-first century.

Unfortunately, an unintended consequence of this conceptual evolution is that it makes astronomy more distant from you. It is much less "hands-on" than it used to be. Astronomy teachers would like to see some of you pursue the subject, in your lives beyond the classroom. They hope that you might, for example, acquire a simple, quality telescope (or binoculars) for your own enjoyment.

Realistically, though, even such a mild commitment is unlikely. It takes money, some skill, and practice time.

This leaves those of you who simply desire to look up at night. Perhaps you can foresee a day when you wish to pass your familiarity about what you see on to your children and grandchildren. You may do this with only a little bit of astronomical knowledge applicable to the naked-eye sky. A class just about naked-eye astronomy is a step toward making this possible.

https://doi.org/10.1515/9783111441245-001

Figure 1.1: Students learn the Quadrivium, a standard curriculum composed of music, arithmetic, geometry, and astronomy. From William Caxton's *Myrrour of the Worlde,* the first illustrated book printed in English (1481).

1.3 Naked-eye astronomy in days gone by

It is ironic that, long ago, the average man or woman was more equipped to interpret what could be seen in the sky than is the college-educated person today. This is true for a variety of reasons.

Notes from yesterday and today

In late 2024, an internet-fueled instance of minor mass hysteria occurred on the northeast coast of the United States. It supposedly involved a horde of mysterious drones in the sky, both organized and nefarious. A local politician provided a photograph and demanded that the government take action. The photograph turned out to be of a famous region in the heavens, populated every autumn for millennia by several bright stars. So unaccustomed to looking up at the night sky was this man that instead of seeing the bespangled, resplendent, and calming, he saw the inexplicable, sinister, and alarming. The episode was farcical but also, in a way, tragic.

There was a time – a very long time! – when the naked-eye sky supplied a practical tool for timekeeping and navigation. The first is vital for assuring a regular food supply. The second makes trade possible.

Moreover, the cycles seen in the sky were reassuring, because they promised a predictability that appeared in very few places in life here below. And there was an aesthetic to the sky, shared by all humans, one sadly lacking in a world below that frequently could be quite ugly.

Perhaps most significantly, the lights in the sky were a clue to **cosmology**, the universal attempt to understand one's world. Indeed, the sky was considered and remains a gateway toward answering those most fundamentally human of questions: "Who am I?" and "What am I doing here?"

Figure 1.2: "Dionysius the Areopagite Converting the Pagan Philosophers," an oil painting by French Antoine Caron, court artist of Catherine de' Medici and Henry II of France (1570s).

Taking a naked-eye astronomy class is one way to join a community of people spread over all the Earth and all era of time.

1.4 Astronomy *versus* astrology

You will explore *astronomy* in this book, not *astrology*. (The exception will be when the latter has a historical or culture connection to naked-eye astronomy.) It does not help that these two words sound so similar; however, what they mean could not be more different. Taken broadly, astronomy is much older than astrology. Astrology – sort of – claims to be based upon astronomy.

1.4.1 What is the difference between astronomy & astrology?

Simply put, **astrology** is a belief system in which people's personalities and future lives are influenced by **celestial** phenomena in the sky. Because it is a *belief*, anybody is free to hold it. However, it is not a science. Astronomy is.

> **Where did that come from?**
> "Celestial" is a Latin adjective referring to the sky. Since the Romans, it has taken on other meanings not germane to this book such as "ethereal" or "delightful."
> For a prefix, astronomers switch to the Greek "urano-." Uranus was god of the sky.

Why can such a thing be said so confidently? It is because astrology does not pass a test that defines science.

1.4.2 Value of theories

Astronomy is based upon what you can see and theories developed based upon what you can see. The word "theory" has an undeserved reputation. There are those who consider a theory to be the same thing as a guess.

A theory is something more than that. You literally cannot survive based just upon guesses. Yet how are you to deal with everything that you cannot see, or else-wise sense, at this very moment in your environment?

While you may agree to call the things that you can see by the word "facts," so much of what you do is not based upon this limited kind of fact. How can you live with only such facts at your disposal? You will be frozen – unable to move forward, either figuratively or literally – if you have no idea what to expect once you leave the safety of what you immediately can see (or hear, or feel, *etc.*) before you.

Pretend that you drove to school and now sit in a windowless classroom. That your car still will be in the parking lot after class is by no means a fact as previously defined!

Theories come to the rescue. While you may not call them so, theories are every-thing that you think you know that is not in front of your face. They are a common tool – like a screwdriver. They help you when you need them. In the case of theories, that is much of the time. Instead of searching aimlessly for your car, your theory advises you that it very likely is in the place where you left it.

Have you noticed that it is difficult to positively prove something? Take Isaac Newton's famous Universal Theory of Gravitation. If it is supposed to be "universal," how can you possibly test it at every place (and during every time) in the Universe?

However, gravity is easily the opposite of provable. It is *falsifiable*. You may hold a pencil right now. Drop it. If it does not fall to the ground, you just proved false Newton's theory. You are famous! It would be so easy, too.

Yet people all over the world have dropped things for a long time. Never once did a thing fall, say, up. The fact that this theory is so simply falsifiable and yet *never is*, advertises that it is a really good, scientific theory. People routinely risk their lives on it.

1.4.3 Astrological theories fail the test

What about astrology? Read an astrological "theory": your **horoscope**. Can you say that it is categorically false? Probably not. Think about it. Can you envision *any* reasonable situation in which it would be categorically false? For that matter, can you imagine any time at which (and for any person, for whom) it would be definitely false? The answer is usually "no."

Astrological statements (if we are uncertain whether to call them theories or not) intentionally are worded so that they are never falsifiable. They may sound good. But are they examples of a scientific theory? No. They fail the principal test for authenticity, potential falsifiability, miserably.

Moreover, the bedrock of science is causation; it is implicit in a good theory. Nobody has ever produced a plausible way in which the sources of remote light in your sky influence you.

The rest of this book discusses observable facts and scientific theories. The hope is that what is written here is as close to the idealized goal of Truth as can be. This cannot be guaranteed. However, it can be assured that astronomical theories presented herein are falsifiable.

1.5 Architecture records naked-eye astronomy before writing

When nearly everything else a society produces disappears, its architecture probably is what remains. This is why you will read in this book about a number of buildings and other monuments in an attempt to convince you that people have interacted with the sky for a very long time.

The study of prehistoric peoples' relationship with the sky of Sun, Moon, planets, stars, *etc.*, is called **archaeoastronomy**. A scholar who does this is an **archeaoastro-**

nomer. (An **ethnoastronomer** studies the astronomies of current cultures; this field is **ethnoastronomy**.)

1.5.1 Orientations & alignments

Two words are used in this book that, at first, sound the same. However, the meaning of each differs.[1] One is **orientation**. An orientation occurs when a line drawn through

Figure 1.3: This canon obviously is oriented with respect to the building at which it is aimed. However, the campus ministry across the street was never a target. It was constructed after the ornamental canon was installed. This does not represent an alignment. Photograph by the author.

---

1 There is not complete agreement among all regular users of these two words; sometimes one is applied with the definition here presented for the other.

two or more objects (or a line **orthogonal** to a plane – like a carpenter's T-square) points in a particular direction. Orientation does not require intent.

The more subtle word "alignment" is used for an intended orientation. An **alignment** is *built*.

A typical two-point orientation or alignment requires both a **backsight** and a **foresight**. A sphere is worthless insofar as determining a preferred direction. There needs to be an asymmetrical object with at least two discernable points (ends, edges, openings). In general terms, the one closest to you is the backsight; the one closest to that which you are looking at is the foresight. The notches on a rifle are effective back- and foresights.

The earliest use of astronomical foresights and backsights known about today is in the Nabta Playa, amidst the Nubian Desert of Africa. Here, sandstone blocks were moved into apparently intentional astronomically based alignments. This was done by pastoral peoples as early as the fifth millennium BC/BCE! (The recently excavated Karahan Tepe site in Turkey eventually may turn out to be a much *earlier* example of astronomical fore- and backsights.)

Figure 1.4: 800 kilometers south of Cairo sits the Nabta Playa "Calendar Circle" (reconstructed). Courtesy of Ray M. Betz.

It used to be that to measure the orientation of architecture, you had to become familiar with the ways and means of the surveyor. Besides direction, slope must be taken into account. (A hillside has a different natural horizon than does a flat plain.) Today, though, the marvelous invention of GPS technology greatly simplifies the task.

1.5.2 Example: Illinois, United States

About the eleventh century, people belonging to the Mississippian Culture built a city called Cahokia on the river whence they are now named. Evidence of the Cahokian people's civilization includes the world's largest platform mound made of earth.

West of the major Cahokian mound, atop which the supreme ruler of Cahokia is thought to have lived and surveyed his kingdom, there was once erected a 120-meter-diameter circle of tree trunks: 48 in all. In the middle was a lone such post.

The Cahokia Sun Circle[2] has been reconstructed, based on excavation of the original post holes. Standing at the interior post, the backsight, it is possible to see the point on the horizon where certain objects rise and set in the sky at particular times of year. Three of the outer poles on the Circle serve as foresights. What is described so far is an orientation, but not necessarily an alignment.

1.5.3 Archeoastronomy in practice

With so many posts in the Sun Circle, it is possible that these events occur by chance. We need further evidence to suggest an alignment. This evidence comes in many forms – from archaeology, other sciences, or somewhere else – and is somewhat subjective. However, it is important to try to distinguish human accomplishments that are alignments from all the orientations that *must* (statistically) occur.

It is easy to skip this warning about mistaking orientations for alignments, because our minds are designed to 1) recognize patterns and 2) pick up on exceptions to normal life: We notice when we see "11:11" light up on the digital alarm clock, forgetting all the times at which we have glanced at the same clock seeing no such interesting pattern. We remark on the coincidence, even though a similar pattern must show up for a minute every hour of every day.

At Cahokia, American archaeologist Warren Wittry suspected a real alignment because the observed orientations do not work if the middle pole were to be at the geometrical center of the Circle, as likely it would be if somebody was merely building a "pretty wheel" of wooden poles. Here we use the scientific method: We hypothesize

---

2 or Cahokia Woodhenge

that the Sun Circle is meant to mark natural phenomena; therefore, we figure out where one must stand for the alignment to work. This turns out to be a couple of meters east of the geometrical center.

We then test our hypothesis by digging for the remains of an ancient post at this spot. It is there. Notice that we did not dig a hole everywhere we could within the Sun Circle, hoping to find something; we dug in one – and only one – predicted location.

It looks like the builders of the Cahokia Sun Circle ignored the strong human urge for symmetry (placing the middle pole equally distant from the circumference poles) and put function over form: The Cahokia Sun Circle may have looked a bit off, but it worked.

Is this enough evidence? Now traditional archaeologists weigh in. There seems to be a connection between one of the outer poles and the Sun. They dig *there*. The excavators find what they know to be, because of their study of Cahokian culture through a myriad of other artifacts, a small ceramic vessel. It is the kind believed used to hold religious offerings. Adorning it is a symbol representing the Sun.

Do we have enough evidence *now*? We never can prove as a mathematical certainty that the Cahokia Sun Circle was aligned based upon the astronomical knowledge and interest of the Mississippian people. Questions remain: What were the other peripheral poles for? Still, there seems to be enough evidence to make it unlikely that this edifice was not aligned. Conclusion: the sky mattered to the Cahokians.

1.6 The typical versus the atypical

The entire Universe seems "painted" upon your sky. Certainly, its overall brightness is not great, considering that the brightness of the objects within it must be averaged with the (effectively) zero brightness of vast spaces in-between. (This subject will be revisited in the last chapter.)

What about color? On a lark, British extragalactic astronomers Karl Glazebrook and Ivan Baldry set about determining the "color" of the Universe. In their algorithm, the black of the Universe played no part. The calculated, combined color of the remaining parts of the Universe that we can see turned out to be a (decidedly) pale tan.

This is ironic. Comparatively few astronomical objects appear, by themselves, to be "tan." The two astronomers later described their universal hue as a very light beige or "cosmic latte."

Figure 1.6: The "color of the Universe" is based upon spectral analysis of more than 200,000 galaxies.

Such a result is not to be taken too seriously but does help illustrate the human mind's interest in seeing the unusual over the commonplace. This textbook quite intentionally falls into that trap. Most of the naked-eye Universe needs limited description and (with only a few major exceptions) consists of so-called garden variety stars and a few planets.

1.7 Commonality of your sky

Let us conclude this chapter in this way: Imagine that you are magically transported to a bar stool in an unfamiliar corner of the world. You do not know where you are. You cannot speak the language. You cannot read the menu. The persons to your left and to your right are strangers. They look, dress, and act differently than you do. At first, you feel very alone.

Then you remember that you have at least one thing in common with your newfound neighbors. It is the sky. Everybody was born, lives, and dies under the same sky. And the sky makes up half of everything anybody can see surrounding them.

Maybe this tale is a silly mind game. Still, there seems to be little doubt that the sky is one thing that brings all of us together. The study of the sky presented in the following pages is as human a story as can be.

1.8 Chapter summary

The subject of this book is not something new. Naked-eye astronomy is very old. It was at the root of our species' hunt for a cosmology. At no time has it been forgotten, though elements of it may be left behind in the closet of astronomical knowledge when new findings made with new methodologies are brought out.

While naked-eye astronomy might (at first glance) resemble astrology, it is different in that astrology encapsulates no useful scientific theories. Perhaps ironically, astrology is not predictive. Astronomy is.

Naked-eye astronomy has been important enough to the world's societies that many incorporated it into their architecture. We still can see it embodied there, even in cases where the people who practiced it have disappeared.

The sky is much the same for all. How it is, and has been, interpreted varies greatly.

Admittedly, this book dwells on the atypical. With that always in mind, it seems safe to proceed ahead.

1.9 Chapter review

1. What are two reasons why naked-eye astronomy is worth studying today?
2. Throughout most of history, naked-eye astronomy had practical uses. Describe one.
3. What is the difference between astronomy and astrology?
4. What is a scientific theory? How does this definition not apply to astrology?
5. What is the difference between an orientation and an alignment?
6. Why is the Cahokia Sun Circle not considered merely decorative?
7. Describe astronomical backsights and foresights. What is the earliest known usage of a backsight and a foresight?
8. What is archaeoastronomy?
9. What are two reasons (that we know of) for which naked-eye astronomy was practiced long ago?

2 Your eyes

What is naked-eye astronomy? It is the intentional observation of astronomical objects in your sky. This is done without optical instruments. (An example of the latter is the telescope, which affects light passing through it.) Instruments of this kind augment our sense of sight but introduce biases.

2.1 Chapter learning outcomes

After reading this chapter, you will be able to:
- explain the purpose of each component of the eye.
- recognize the limitations of human vision.

2.2 Your field of view

Naked-eye astronomy obviously concerns the eyes. With two eyes, how much of the sky can you see at any given time? No eye movement is "allowed." Typically, your vision spans 210° left to right. Up and down is less: about 120°. This includes the so-called peripheral vision.

Binary vision produces a "landscape format." Along with all humans, you spend most of your time watching your surroundings, instead of looking above or below you. This genetic trait makes survival sense.

2.3 Bright or dim?

Each of your eyes consists of an outer, transparent covering called the cornea. Beneath the **cornea**, the **pupil** dilates and contracts. It regulates the amount of light admitted to the inside of your eye.

Finally, light arrives on a surface named the **retina**, upon which **photoreceptive** (light receptive) cells are situated. These send information via nerves to your brain in order to construct what you "see." There are more than one hundred million of these cells!

All of the astronomical objects discussed in this textbook are visible to your naked eye. Admittedly, a great many are very faint. Astronomy, without the aid of instruments, obviously is limited by the ability of your "detector," the human eye. Your eye is only capable of seeing objects brighter than a minimum brightness, which varies from person to person.

Your eye is only capable of seeing objects with a minimum brightness.

https://doi.org/10.1515/9783111441245-002

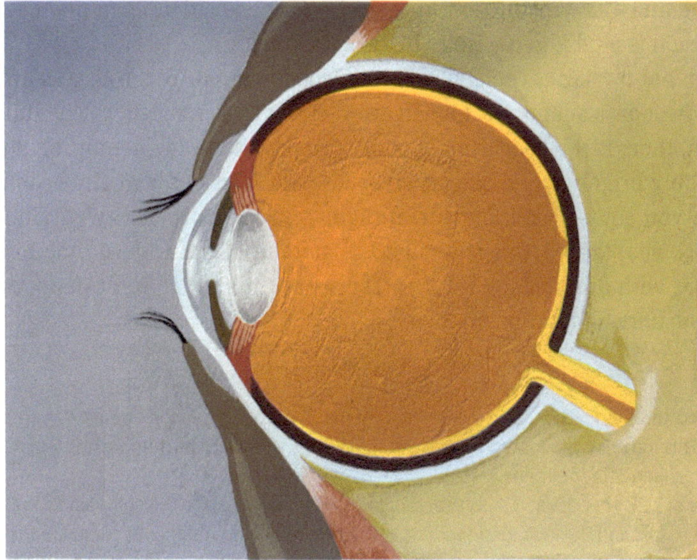

Figure 2.1: Cross section of the human eye. Art by Christine Tarte.

Interestingly, the colored part of the eye, the **iris**, appears to correlate with sensitivity to light. Those with lighter-colored irises tend to have more light sensitivity in their eyes than those with darker colored irises.

2.3.1 Dark adaptation

You have to be patient to really see the night sky. Long ago, people automatically spent more time outside than they do now. The biblical "shepherds abiding by their flock at night" come to mind.

Plus, the dark[1] was then more familiar. Today, we rush from our interior-lit car to our fully lit home. In pre-history, without the invention of so much as a candle, it easily could be just as dark "indoors" as it was outdoors.

With the intent to see the night sky for what it really is, it is desirable for as much light as possible to strike and affect your retina cells. You want your iris to make your pupil as large as possible. You also want to **dark adapt**. If you just came outdoors from a brightly illuminated building or vehicle, you are using your **photopic** vision (daytime vision), designed for resplendent lighting. You must transition to **scotopic** vision (nighttime vision). Without these two different systems, your vision would satu-

---

1 Humans cannot see in the dark. Here, "dark" is used in the vernacular as a synonym for "low light levels."

rate. There would be much less distinction in intensity. The transition from photopic vision to scotopic vision is vital if you wish to discern faint stars.

Your eyes adjust, but it takes a while; first comes blue, then red. This trait is called the **Purkinje Effect**. As it dark adapts, the human eye's sensitivity to red light catches up.

Dark adaptation, then, is the triggering of your scotopic vision. Light-emitting objects become noticeably brighter. And objects otherwise too faint to see at all become visible. However, if you simply walk outside at night, you need *at least* twenty minutes to achieve dark adaptation. This is an average amount of time. More time furthers the process, but with diminishing returns. There are things that you can do (or avoid) to speed up the dark adaptation, though.

> **Misconception alert**
>
> It is commonly assumed that dark adaptation means that the pupil of the eye widens to allow more light to enter the eye. This indeed happens. You can see it in a mirror. However, much more important is what you cannot see: changes in the retina.
>
> There are two kinds of cells that act as photoreceptors in the retina. These are the so-called *cones*, designed for high light levels, and the *rods*, designed for low light levels. During dark adaptation, both kinds of cells become more light-sensitive. This especially is true of the rods, which are at a low level of light sensitivity when surroundings are bright. It is the cones that adapt more quickly, but the efficiency of the rods is so much greater that they eventually take over and become the major contributor to dark adaptation.

2.3.2 Impediments to dark adaptation

Perhaps not surprisingly, the higher the level of illumination in the space you occupy before you go outside at night negatively affects both degree of dark adaptation and the time needed to achieve dark adaptation. Astronomers have gone so far as to place themselves in a black box prior to observing the sky, or (less dramatically) they wear eye patches.

Exposure to rapidly pulsing light sources, such as fluorescent tubes that flash on and off at 100–120 times per second, significantly prolong adaptation time. This is as opposed to (for example) steady, incandescent lamps providing the same average brightness. On the other hand, there is some evidence that if you frequently change between constant high and low light levels, such practice helps you develop an ability to dark adapt more quickly in the long term.

Smoking is bad for dark adaptation. Vitamin A is known to aid vision in general and dark adaptation in particular. Large quantities of consumed quinine or carbon-sulfide compounds impede dark adaptation.

As a rule, a fully dark-adapted child sees fainter stars than a fully dark-adapted senior citizen. Throughout life, the human pupil shrinks, though how much and when is specific to the individual. Compounding this deterioration is that the lens in the human eye becomes less transparent with age.

These effects cut down the light that will make it to your retina in the first place. Your lens also becomes yellower with time, introducing a color bias. As for the cornea, it may dry and eventually become irregular, thereby increasing annoying internal light scatter that also gets in the way of your sky vision.

Finally, your genes include the characteristic of good (or poor) dark adaptation. Not much can be done about that. There are even those who suffer from **nyctalopia** (night blindness).

2.3.3 A trick to improve your sensitivity to light

There is a technique that you can use to see stars and other astronomical objects at the limit of your sight, even when fully dark-adapted. It is called **averted vision** and is described below.

The eye's retina provides its greatest **resolution**, the ability to distinguish detail, at its center. This was evolutionarily helpful: If early humans stared across the steppe and distinguished a charging saber-toothed cat from the background, it would have been best to be able to do so while the cat was still far away. At the same time, the middle of the retina is less sensitive to brightness than is the retina's periphery.

By glancing slightly away from, say, a faint star, you can use the more light-sensitive outer retina to make a dim object look brighter. Try it. It may only work for a moment, until your brain overrides your instruction to the eye to avoid looking straight ahead in a normal fashion. (Practice extends the ability to avert vision.)

The instant that you avert, the star, now more easily visible because it appears brighter, "jumps out" at you. It is as if somebody turned up a dimmer switch. This experience almost is startling.

2.3.4 Limiting magnitude

On a given night of sky watching, it might be useful to begin by establishing how faint an object you are able to see. This is quantified as your limiting brightness. Establishing your personal **limiting brightness** usually is done by determining the faintest stars that you can make out among a set of stars of known but differing brightness. (It helps if the members of this set are near each other in the sky.)

The stars together called the Little Bear are a good test for limiting brightness. This group of stars is compact, easy to find, and in the sky of most people most of the time.

Each of the stars that make up the "Little Dipper" part of the pattern happens to have a noticeably different brightness. If you can see all seven stars, you are in great shape. (See Chapter 5.) However, there may be some that you cannot see – at that time from a given place. The faintest of the "Little Dipper" stars that you can see is your limiting brightness. It is therefore unnecessary to try to find a celestial object that is known to be fainter than your limiting brightness. However, it has been shown that your limiting brightness can be improved with practice.

Do not test your limiting brightness in the part of the sky occupied by the Milky Way. (See Chapter 5.) The light of the numerous very faint stars found there seems to merge together into a white background. This tends to cause one to assign a pessimistic limiting brightness.

2.4 Special precautions for observing the Sun

Naked-eye astronomy is about light. Normally, the more light the better. An exception is viewing the Sun. The Sun is *too* bright. Its light is equivalent to sharing a room with a trillion, trillion 100-watt light bulbs! Not only does the glare from its disk make it difficult to see any features associated with the Sun, too much light can harm the eye.

There is no documented case of any animal (save among us) ever injuring its eyes by staring at the Sun. The natural predilection, after peering at the Sun briefly, is to look away. Only humans will deliberately overcome this response and stare at the Sun for too long.

Whether it is in the pursuit of astronomy or something else, this is a bad idea. Your retina may be damaged by doing so.

For most people, there ordinarily are few compelling reasons to gaze at the Sun. Without attention to detail, it is assumed to look pretty much the same all the time. A beautiful sunrise or sunset is heavily filtered by the Earth's atmosphere and presents far less risk than a noontime Sun. (See Chapter 3.) Similarly, sunlight diminished by thin clouds, smog, fog, dust, or reflection in water may be tolerable briefly.

One situation that draws people to the Sun is the unusual circumstance of a solar[2] eclipse. (See Chapter 16.) Ironically, a totally eclipsed Sun is safe to view. However, the partial eclipse before, afterward, and nearby remains a danger.

The Sun is almost uniformly bright. That brightness is so overwhelming that even a small area of the whole solar disk is enough to affect your eyes. A quick glimpse might or might not be safe. A longer fix definitely is not. Why risk it?

Notes from yesterday & today
Normal sunglasses, no matter their tint, coating, polarization, or prescription will *not* protect your eyes from the Sun. The same goes for folk filters such as smoked glass, exposed photographic film, or old computer floppy disks. All that these attempts at filtering may do is to lull you into a false sense of security and cause you to look at the Sun for more dangerously long than you would without any of them.

---

2 adjectival form used for the Sun

The most harmful portion of the Sun's light is not within the visible spectrum. It is the Sun's invisible ultraviolet rays. These pass through many visually transparent materials, including your cornea, and arrive upon your retina at nearly full strength. Here they produce "burns." The most harmful portion of the Sun's light is not within the visible spectrum. It is the Sun's invisible ultraviolet rays. These can give your cornea the equivalent of a sunburn.

Destruction of the retinal cells may result in partial or full blindness. There is no guarantee that this blindness is temporary. If that were not bad enough, solar infrared radiation contributes long-term to cataracts. Solar infrared rays can pass through your eye's cornea and burn its retina. Damage to retinal cells may result in partial or full blindness. There is no guarantee that this blindness is temporary. Such burns also contribute to your risk of cataracts.

Were you to feel this attack on the retina, you would defensively blink. Unfortunately, the insidious thing about retinal damage is that you may have no awareness of what is happening. The retina has no pain-producing nerves.

The good news is that there are filters you can use to safely look at the Sun. Out of these, those marketed explicitly for observing the Sun should be labeled as approved by the International Organization for Standards. Look for their eye-safe code, ISO 12312-2.

Figure 2.2: Safe filters for direct solar viewing. From left-to-right: high-grade mylar plastic, #14 welders' glass, optical-quality silvered glass, and polymer film (carbon black powder embedded into a resin). Photograph by the author.

> **Tools of the naked-eye observer**
>
> It is not necessary to observe the Sun directly. Consider this simply made apparatus. Put a pinhole in one side of a topless box. Place a piece of high-contrast paper on the opposite side. Aim the hole at the Sun. An image of the Sun will be projected onto the paper. (The box is really just there to shade the paper.) You will be able to see any naked-eye detail on the Sun. (See Chapter 19.)
>
> Alternately, place a pinhole in a piece of paper. Use this paper to cover a flat mirror. With the mirror aimed at the Sun, a solar image will be reflected onto any desired surface. Such a technique allows many people to safely observe the same solar image at the same time.

2.5 Warning

Never use an unfiltered telescope or binoculars to look at the Sun. These optical aids gather more light than your naked eye does. The increased amount of light that they funnel into your pupil may harm your eye severely. In the case of a large-aperture telescope, the damage may occur too quickly for you to react.

2.6 Angular resolution

Earlier, resolution was introduced. Recall that this is the ability to distinguish detail. You can think of it as equivalent to pixel size in a digital camera.

Angular resolution is quantified as the minimum angle through which the separation between two objects is visible. For example, consider a tall fence in which the slats are fifteen millimeters apart. You view this fence from one hundred meters away.

This is absolutely as good as it gets for the resolution of the human eye: 20–30 seconds of arc. You may or may not see the fence as continuous.

Yet interestingly, when you reach the limit of your eyes' resolution, your brain takes over. It unconsciously plays "connect the dot" and provides assumed detail based upon experience.

Most of the time, the brain gets it right: You perceive more detail than you optically have a right to. Knowing the fence to be picket, it is possible that you "see" it as such. However, sometimes the brain makes a mistake, especially when you are looking at something unfamiliar.

Here is a test of your resolution.

Two stars coincidentally near each other in the sky are called **optical doubles**. Sometimes, one of these stars is much further from you than the other. If you were to observe such a pair from a very different viewing location, say from a planet orbiting another star other than the Sun, their association in your sky would no longer exist. But this is irrelevant to our immediate purpose.

A famous dyad visible from the Earth is that of Castor and Pollux, the "twin" stars of the constellation Gemini. (See Chapter 5.) This optical double is almost five degrees in separation. This is a fairly large one as optical doubles go. The fact that the two stars are of roughly equal brightness and both are among the brightest in the sky, makes the pair even more unique.

However, to assess resolution, you want to try to separate doubles that appear *close* to each other. An example is the twosome, Alcor and Mizar, the "Horse-and-Rider Double" in the constellation of Ursa Major. (Again, see Chapter 5.) They are separated by less than one-quarter of one degree. The ability to see them as separate stars has been used as a vision test since at least the days of the Roman Army.

The best resolution test is a set of two stars near to each other and also of approximately the same brightness and color. The minimum angular separation in the sky between two stars, with spacing between them that you can make out, is the limit of your resolution. Still closer pairs merge into a single "star" as seen with your eyes alone.

Figure 2.3: In the optical double 11 and 12 Camelpardalis, the angular separation equals three arc minutes. This pair also happens to be a physical binary. (See Chapter 19.) Note the very different colors of the components. Courtesy of David Ritter.

2.7 Color

Do not be too disappointed if you do *not* see color in objects brought up in this book. Individuals' color perception varies markedly. If you cannot detect color in the stars, you are unlikely to see it elsewhere in your night sky, either. The eye needs a certain amount of light to trigger color vision at all.

Recall the Purkinje Effect. This is why, when you wake up in a dark room at night, everything looks grayish blue. Even bright stars may not be enough to properly trigger your particular color vision.

In truth, it is unclear whether color is an objective property of the Universe as is brightness. You can measure brightness with a meter; color is a more daunting task. Whether color is real is an ongoing debate among philosophers. Yet, color adds to the sky-watcher's experience. This book will adopt its colloquial meaning without further comment.

2.7.1 Your visible spectrum

The colors that you see are centered upon the visible rainbow-like spectrum: red, orange, yellow, green, blue, "indigo," and violet. The list is somewhat fluid. In a sense, each of these individual color "bins" is arbitrary. Scientists estimate that humans (unless they are color blind) can distinguish a million subtle variations in color!

> **Notes from long ago**
> In the Middle Ages, there were seven colors recognized: white, yellow, red, green, blue, purple and black. Imagine – no orange! The inclusion of white and black demonstrates a primitive understanding of light.
>
> The autodidact physicist Isaac Newton conducted experiments that led to recognizing black as the absence of radiated light and white as a mixture of radiated light of different colors. He inserted orange and indigo into the list of colors because of the supposed mystical nature of the number seven.
>
> The human eye has more difficulty distinguishing colors at short wavelengths. These correspond to the spectral color blue and beyond (away from red, in rainbow order). For this reason, the definition of indigo is elusive. It is a term no longer used by physicists.

Keep in mind that, even though a light source may look one particular color, this merely may be the color at which it produces the most light. It may generate light of other colors at the same time. In fact, this is true for any dense, glowing object.

In the middle of the range of colors that you can see, your brain mixes them together, and you see white. (Your perception is of the colors evenly blended.) It is not surprising that your vision is centered upon these particular colors: the human eye evolved under sunlight that produces just this spectrum. Picture the whitish-yellow Sun.

However, your exact range could have been different, slid to one end of the Sun's spectrum or the other. For example, dogs have more difficulty seeing red than do humans, but your pet can see colors past violet that are invisible to you.

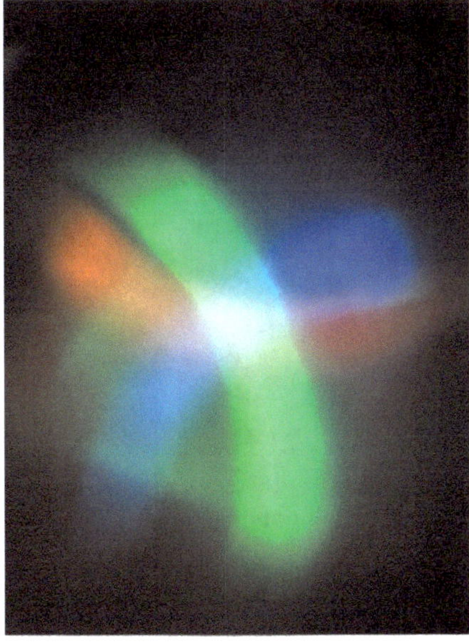

Figure 2.4: Colors of light (different wavelengths) combine to form white. Photograph by the author.

2.7.2 Not all color vision is alike

Unfortunately, at least one in ten – mostly men – suffer from one of several forms of **color blindness** that skew color perception in some way. Some people do not possess all the three different color-receptors in their retinas; they are at least partially color blind. Because of this, not everybody will agree on the colors seen in the night sky. Few of the color-vision impaired, though, are completely color blind.

There are many physiological factors that influence what you may be able to see in your night sky. Other factors are described in the next Chapter.

2.8 You can see a lot – but not everything

To be sure, only a tiny fraction of the Universe is available to the human eye alone, limited as that organ is by finite brightness sensitivity and restricted as it is to a narrow range of colors. You cannot see X-rays. You cannot see radio. You cannot see gamma rays. The Universe emits all of these, and more, besides visible light.

Quite clearly, the observable universe is profoundly dark to the naked-eye observer. Those bits that you are able to see, hint at the variety among objects that exist in your Universe.

Figure 2.5: Different kinds of colorblindness. Courtesy of Johannes Ahlmann.

Perhaps the glass is half-full. By the time you reach the end of this book, you may ask yourself, "Is it not amazing that I can see as much as I can?"

2.9 Time to get started!

Notwithstanding the caveat about the Sun, naked-eye astronomy is both safe and inspiring. You live in a world where much of what you see is processed for your consumption: video, audio, artificial scents, and food additives. It can be exhilarating: the fact that no instrument, indeed, *nothing* stands between you and that which you are observing (save for the air). It also is highly personal.

Whether anybody truly sees a thing in the same way as you, is a question to be left unanswered. Notwithstanding, it is a good real-world guess that unless you are part of a party, no one else sees what you see at exactly the same time and under exactly the same conditions.

You watch your Universe "live."

2.10 Chapter summary

Naked-eye astronomy allows you to "travel" lightly. It literally involves you and your own sense of sight only.

Your eyes are your portals to the Universe. They, and their signals as interpreted by your brain, determine your field of view, resolution, photo-receptivity, the colors that you see, and even how you appreciate those colors. Different people have varying abilities with regard to each of these parameters.

The anatomy of the eye makes possible different two-stage processes (involving your iris and cornea), allowing you to see in a great range of illumination. One operates for the very bright; another operates for the very faint. There exist ways in which you can aid the transition from your day vision to that useful for investigating the night sky.

However, there are limits. Too much light, to which you are exposed by an overly, long direct (unfiltered), view of the Sun, can injure or damage your eyes.

2.11 Chapter review

1. Why does it make sense that our field of view spans a greater left and right angle than up and down?
2. What is the function of each component of the eye?
    Retina
    Pupil
    Cornea
    Iris
3. What are two things that impede your dark adaptation? What are two things you can do to improve your dark adaptation?
4. What is the Purkinje effect? How might smart phones, which largely emit blue light, impact your dark adaptation?
5. What is limiting magnitude? How can you determine your limiting magnitude?
6. What is angular resolution? How did the Roman Army determine people's angular resolution?
7. Why is the Sun dangerous to observe with your naked eye?
8. White is not a color. Why do you see some things as white?

3 The atmosphere

You observe the sky above through a veil of atmosphere. While transparent to light for the most part, the Earth's atmosphere contains opaque materials (dust, ice crystals) that dim starlight. In addition, it **airglows** (ever so slightly) due to chemical reactions high up; this background glimmer can interfere with the view of very faint light sources beyond. The more atmosphere you look through, the brighter a – say – competing star must be in order for it to remain perceptible.

3.1 Chapter learning outcomes

After reading this chapter, you will be able to:
- adjust for changing effects of the atmosphere that affect your skyward view.
- time sky observing to the twilight you seek.
- take note of, and avoid some of the pernicious effect of, light pollution.

3.2 Air mass

Your planet's atmosphere is a layer of gas surrounding the solid globe. It is only about 100-km thick (though at the top it is much rarified). Thus, it is thin compared to the size of the Earth.

Your geography books correctly teach you that our planet is "round." Yet, because we are so small in comparison to the Earth, it is easy to imagine yourself as standing at the bottom of a flat dish of atmosphere. (You would need to be at an elevation of more than 10 km to see the curvature of the Earth; the highest mountain on the planet is less than nine kilometers at its peak.)

The "dish" is wider and longer than it is tall. You want to look upward through as short a column of obscuring air as possible. The atmosphere is thinnest when you look directly overhead.

Notes From yesterday and today
By at least the fifteenth century, people the world over recognized that the Earth is a sphere. The modern, fringe belief that it is flat dates to the nineteenth century. It relies on conspiracy "theories" that supposedly debunk direct visual evidence of the Earth's sphericity, such as photographs taken from space.

Think of standing (or floating) at the bottom of an Olympic size swimming pool. Soon you run out of air to breathe – but you are at the *middle* of the large pool. What do you do? Do you swim for the nearest poolside? Likely not. You bob straight up. Intuitively, you know that it is the direction in which you have the least water to pass through before surfacing. This goes for looking through the atmosphere, too.

https://doi.org/10.1515/9783111441245-003

Astronomers call the length of the air column, which you look through in order to see the sky, the **air mass**. Directly overhead is defined as 1 air mass. Observing a celestial object at any increasing **zenith angle** with respect to the point directly overhead, means that you are peering through a greater and greater air mass. Your view is dimmed. As you get closer to the horizon,[1] that view deteriorates quickly. Indeed, it is feasible to see only the brightest astronomical objects (*e.g.*, the Sun and Moon) *at* the horizon. Fainter objects face **extinction** before reaching it; the air mass becomes high enough to dim them below naked-eye visibility.

The first 10 km of the atmosphere above sea level are most important to astronomers. Here it is densest, and here almost all weather takes place. So let us assume that the atmosphere is effectively 10-km thick.

You observe a star from sea level. The star makes a 45° angle with the horizon (locally assumed to be parallel to the effective top of the atmosphere). This is one half of the way "up" the sky. Because the two angles are complementary,[2] the zenith angle of such a star is 90°−45° = 45°. At this angle, you are observing the star through almost twice as much atmosphere as you would if it were directly above you.

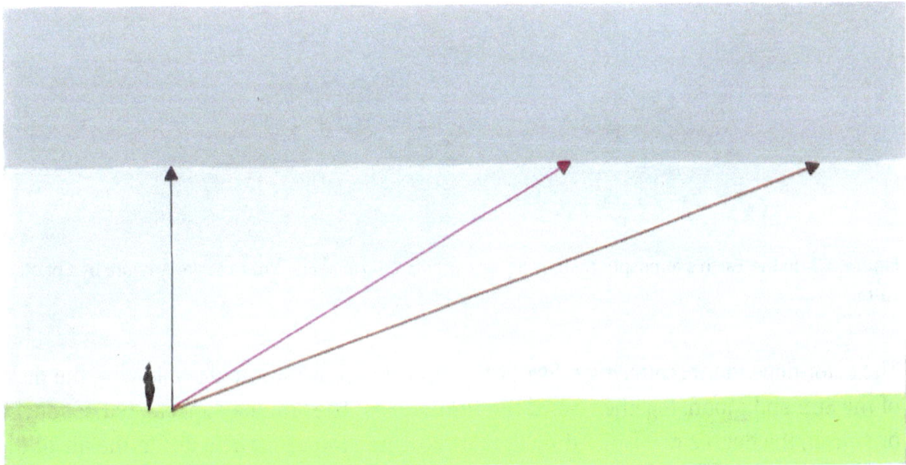

Figure 3.1: Lines of sight through the atmosphere representing (from left to right) one, two, and three air masses. Not to scale. Artwork by Christine Tarte.

3.3 Preferential transmission of light

It is the atmosphere that makes celestial objects near the horizon (*i.e.*, seen through high air mass) look redder than they appear at other times. The blue component of,

---

1 Horizon is formally defined in the next chapter; for now, it is used with its popular meaning.
2 from the Latin for "completed"

for instance, direct sunlight is scattered by the atmosphere so as to appear to come from all directions above the horizon; this is the cause of daytime's blue sky. However, the red component maintains a more line-of-sight path with the Sun. The greater the air mass, the greater is this effect.

Figure 3.2: In the Earth's atmosphere, blue light scatters, red transmits. Not to scale. Artwork by Christine Tarte.

The color-dependent scattering effect of the atmosphere is most noticeable with the disks of the Sun and Moon, together called the **luminaries**. The Sun may appear red at sunrise or sunset, the degree determined by the density and kind of particulates in the air at the time (Homer's "rosy-fingered dawn"). The same thing happens to the Moon and, to a less noticeable degree, other celestial objects that can be seen at least near the horizon.

When the Sun is "up," what looks like background illumination of the blue sky makes it difficult to see other celestial objects. However, that you must wait for nightfall to see anything else is not entirely true.

It is sometimes possible to see the Moon, certain planets, and the brightest star – it is named Sirius[3] – against daytime's blue sky. The brightness of these bodies is, of course, much diminished in these circumstances. To spot them, you likely will have to know just where to look ahead of time.

---

3 from the Greek Seirios, meaning "sparkling"

The atmosphere can have an additional, noticeable effect on – in particular – the luminaries. This phenomenon is in addition to dimming them and changing their apparent color. It is most striking in regard to the Sun.

The atmosphere acts as a natural prism. Were there to be no atmosphere, the Sun would be white. At any zenith angle other than zero, the atmosphere introduces **dispersion** of sunlight. This means that the atmosphere causes component colors in a ray of sunlight to "bend" – change direction slightly. The result is that different colors appear to come from slightly different directions.

This modification in the direction traveled by a ray of light is a well-known optical property. It is called **refraction** and occurs anytime light passes from one transparent material into another. Artificial lenses, such as those in eyeglasses, send light rays in the proper direction – unlike your faulty natural lenses. They work because of refraction. The natural version, when light from a celestial object traveling through space passes through the layers of the Earth's blanket of air, is atmospheric refraction. The overall effect of atmospheric refraction is to direct light away from your horizon.

The resulting angle of deflection caused by atmospheric refraction is zero in the "middle" of your sky. It is at its maximum when the celestial source of light is farthest from this point. At least, this is the case theoretically. Its real severity changes with atmospheric conditions, for instance, the difference in density between different layers of the atmosphere, the interface between which each offers another contribution to atmospheric refraction.

Let us continue to use the Sun as an example. The total (time-dependent) result of atmospheric refraction will be this: At sunrise or sunset, light from the "top" of the Sun's disk, and light from the "bottom" of the Sun's disk, appear to be slightly different colors and coming from directions other than what you would anticipate. For instance, you expect the Sun's disk to be circular. Yet atmospheric refraction may cause it to appear elliptical (up to twenty percent) with the long axis parallel to the horizon. Or it may appear otherwise distorted – even as if its disk is broken!

Atmospheric refraction has a less obvious effect on the diskless stars. Yet it can change *where* stars appear in the sky. Starting with the Hellenistic star mapper Posidonius, **uranographers** have tried to take it into account when describing star positions.

An unusual atmospheric-refraction-caused phenomenon is difficult to catch because it requires a smooth "rim" to your sky. Such a view is available at sea. As the last sliver of Sun disappears at sunset (or first appears at sunrise), its color will flash *green*. (The word "flash" is not an exaggeration; the dispersion effect lasts a second or less.) Sailors call this the **green flash**.[4] If the air is especially clean, there may even be a "blue flash."

---

[4] or emerald flash

Figure 3.3: The effect of atmospheric refraction on the Sun as it sets over the Wadden Sea, off the coast of Germany. Courtesy of Hans Juergen Heyen.

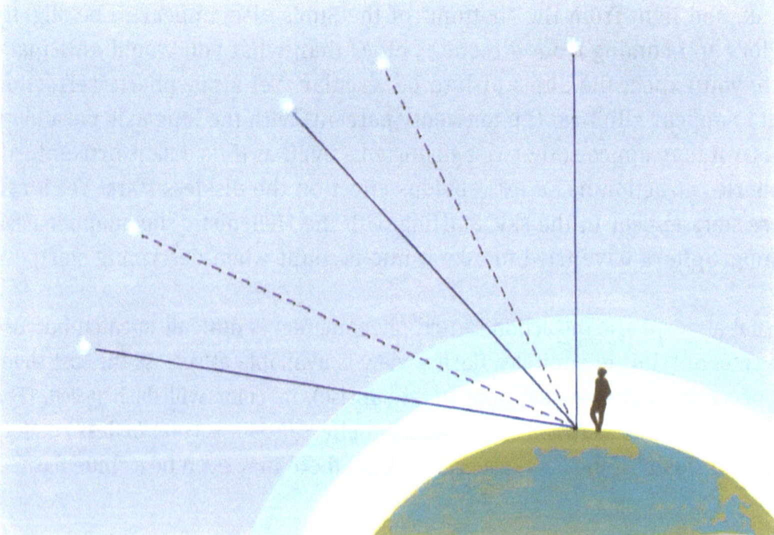

Figure 3.4: Atmospheric refraction causes a star's position to appear closer to the horizon; the lower the star is in the sky, the greater the refraction. Angles exaggerated. Artwork by Christine Tarte.

Figure 3.5: The green flash seen from the shore on the Gulf of Mexico.[5] Courtesy of Kai Schreiber.

3.4 The quality of the sky for viewing varies

Let us return to the night sky. Even if you happen to look through the same air mass, the quality of your sky view still can vary.

3.4.1 Transparency

The **transparency** of the atmosphere may change from night to night (or hour to hour). By "transparency," it is not meant whether it is clear or cloudy. Even on a technically clear night, the amount of light-absorbing material in the atmosphere can be greater or less than on another.

Volcanic eruptions and forest fires can dramatically affect transparency adversely and for a long time. They may even reduce transparency globally. But nearby phenomena, such as smoke from wood stoves and fireplaces in winter, can deteriorate the transparency that you encounter, too. After a weather front passes through, your local atmosphere often is cleansed, improving whatever the transparency was, earlier.

---

5 The author considers the term "Gulf of America" to be doltish; it is not used in this book.

3.4.2 Seeing

A second factor that affects clarity of view is called simply "**seeing**" by astronomers. It is what makes stars twinkle and therefore, on average, appear dimmer. Your seeing may be affected adversely by observing through unstable air convecting off of, for example, a hot driveway. Moving can improve it. More often seeing is regulated by what is taking place in the atmosphere much higher above you. In this case, there is little that you can do to change it.

The air is never static. Twinkling is produced by starlight passing through the Earth's undulating and translating atmospheric cells (defined by their common density), which make up a given layer of atmosphere.

Stars appear point-like. As horizontally moving atmospheric cells pass between you and a star in succession, they refract its light a bit more this way or that way, compared to the atmospheric average. It is as if the starlight appears to be coming from slightly different directions all the time. You see it when this direction is toward you.

Your eyeball may try unsuccessfully to track the starlight. Notwithstanding, when the alignment between your direction of gaze and the star is disturbed, the star's brightness decreases. When the alignment is restored, the earlier brightness of the star is (perhaps momentarily) restored, too. This happens again and again. You perceive this rapid and irregular cycle as twinkling.

Figure 3.6: On most nights, layered atmospheric cells moving this way and that way redirect light on a fast timescale; the result is some degree of twinkling. Angles exaggerated. Artwork by Christine Tarte.

The degree of twinkling depends upon temperature variations and how fast upper-air cells are moving tangentially to your line of sight. It is also dependent upon the size of the cells, the difference in density of the cells, and their elevation. Nights on which the star view is steady are called nights of good seeing. Nights on which the view is less so are called nights of bad seeing. However, seeing can vary throughout the night and, indeed, on a night of average bad seeing it may be possible to catch a few moments of good seeing and vice versa.

Figure 3.7: Bad and good seeing. Courtesy of Emma Smith.

Seeing affects extended objects in the sky less. For example, planets are actually tiny disks, though ones too small to resolve with your unaided eye. Nonetheless, their light does not act as if it is coming from a single point. If poor seeing causes the light from, say, the top of the planetary disk to be refracted away from you for a moment, the light coming from the bottom of the disk may be refracted toward you. On average, the total amount of light you see from the planet is constant.

Twinkling, or the absence thereof, can be used to distinguish planets from stars. However, high air mass may amplify the effects of bad seeing. Even planets may twinkle "low" in the sky.

3.5 Twilight

If it was not for the atmosphere, once the Sun sets, the sky would become instantly dark – until the Sun (without warning) rises again. Instead, you experience the brightening and diminution you know as twilight. Twilight is caused by the Sun's continued scattering of light into your sky, even when it is no longer (or not yet) itself visible.

When it is late enough (or early enough) such that the Sun is far from your sky, this illumination ceases. Twilight ends at dusk and commences at dawn.

Obviously, astronomical viewing is best without background light of any kind. When does twilight begin and end?

Three definitions of twilight are used by three more or less separate communities. Confusion takes place when different people use different definitions.

3.5.1 Civil twilight

Civil twilight is that time before sunrise or after sunset when there remains (or begins to be) enough light to see common objects outdoors. Indoors, though, we set about "turning on the lights." However, only the brightest stars are visible during civil twilight. Bright planets may be visible during civil twilight.

Formally, civil twilight is defined to exist after sunset and before sunrise, while the Sun disk's geometric center is up to 6° below the limit of the sky (the point at which this midpoint set or rose). Civil twilight ends or starts at **civil dusk** and **civil dawn**, respectively.

3.5.2 Nautical twilight

Sailors use a different standard. After **nautical twilight** in the evening (or before nautical twilight in the morning), it is not possible – absent moonlight – for a mariner to discern the sea from the sky, and this makes navigation more difficult. Sky and sea appear to merge. On land, outside activities begin or cease, if artificial outdoor illumination is not available. Many stars are visible during nautical twilight.

The formal definition of nautical twilight requires that the geometric center of the solar disk is a further 6° below its sunset (or sunrise) position. This equals a total of 12°. Nautical twilight ends or starts at **nautical dusk** and **nautical dawn**, respectively.

3.5.3 Astronomical twilight

Professional astronomers use an even stricter criterion. They want the sky to be as dark as it is going to be at any other time of the night. **Astronomical twilight** begins and ends at the earliest (morning) and the latest (evening) time, out of the three standards for twilight.

To avoid astronomical twilight, the Sun must be another 6° below its sunset (or sunrise) position. This equals a total of 18°. By then, the contribution of the Sun to the brightness of the sky is equivalent to or less than that of the Moon, stars, and the natural faint glow of the air. Astronomical twilight ends or starts at **astronomical dusk** and **astronomical dawn**, respectively. Afterward/before, it is as dark as it is going to get!

Twilight often implies sky illumination near your horizon. Once the mid-solar disk has set about 9 ½°, overhead already is dark.

In later chapters you will see why the lengths of time during which these different definitions of twilight hold true are not always equal.

Figure 3.8: Civil twilight, nautical twilight, and astronomical twilight. Photographs by the author.

Notes from cultures around the world

Different formalizations of twilight have been and are used. For instance, a Jewish tradition holds that the Sabbath begins and ends, on successive nights, once three stars can be made out in the evening sky. The stars must not include one or more particularly bright stars. Notice that, defined this way, the Sabbath can last more than 24 h.

Another way to establish the beginning and ending of the Sabbath is to base it upon the position of the Sun, similarly to the angular definitions already introduced. However, the corresponding relevant Sun angle for the Sabbath usually is somewhere between 5° and 9°, depending upon the textual source.

The three definitions of twilight assume normal conditions. When the amount of particulate matter in the atmosphere is especially great, it can extend twilight by a negative solar altitude of several degrees. (See the next chapter.)

3.5.4 The "Belt of Venus" (a misnomer)

A phenomenon that may confuse establishing twilight is the **anti-twilight arch**.[6] After sunset, the sunlight that reaches your sky does so at a very small angle with respect to your line of sight (the direction in which the Sun has disappeared). In this situation, the atmosphere can **backscatter**. Atmospheric aerosols illuminated by the vestiges of sunshine in the West backscatter in the East. Starting when the solar altitude is about −5°, a band of (usually) pink from the now-reddened sunlight appears. The anti-twilight arch appears 10–20° above the limit of your line of sight on the side

---

6 or Belt of Venus (perhaps so nicknamed due to some imagined resemblance to the girdle of the goddess)

of the sky opposite the Sun. Eventually a bluish and quickly darkening band appears below the arch. This is the shadow of the Earth.

Figure 3.9: Anti-twilight arch. While its alternate name invokes Venus, the only connection this phenomenon has with the planet is its appearance in the evening and morning skies. Courtesy of Kent Duryee.

The "arch" is barely curved; you may not see it as such. The appearance changes within, and the disappearance of the anti-twilight arch occurs quickly – faster than the rate at which the Sun itself rises and sets. (Of course, directions described are reversed when encountering the anti-twilight arch before sunrise.)

A last word on twilight. The formal definitions of twilight do not hold the status afforded by the rule of law. Ultimately, *you* get to decide when it is "dark enough" to consider nighttime to have arrived. Similarly, you get to make the call in regard to when *you* consider it to be close enough to sunrise to conclude that daytime has arrived. For these subjective times, the English language provides two words that are purposely vague: "daybreak" and "nightfall."

Misconception alert

Crepuscular rays that appear to emanate as the Sun prepares to set (or has just risen) have little to do with the Sun itself. These are formed by sunlight broken into beams by holes of clear sky between clouds.

Figure 3.10: Geometry of an anti-twilight arch. Not to scale. Artwork by Christine Tarte.

3.6 Light pollution

In this chapter, you have been introduced to how the atmosphere may impede or alter your view of the astronomical sky. There is a final such impediment but one that is mostly created artificially.

3.6.1 Wasted light

Your grandparents or great-grandparents had an advantage over you when it came to sky-watching. All over the world, night skies were darker a century ago. Outdoor lighting in this age was in its infancy. Ours was a much more a rural civilization. It was easier to find a dark place back then; it might even have been your own front yard.

Today, you are losing the dark of night. For most urban dwellers, even the brightest stars have all but disappeared from view. They have not gone anywhere – the fault lies in streetlamps, spotlights, glowing fast-food signs, *etc.* Astronomers call this nuisance **light pollution**.

Light that lets you find your way at night is a good thing. Nobody wishes to return to the days when all of us were virtual prisoners in our homes, once the Sun had set.

Everything is fine as long as artificial illumination is directed where it is intended: down onto the ground. However, too much of this lighting is carelessly directed upward. This includes lights intentionally beamed up for advertising purposes but, more commonly, improperly aimed street and yard lights.

The problem is once again facilitated by the atmosphere. This escaped light does not just disappear as it shines upward. It scatters in the air and makes the sky glow, thereby ruining the contrast needed to see faint astronomical objects.

Photographs of the Earth at night, taken from space, show the effect clearly. Centers of brightness in such photographs are caused by misdirected artificial lighting or, in some cases, rural, anthropogenic forest fires.

The view from above shows that North America and Europe are particularly ablaze with light. (Shiny snow on the ground amplifies light pollution further.) The location of nearly every city and village can be pinpointed. In North America, the paths of Interstate Highways can be traced just by the glow of roadside rest stops.

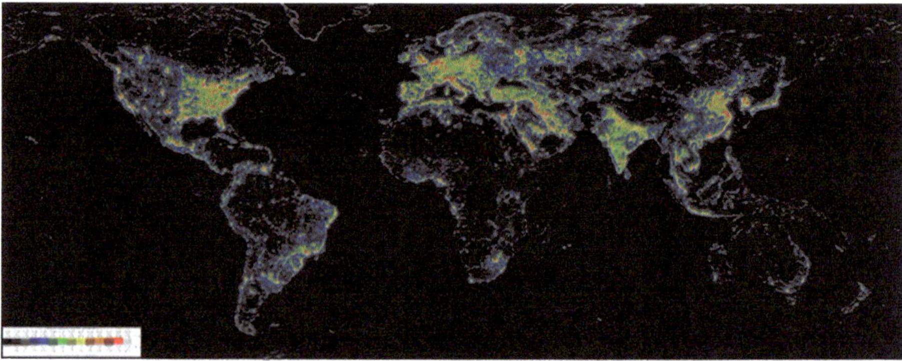

Figure 3.11: Light pollution as seen from space. Courtesy of David Lorenz/NASA.

This escaping light is not doing anyone any good. It represents wasted energy and the associated cost. A great deal of the latter is at public expense.

Proper security lighting makes private and communal spaces safe. Yet more light is not synonymous with better light. Efficient illumination involves putting enough light where it will be useful, and not where it will fail to be. Indeed, improperly designed artificial lighting can produce glare that makes objects and people near the source more *difficult* to make out.

3.6.2 What can you do about light pollution?

Unless you are in the middle of a desert or on a ship on the ocean, light pollution is difficult to escape entirely. You can relocate to a mountaintop; less atmosphere above you means less atmospheric scattering. This is seldom practical.

Truth be told, a little of the artificial light scattering we call light pollution occurs within the interocular fluid (inside your eyeball.) There is nothing at all that can be done about that. Thus, light pollution must be dealt with at the source.

The irony of light pollution is that, unlike many other forms of pollution, the problem is so easy to cure. Often, that solution is as simple as placing a reflector over a lamp bulb to bounce light back down instead of beaming it up.

Figure 3.12: Are these four streetlamps where one would do? Courtesy of James Lowenthal.

It may be possible to narrow the beam of a light, too. Are two lamps shining on the same spot, when just one would do?

Meanwhile, light pollution continues to rob us of a birthright no less legitimate than clean air and water: the right to enjoy a truly dark and beautiful night sky.

Physicist Christopher Kyba and others have determined that the night sky is brightening by almost 10% each year. Imagine that you are a toddler, just capable of looking out of your bedroom window. Suppose you see 250 stars. Before heading off to college, you look at your sky through that same window. You will see less than 100 stars!

Luckily, there still are places you can go where there is *less* light pollution, though citizens of major cities may have a fairly long trip. You may be amazed how many stars you can detect at a non-light-polluted site. (An added benefit of this detour is

Figure 3.13: The effect of light pollution: 1970 (left) and 2012 (right) views taken at roughly the same location. Courtesy of Joe Roberts.

that it may get you away from industrial and transportation-produced smog, which adversely affects the transparency of the sky.)

A convenient test of your chosen observing site is the Milky Way. (See Chapter 5.) A third of all people cannot see it from home due to light pollution; the fraction is 80% in the USA. If you can spot this band of light, you are seeing the sky as the Ancients did – after they had extinguished their campfires.

If you cannot avoid light pollution entirely, there is a certain practical advantage to observing after midnight. (This assumes, of course, that your target object is above the horizon at this time.) The reason is that more outdoor lighting gets turned off as you head into the early hours of the morning, and your sky will be a little darker.

Tools of the naked-eye observer

If you require a *little* illumination while observing (for instance, to consult a sky chart), try putting a red filter over your flashlight. Red light does not disturb your night vision as much as other colors do. (This is the same strategy used in lighting airplane cockpits.) While you can buy red incandescent light bulbs – even special red LCD flashlights – a piece of red cellophane over the front of a torch[7] may do. Another inexpensive option is to paint the bulb of a penlight with red fingernail polish.

3.7 Chapter summary

Locally, the Earth seems flat, and therefore, too, does its blanket of atmosphere. This air layer comes between you and the astronomical objects that you want to see, though mainly only that portion very near the ground.

---

7 flashlight

Air mass (as a function of zenith angle), scattering, bad transparency, bad seeing, and twilight may all impede an optimum view of the night sky. Indeed, light from objects that should be visible in your sky can become extinguished. Additionally, atmospheric refraction can fool you. Of course, we are indebted to the atmosphere for other reasons.

These are natural impediments to naked-eye astronomy. Light pollution is anthropogenic, and it is getting worse. Just as we created light pollution, we can take steps to lessen it or avoid it.

Dusk and dawn occur at three different times, depending upon the definition used. The result is three different kinds of twilight. Astronomical dusk to dawn represents the shortest and darkest period of the night.

3.8 Chapter review

1. Why is the daytime sky blue?
2. Why does the Sun appear more orange/red during sunrise and sunset than during the rest of the daytime?
3. Why do stars and planets often appear to "disappear" when they are near the horizon?
4. Why do stars appear to twinkle?
5. Why do stars and planets appear to be in a slightly different spot in your sky than they actually are?
6. Explain the "Belt of Venus."
7. What are major causes of light pollution? How can we reduce light pollution?
8. Why might a mountaintop be a good place to observe the stars?
9. There is almost no atmosphere on the Moon. Describe what dawn would be like on the Moon.
10. Describe the different types of twilight below.
    Civil
    Nautical
    Astronomical

4 The dome of your sky

Before discussing what is *in* the sky, what about the sky itself? You experience the sky as an imaginary inverted dome, the center of which is directly over your head. Its rim is defined by the **horizon**. While the word is so familiar that it already has been used, its formal definition is the intersection of sky and ground.

4.1 Chapter learning outcomes

After reading this chapter, you will be able to:
– direct another person to a location in your sky.
– make angular measurements in your sky.
– know when and where to look for stellar[1] phases.

4.2 Significance of and direction in your sky dome

Your impression of the sky is likely of a distorted half-hemisphere, squashed at the top. This may be because, when asked to look straight up, many of us do not bend our necks back a full ninety degrees. Other information-processing factors in the brain likely are at work, too.

The concept of the sky as some sort of dome centered upon you is a practical way of thinking about your sky. It really looks like that.

Moreover, the sky as a dome is somehow comforting. Look up at the night sky and try to force your eyes (and imagination) to consider it for what it really is: a black expanse as near to infinity as you ever will see. It is punctuated by stars at such vastly different distances from you that obscure stars are obvious, because they are comparatively near, and resplendent stars are difficult to detect, because they are so far away. What do you feel in that instant before your brain again overtakes your vision? A bit of panic? Maybe a tinge of vertigo?

Thinking of the sky in "3-D" is difficult. It is contrary to any other human experience, where our environment is bounded. It can be exciting. It also can be scary. Your mind tells you that a dome is a much more palatable construct.

There is nothing wrong with this. The idea of a black opaque dome over your head is a useful fiction for describing things in your sky at any moment in time.

What is more, in popular culture, the idea of the sky as a physical thing has had great traction. "It's like the sky is falling" remains a simile for calamity.

---

1 adjectival form used for stars

https://doi.org/10.1515/9783111441245-004

The sky dome also helps you communicate with others. You can tell another person where to look for a sight-to-be-shared in the sky by providing them with just two pieces of information. One is how high to look on the dome, the other is where (in which direction) to look.

4.3 "Distance" & "Place"

In using "height" and "where" in this context, the reference is not to a *physical* distance or *physical* place. It refers to how much you must tip your head, and how to turn it, in order to see (for instance) a particular star.

Without prior knowledge, it is impossible to know how far away a star – or anything, for that matter – is. Once an object is too high to reach (*e.g.*, an orange in a tree), would you know how far away it is without first knowing how *big* an orange is?

Similarly, how far are the stars if you do not know how big a star is? Finding alternate means of determining the distance to stars has been one *raison d'etre* of astronomy, but none can be accomplished using the naked eye.

You never pick up and examine a star. You may read in other books that a star is exceedingly large, but would you necessarily assume that without being told?

On the Earth, you might get lucky: There are clues to distance that are familiar to artists who deal in perspective. A landscape feature nearly lost in mist is likely farther away than a similar one that is in clear view. But such clues are not useful when you consider the sky.

4.4 Altitude

Your sky is limited by the horizon. Astronomers define the positive angle between a celestial object (as opposed to, say, a skyscraper) and the horizon as its **altitude**. Altitude is an angle. It is measured in degrees of arc, a system in which a circle measures 360°, with zero degrees at the horizon and ninety degrees directly overhead.

Misconception alert

A more common use of the word "altitude" is as a synonym for "elevation." Airline pilots share with you the altitude at which your airliner is flying. However, this kind of altitude is measured in units of length. In Latin, *altus* simply means "high."

The Earth is large, 40,000 kilometers in circumference, and we live within a very thin layer upon its surface. Standing on the (sea-level) ground, and contrary to what you might think, you cannot peer terrestrially very far before encountering the horizon. An average-height person sees it at a distance of about five kilometers.

Currently, the world's tallest building is the 163-floor Burj Khalifa in Dubai. Your horizon viewed from the top? It is only about one-hundred kilometers way.

A circle in the sky of constant altitude is an **almucantar**. The points on an almucantar include one in every direction, but all have the same altitude. The almucantar of the greatest apparent diameter is the horizon itself. The almucantar at altitude ninety degrees is, instead of a circle, a mere point. This almucantar has no circumference at all.

Admittedly, our horizon is often interrupted by trees, buildings, hills, and billboards. The **astronomical horizon**, implicit hereafter unless an explicit distinction is made, is the one that would exist if you stood on a featureless plain.

Figure 4.1: The yellow angle indicates the altitude of an object in the sky above the astronomical horizon. Notice that anything on the same almucantar will have the same altitude. Artwork by Christine Tarte.

4.5 Azimuth

Besides altitude, there is second piece of (angular) information necessary to convey, so that you and somebody else are looking at the same thing in the sky. It is azimuth. By itself, altitude tells your friend whether to look near or far from the horizon. But in which direction? Actually, azimuth is a more familiar concept than altitude. It is an angular equivalent to compass direction.

You may be used to giving directions here on the Earth. You tell somebody to look north, south, east, or west. Or perhaps northeast, southwest, southeast, northeast, *etc.* The **azimuth** angle is measured starting in the North; due north is azimuth zero degrees. Azimuth increases as you turn clockwise (eastward) to 90° at due east, 180° at due south, and 270° at due west. It nears 360° as you approach due north again.

Notes from yesterday and today
Those numerical signs on airport tarmacs tell aviators the names of runways. They are based on runway direction; the signage has just dropped the last digit. A "9" runway points eastward (90° in azimuth). Approaching the same runway from the other end, the pilot will see "27" (270° in azimuth).

A great semi-circle on the sky of constant (and constant supplementary) azimuth is a **vertical circle**. Because altitude is measured with respect to the nearest horizon, every altitude is represented on a vertical circle twice.

Notice that it is called a vertical *circle*. Technically, it includes every *negative* altitude (zero to ninety degrees) twice, too. A negative altitude is below your horizon. Obviously, you cannot expect to see anything in the sky at a negative altitude!

Figure 4.2: The yellow angle indicates the azimuth of an object in the sky along the astronomical horizon. Notice that anything on the same vertical circle will have the same azimuth. Artwork by Christine Tarte.

4.6 Measuring angles in your sky

Astronomers are not obsessed with angles. It is simply that the sky is best described this way. Degrees are conventionally used as a unit of measurement, not radians.

Here is a simple and approximate way of measuring both altitude and azimuth in degrees: Extend your hand so that it is at arm's length from your face. Different hand gestures subtend different angles!

Figure 4.3: From left to right: one degree, two degrees, ten degrees, fifteen degrees, and twenty-five degrees. Photograph by the author.

Does this really work? Some people have small hands, others large hands. Some people have short arms, others long arms. Hand measures of angles work because those with large hands usually are those with long arms, and those with small hands usually are those with short arms! It is "handy" that the ratio of hand size-to-arm length is more-or-less constant among humans. The angle measured in this way is more or less the same from person to person.

Your hand is not the size of a constellation. (See Chapter 5.) Your arm length is not the distance to a star. However, due to the geometric property of similar triangles, this method functions for measuring altitude and azimuth.

Tools of the naked-eye observer

A more accurate means of measuring angles, while still using that naked eye, was part of the accoutrements belonging to Chinese navigators. It consists of a set of wooden squares, each of a different area. These squares are cut to sizes such that holding one up at arm's length represents a predetermined angle in the sky.

Medieval Arabic sailors of the Indian Ocean used to carry with them a *kamal*.[2] This is a piece of wood with a hole in the center. Through the hole is threaded a string. While holding the string's end by mouth, the user slides the wood piece along the taut string. Its apparent size changes. The navigator moves the wood piece so that it covers an angle of interest. Perhaps a star is at the top of the wood piece, and the horizon is at the bottom.

Knots in the string mark locations where the wood piece takes up a given angle. The number of knots passed by the wooden bar indicates the measure of the angle in question.

---

2 "guide" in Arabic, or Khashaba

Yet another method for measuring sky angles, dating back to at least the eleventh-century Persians but claimed as a Portuguese invention, is using a *cross staff*.[3] A short (normally wooden) bar, called the *transom,* slides along another, longer wooden track at right angles to it. The user looks down the track at the transom, sliding this bar until its apparent angular extent overlaps that between the objects (or horizon and object) being measured. The angle is then read off a scale on the track.

Notice that all of these methods require a challenging act: Essentially, the user must look at two things (edges) at once!

Altitude and azimuth. That is it. Two angles define a unique spot on the dome of the sky. They are all that you need to hop among the stars in your sky – *if* your sky did not appear to change with time.

4.7 Cardinality

Cardinality is perhaps the most fundamental gift imposed by the sky upon Earthly inhabitants below. The concept of a four-quadrant grid superimposed over topography pervades a great many cultures, both old and new. In writing, it is as ancient as Homer, and as modern as GPS navigational instructions in your car: north, south, east, and west.

Notes from long ago

Near Xian, China, is the mausoleum of Emperor Qin Shi Huang. He was the ruler who first united the country, both politically and culturally. It was the third century BC/BCE.

Qin's elaborate, hill-sized tomb took hundreds of thousands of laborers decades to construct. Today, the site is an archaeological work-in-progress; the tomb has not yet been excavated. We do know that it is cardinally aligned. Moreover, we have a hint as to what we might find inside.

In 1974, local farmers dug a new well. To their surprise, they struck an earthen head! It turns out that here, near the tomb of Qin, is an entire buried Chinese army in replica. There are thousands of "soldiers." Their weapons, vehicles, and horses – are all life size.

This eternal infantry guards Qin's tomb, facing the East, the direction from which Qin's enemies came. Most of the figures are made of terracotta. The Terracotta Warriors are one of the great archaeological finds of the last century.

Human ideas often are long-lived, even longer than the societies that create them. The symbolic seat of government in modern China, the famous Forbidden City, also is aligned cardinally.

Cardinal directions are so natural to language that this book already has begun to use such terms without naming them as a group. Almost as familiar are the **intercardinal points** half-way between them. These can be further divided, divided again, and so forth. The result is the 32-, 64-, or 128-point graphical depiction, called the **compass rose**,[4] which often appears on maps. Why do you suppose it is called a rose?

———————

3 or Jacob's Staff; fore-staff; ballastella/ballestilla; "balestilha," in Portuguese

4 sometimes called the "wind rose" or "compass star"

Figure 4.4: Compass rose. Courtesy of Tau'olunga.

For instance, there are 11 ¼° between north and north by east, 11 ¼° between north by east and north-northeast, 11 ¼° between north-northeast and northeast by north, and 11 ¼° between northeast by north and northeast. And that just takes us one-eighth of the away around the compass rose!

Outside of astronomy, these words are used more often than azimuth. To relate the two concepts, think of cardinality as alignment to specific azimuths.

Human architecture based on cardinal alignment exhausts cataloging. It includes many a town plot, where, disregarding the natural topography of rivers, mountains, and seashores, numbered streets align east-west, and Main Street runs north-south or vice versa. Cities, larger political divisions (*e.g.,* in the United States, counties), whole regions (*e.g.,* in Canada, provinces), and even nation states form a whole or partial grid work with east-west- and north-south-running borders.

There is a continuous, cardinal progression on the map. It extends from many exactly east-facing houses (perhaps including your own) to the **axis mundi,**[5] the Earth's axis of rotation. It is the axis mundi that universally defines north-south for everybody on our planet.

---

5 also cosmic axis, world axis, or world pillar

4.7.1 The compass

The magnetic compass is a useful device – try finding north on a cloudy night without it! In such a compass, a magnetized needle is free to swing toward the natural magnetic north of the Earth. But magnetic north is not the same thing as geographic north, which is defined by the axis mundi.

It is correct now that, for those at mid-latitudes and equatorward, the magnetic north pole is close to true north; however, one can find oneself at high or low latitudes where there is a major discrepancy in direction between the two. Moreover, magnetic north is moving all the time due to changes in the Earth's magnetic field. Tables showing the difference must be updated constantly. A further complication is that every object near you with magnetic properties can affect the direction in which a compass needle points. Two compasses, used at two different places on the globe or at two different times, may not indicate the exact same "north."

When talking about the sky, geographic (which we are using interchangeably with "true") north is always meant, not magnetic north. For instance, when an astronomer says "south," they always refer to the direction 180° opposite geographic north. In other words, they mean geographic south.

> **Notes from cultures around the world**
> The concept of the North seems so fundamental to most of us that it is difficult to imagine its absence. Nevertheless, Canadian ethnologist John MacDonald points out something interesting.
> For the Arctic-dwelling Inuit, the North has no special meaning. They do use a sort of cardinality of their own; it is established by the direction of winds, that of sea currents, and the orientation of snow drifts!

4.7.2 Alternatives to cardinality

It may seem reasonable to divide the horizon into four sectors, between equally spaced azimuths. You call these the North, the South, the East, and the West. However, if you live at a high latitude, there is another way to divide the horizon that may make just as much sense to the indigenous peoples living there.

Imagine this: There is a sector – let us call it "the East" – in which the Sun rises. Its limiting azimuths are the most northerly and southerly points on the horizon at which the Sun emerges, sometime during the year. (See Chapter 10.) Likewise, there is a sector – let us call it "the West" – in which the Sun sets. Its limiting azimuths are the most northerly and southerly points on the horizon at which the Sun disappears, sometime during the year. Then there is a sector – let us call it "the South" – in which the Sun *never* rises nor sets, but instead, always traverses above the horizon. Finally, there is a sector – let us call it "the North" – in which the Sun *never* appears.

The angular sizes of these sectors vary with latitude: "The East" and "the West" take up more and more of the horizon as you travel farther northward. (Again, see Chapter 10.)

"The North" and "the South" take up more and more of the horizon as one gets closer to the equator. Before you get to the equator, the system breaks down. Still, for certain latitudes, these practical definitions of "the North," "the South," "the East," and "the West" might make as much or more sense as arbitrary sectors of ninety degrees each.

4.8 The "Heavenly Firmament"

The dome of the sky has been an important philosophical concept throughout history. In portions of the *Bible,* borrowed from the civilizations of Mesopotamia, its physical manifestation is the "Heavenly Firmament." It served to separate the Earth from the heavenly waters.

It is not outrageous to think that there is water above you. What color is the sea? Blue. And, of course, water *does* sometimes fall to the Earth – in the form of rain.

Figure 4.5: The Heavenly Firmament, as envisioned by Cosmas Indicopleustes. Published in Raymond Beazley's *The Dawn of Modern Geography* (1897).

Unfortunately, the view that the heavens are separated, and different, from the Earth (that they literally are divided by a barrier) obscures the fact that all of us on the Earth have a great deal in common with the rest of the Universe. Indeed, we are intimately related to it. The suggestion that the rules (working theories) are different "up there" and "down here" held back western science for millennia.

4.9 Horizon events

Air mass, the quality of the atmosphere, and twilight all conspire to create a set of horizon phenomena. The terms used for them, "heliacal" and "acronical," probably are unfamiliar. This is ironic, because in long-ago-but-still-historical times, the meaning of a phrase such as "heliacal rising" would have been commonplace knowledge. It refers to a concept that has dropped out of our popular, cultural usage. Its early introduction here, as our first sky *event*, is intended to make this point.

4.9.1 Heliacal rising & setting

The **heliacal rising** of a star occurs in the early morning of the day on which that star is first seen in the remaining hours of twilight. Previous to that date, the star was lost in the glare of the Sun. Before that, it was an evening-sky star that rose after the Sun and was invisible at any time during the morning. (See Chapter 5.)

To see a heliacal rise, there must be a certain minimum angle between the center of the Sun and the star. This star-specific angle is called the **arcus visionis**.

> **Where did That come from?**
> Helios was a classical Greek titan and god of the Sun. The prefix "helio-" or "helia-" is used several times in this book as part of a noun or adjective for something that has to do with the Sun. In the Latin of the Roman Empire, Helios became Sol (as in "solar").

Imagine watching morning after morning for a favorite bright star. Then, one special morning, you see it, but only for an instant. (Again, see Chapter 5.) This is because, soon thereafter as the Sun rises, twilight gives way to daytime and the star disappears from view. You have witnessed the heliacal rising of that star. A star that rises heliacally appears to pop out of nowhere, and then to vanish just as quickly.

You may never have heard of a heliacal rising before. Nevertheless, there are plenty of relevant written records, starting with the sixth-century BC/BCE Mesopotamians.[6] These document ancient peoples, the world over, paying attention to the heliacal rising of stars. The Egyptian year started with the heliacal rise of Sirius; for millennia, it happened to mark the time of the **annual**[7] Nile River flood.

A **heliacal set** is a similar phenomenon. It occurs just after sunset. Together, heliacal rise and heliacal set are called **heliacal phases**. (These and similar "phases" are unrelated to the phases of the Moon.)

---

6 The multitude of city states that occupied the Tigris and Euphrates River basin since the beginning of civilization shared a common astronomy. Therefore, you are not asked to keep track of them. Instead, the generic term Mesopotamians is used.

7 Latin for "happening once per year"

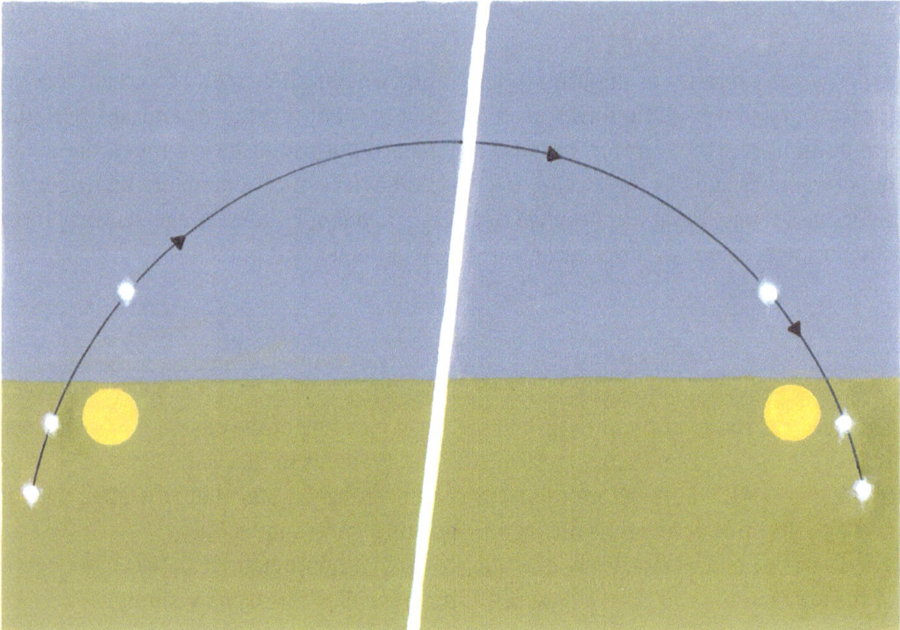

Figure 4.6: Left: A star rises in the East, just before, with, and just after the Sun; the first configuration is an example of a heliacal rise. Right: A star sets in the West, just before, with, and just after the Sun; the last configuration is an example of a heliacal set. Artwork by Christine Tarte.

Figure 4.7: Heliacal rise of Sirius. Courtesy of H. Rabb.

Notice that it is difficult to predict the *exact* date of a heliacal phase. Visibility of the phenomenon depends upon the brightness of the object (*e.g.*, star), the arcus visionis, your latitude on the Earth, the time of year, the presence or absence of the Moon above your horizon, its location in the sky, and its phase!

Most everybody has heard of the Dog Days of Summer and knows that the expression refers to the hottest, muggiest time of year. But what do dogs have to do with the seasons? In ancient Rome, Sirius was nicknamed the Dog Star. It rose heliacally in late summer, the **Dies Caniculares** ("days of the dog").

4.9.2 Example: Wyoming, United States

The indigenous people of the North American plains – perhaps they were the prehistoric Crow or Shoshone – left no writing and little in the way of indestructible architecture. Stones are one of the few construction materials available on the high, dry steppe that would survive time.

When non-indigenous settlers arrived, they discovered dozens of large stone rings. The bigger rings had "spokes" of stones radiating from the center. Certain

Figure 4.8: The Bighorn Medicine Wheel cannot be precisely dated, as it was disrupted prior to archeological investigation. Courtesy of Ray Gannett.

spokes, as well as the circle centers, were marked by rock-pile **cairns** containing (in some cases) tons of rock. These edifices were named Medicine Wheels, by the newcomers, for no other reason than that they suggested something mysterious; nobody knew why they were built the way they were.

Medicine Wheels appear in unlikely places. The best example is the Bighorn Medicine Wheel. It is located on a flat area high in the mountains, at nearly 3,000 meters in elevation, and has a clear view of the horizon. From it, you easily can see for a hundred kilometers. However, it is only accessible during the summer due to weather.

Jack Eddy, an American solar astronomer with NASA, heard of the Bighorn Medicine Wheel in the 1970s. He visited it and found it to be a 24-meter-diameter "circle" with a central cairn four meters in diameter and one meter high. There are 28 spokes. He concluded that most of the spokes having significant cairns at their ends were astronomically aligned.

Among these are pointers to the heliacal rising of named stars Aldebaran, Rigel, and Sirius. These are among the brightest in the sky, and roughly evenly spaced about the sky.

About three centuries ago, when the Bighorn Medicine Wheel is thought to have been built, Aldebaran rose heliacally at the time of the summer Solstice; about a month after Aldebaran, Rigel rose heliacally; and about another month later, Sirius rose heliacally. (See Chapter 10.)

While Bighorn is first accessible only just before the first of these heliacal rises, two months afterward marks the end of the time when it can be visited safely. The Bighorn Medicine Wheel, and its location, suggest some sort of shamanic site. (It is not big enough for, nor can it support, a great many people.) Once you arrive for the Solstice, the star alignments tell you when to leave.

4.9.3 Acronical rising & setting

The opposite of a heliacal rise is an **acronical**[8] **rise**, that is, the star rises on the opposite side of the sky as the Sun sets. A star can set just as the Sun rises as well. This is an **acronical set**. Together, acronical rise and acronical set are the **acronical phases**.[9]

Acronical events have not been as historically important as heliacal events. They are not as exactly definable: When is "at" sunset and "at" sunrise?

Collectively, heliacal and acronical risings and settings are called **stellar phases**. Planets have similar phases, too. (See Chapter 18.)

---

8 or acronycal
9 or cosmical rise, cosmical set, and cosmical phase

The significance of stellar and planetary phases to past societies, and how they were used calendrically, was largely passed down generation to generation in oral traditions. Sadly, we suspect that most of these are lost.

4.10 Chapter summary

The apparent dome of the sky was considered physically real in the past. Today, it is a useful fiction.

To indicate a location in your sky, you must use angular measurements. These are altitude and azimuth. Humans always have given special attention to the set of azimuths we call the North, East, South, and West. More obviously, we give special attention to the circle of equal altitude called the horizon (and to the zenith). Stellar phases are evidence of the latter.

The Big Horn Medicine Wheel is but one example of a prehistoric artefact, for which its purpose is not known with certainty. Yet, potential celestial alignments within it provide a clue.

4.11 Chapter review

1. Why is the distance of objects in the sky difficult to determine?
2. What two things would you need to have to be able to describe where something is in your sky to someone else?
3. What is altitude? How is this different in an astronomy context compared to the everyday usage of the word?
4. How would you describe azimuth in your own words?
5. How can you measure angles in your sky?
6. What is the difference between geographic and magnetic north? Why are they not in the same place on the Earth?
7. Match the appearance of a star on the left with the name of this appearance on the right.

First appearance in morning sky	heliacal set
First appearance in evening sky	acronical set
Last appearance in morning sky	acronical rise
Last appearance in evening sky	heliacal rise

8. Where did the phrase "the dog days of summer" originate?
9. Why do you suppose it is called a compass rose?
10. Where might you go so as to make your horizon as close to the astronomical horizon as possible?
11. Why might a stellar phase be useful?

5 Stars

It is time to properly introduce the population of your sky dome. When you look up at the night sky, the Moon may or may not be present; the same is true for the planets. You *might* see a meteor, comet, or some other rarer astronomical phenomenon. (See Chapter 19.)

However, you are guaranteed to see stars on a clear night. Under optimal conditions, the naked eye perceives thousands of individual stars above the horizon at one time. Turn-of-the-last-century English amateur astronomer Thomas Backhouse published a catalog with the title, *9,842 Stars Visible to the Naked Eye* (1911)!

5.1 Chapter learning outcomes

After reading this chapter, you will be able to:
- trace constellations and asterisms.
- identify and name individual stars using brightness and color.
- develop strategies for observing unresolvable groupings in the sky.

5.2 Backdrop of your sky

Sky watchers group the stars together into patterns called **constellations**. This is an old human habit. Some maintain that certain constellation patterns are tens of thousands of years old and survive until today largely by word of mouth, passed from generation to generation.

5.2.1 Constellations

Constellations are patterns only, sort of a connect-the-dots game in the sky. They have no physical significance. Some stars in a given constellation may be vastly farther away than others in the same pattern. Thus, viewed from anywhere other than our **Solar System**, the vicinity of our Sun, the pattern would change.

Have you ever seen a rock or mountain with a profile that reminded you of a familiar shape, perhaps a human face? The resemblance disappears when you step away from your particular viewing spot to another. Still, constellations are handy "landmarks" in the sky for someone standing on the Earth.

At one time constellations were named after creatures, heroes, and stories from popular mythology. Only the brighter stars were named. Some of the figures are obvious even today, though most take a great deal of imagination.

https://doi.org/10.1515/9783111441245-005

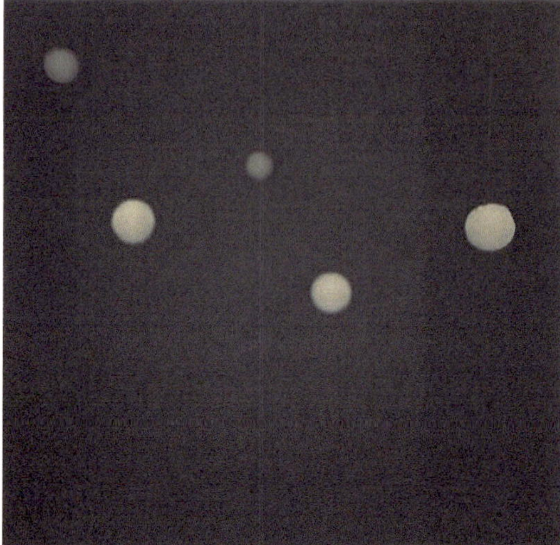

Figure 5.1: Simplified model of the familiar constellation Cassiopeia, which looks like the mythological queen's throne or the letter "W" to most people. Photograph by the author.

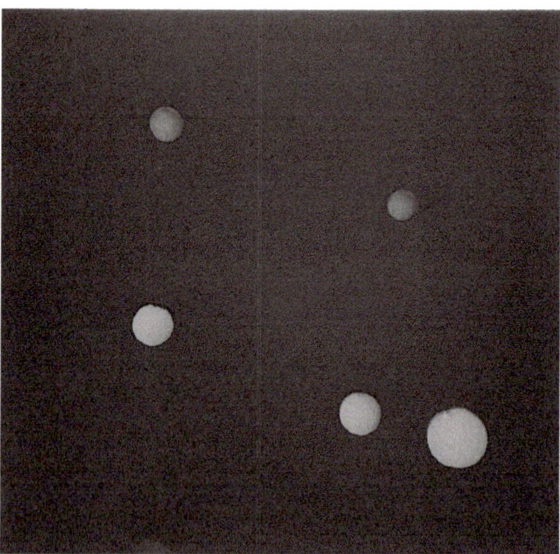

Figure 5.2: However, seen from a different point of view, representing one far from the Solar System, the model of Cassiopeia looks quite different. Photograph by the author.

Yet this was not a pastime such as that indulged in by a child while lying on their back and identifying elephants and candy among the clouds. Remember that the mythological creatures, heroes, and stories existed first.

People assigned star patterns to *honor* these characters in their culture. It was not considered absolutely necessary for a pattern to look a great deal like that after which

Figure 5.3: Eyemouth Point, Scotland. What does it look like to you? Courtesy of Walter Baxter.

it was named. In other words, the constellations may resemble their namesakes only as much as a map of the United States' capital of Washington resembles the nation's first President!

The constellations, as thusly established, were more or less "stick figures." They had no agreed-upon boundaries. Moreover, different civilizations recognized different sets (and names) of constellations.

The famous artwork preserved in the French Lascaux Caves may include contemporary constellation illustrations. They are approximately 17,000 years old! The western constellations, those with which you are most likely to be most familiar, start appearing in Mesopotamian writing from millennia ago.

5.2.2 Constellations made official

Modern astronomers formalize the constellations, fixing their number at 88, and assigning them definite boundaries. The result is that every star is now a member of some particular constellation.

You often can recognize the new constellations: If you come across Antlia (the Air Pump), you rightly suspect that it is not of ancient origin.

A map of the constellations today looks a bit like that of a crossword puzzle: lots of east-west and north-south running borders. Constellations can provide a general description of a location in the sky, without resorting to a numerical coordinate system.

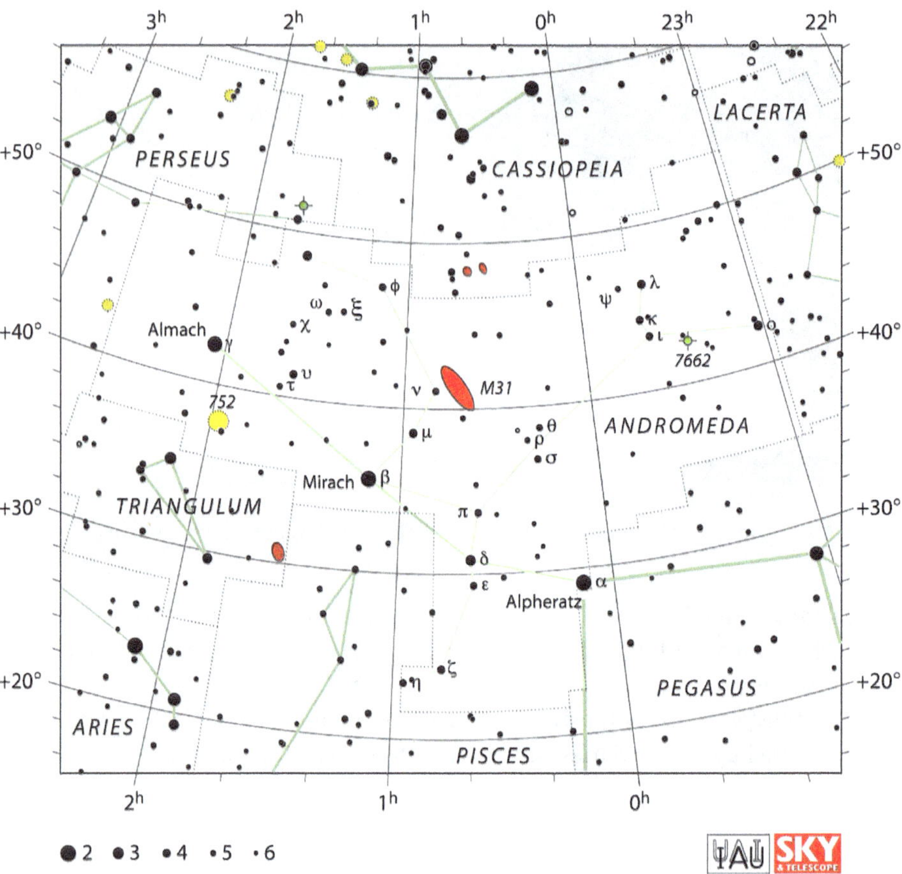

Figure 5.4: The celestial region now assigned to the constellation Andromeda. Star maps routinely use dot size as a convention for indicating brightness, even though all stars are really points of light. Courtesy of the IAU.

Often this is sufficient. If you are driving to a new city, knowledge of its exact latitude and longitude may not be called for. Knowing what canton, constituency, or county it is in might be all that you require to navigate your way there. If the city is big enough, you will *see* it once you get close.

Knowing the constellation an astronomical object occupies may be sufficient to find a celestial object if it is reasonably bright. The only requirement may be a chart of the constellations.

Of course, you also must know which constellations will be visible in the night sky at specific times, in order to know when and where to look for an object. Soon you will be introduced to the role of time in our discussion of the sky.

Misconception alert

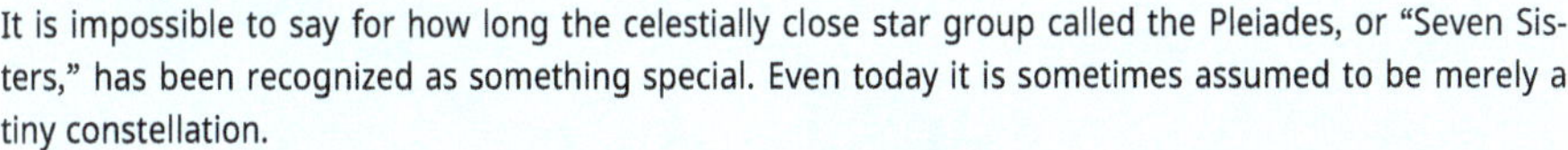

It is impossible to say for how long the celestially close star group called the Pleiades, or "Seven Sisters," has been recognized as something special. Even today it is sometimes assumed to be merely a tiny constellation.

This is technically not true. The reality is that these stars belong to a physical cluster of stars. (See Chapter 20.) They are all at more or less the same distance from you, and not an accidental pattern due to projection onto our sky. The Pleiades are considered to reside within the constellation of Taurus.

Most people see six stars in the Pleiades, instead of seven. Two Pleiads[1] are so close together that it is hard to resolve them into separate stars. Somewhat mysteriously, individual cultures all over the world have their own stories about what happened to the "missing" seventh Pleiad.

5.2.3 Asterisms

Besides constellations, there are unofficial patterns of stars in the sky used colloquially. These are **asterisms**.

Sometimes asterisms are part of a constellation. The Big Dipper (or The Plow) is made up of the brightest stars belonging to the larger constellation Ursa Major, is an example. The motif appears in design as early as a tomb painting from around 1280 BC/BCE Egypt.

Some asterisms are made up of stars from multiple constellations. The Square of (the constellation named) Pegasus must borrow a star from the constellation Andromeda in order to complete its geometric shape. The so-called Summer Triangle and Winter Hexagon[2] each include the brightest star from a separate constellation.

There are asterisms that are designed to guide you. Consider just the handle of the Big Dipper. If you follow this curve away from the Dipper's bowl, you encounter the bright red star Arcturus. The mnemonic is, "arc to Arcturus."

The three conspicuous "belt stars" in the constellation Orion point south in the direction of the famous star Sirius. The Southern Cross in Crux is an important, similar guide, because it directs you to the otherwise unmarked South Celestial Pole. (See Chapter 7.)

---

1 singular form of Pleiades
2 or Winter Circle

Figure 5.5: The asterism of the "Big Dipper" is only a small portion of the constellation Ursa Major (supposedly a bear with an unusually long tail). From *Urania's Mirror* (1825). Edited by the author.

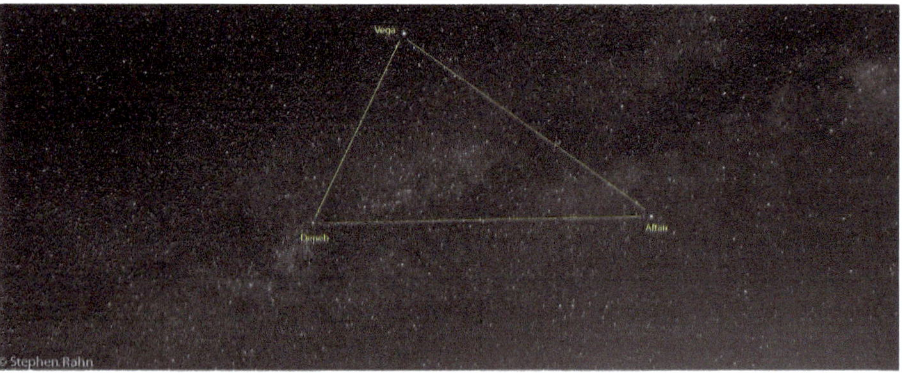

Figure 5.6: The "points" of the Summer Triangle are the stars Vega (in Lyra), Deneb (in Cygnus), and Altair (in Aquila). Each is the brightest star in a different constellation. Artwork by Christine Tarte.

The "horns of the bull" in the constellation Taurus can be thought of as an asterism, though in fact they are made up of a star cluster called the Hyades. Like the Pleiades (box above), they are an **open (star) cluster** and represent a group of stars the members of which are, in fact, physically close to each other. (See Chapter 20.)

Figure 5.7: The Winter Hexagon is formed by the stars Rigel (in Orion), Aldebaran (in Taurus), Capella (in Auriga), Castor and Pollux (in Gemini), Procyon (in Canis Minor), and Sirius (in Canis Major). Courtesy of P. K. Chen.

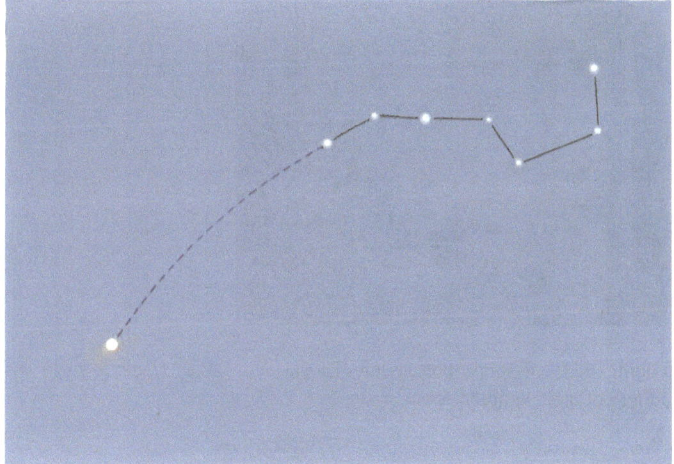

Figure 5.8: The mnemonic "arc to Arcturus" is illustrated. Artwork by Christine Tarte.

Figure 5.9: The Hyades (top right) and the Pleiades (bottom left). The distance to the Hyades is much less than that to the Pleaides. Courtesy of Brian Ventudo/NASA.

The constellation Coma Berenices might better be classified as an asterism. The entire constellation is an open cluster of stars and, therefore, arguably should not be considered a constellation at all.

Asterisms need not be ancient. Part of Sagittarius forms the asterism of the Teapot.

Figure 5.10: Coma Berenices is "Berenice's Hair." Courtesy of A. Fujii/NASA.

5.3 Star names

Bright stars have proper names. The preponderance of such stars with names that begin with "A" hint at the origin of these: In Arabic, the word "the" is made up of the letters *alif* and *lem*, or "al." Other star names come to us from the Greek.

However, most stars do not have proper names, and it would be daunting to come up with such names for all the stars visible to the naked eye (though there is a modern movement to do just that). The constellations also provide a naming convention for the stars.

In 1603, the German attorney and uranographer Johann Bayer invented a star cataloging convention to use as an alternative to a proper name. It is especially useful in cases where no such name is universally agreed upon or simply does not exist. It is called the **Bayer system.**

The Bayer system works this way: The brightest star within a given constellation is given the first letter (out of 24) of the Greek alphabet. (The Chinese tried a similar, placement-ordered system at least five centuries earlier.)

After he ran out of Greek letters, Bayer used the 26 lowercase Latin letters (skipping j and v), followed by the 26 uppercase letters (though he never got past Q). That proved confusing, and now numbers are used once Greek letters are exhausted. Only a few Latin letters remain in catalogs.

That is how it was *supposed* to work. In the case of a few stars, Bayer misjudged stellar brightness order, chose to label them instead in east-to-west order, or adopted an order that is by no means obvious now! He even assigned some stars to constellations different from those we do now.

Nonetheless, the Bayer system, with some adjustments, is still officialized today by members of the International Astronomical Union. For instance, they cleaned up situations where Bayer had assigned the same star two different designations.

The Bayer letter (or number) is followed by the name[3] of the constellation in which it resides. So, the brightest star in the constellation Eridanus is Alpha Eridani. While you can use the alphabetic letters themselves, it is common to write them out. However, a formalized style for abbreviation does exist: α Eri.

The result of the Bayer system is a two-part appellation that allows a tremendous number of stars to be uniquely identified. Just using the constellations known to Bayer and applying only the alphabet, his original system could distinguish 1,564 stars!

5.4 Stellar brightness

Besides their apparent position with respect to each other (*e.g.*, place in a constellation pattern), do stars all look the same? At *first* blush, they do. All are points of light in the sky. The "starburst" pattern of rays you might think you see emanating from a bright star is actually due to physiologically based optical effects within your own eyes.

5.4.1 Can stellar size be seen?

For centuries people have wrestled with the apparent area of stars. Thinking that stars must have *some* apparent diameter, there is a tendency to equate brightness with size. This impression proved wrong when the telescope was invented and revealed that, no matter how much a star is magnified, it remains point-like. Stars are so far away that you cannot see their apparent diameter even using a huge telescope (without some optical tricks available only since the last century).

The fact that stellar brightness does not correlate with size – that stars do not have apparent diameter at all – comes as a bit of a shock. It *looks* like the brightest stars are tiny disks. However, this is an optical illusion.

---

3 The genitive form of the noun is used

Stare at those stars. Do they really have a shape? Does that sense of round remain constant? No. Logically, something with an apparent size must have some constant shape. The twinkling stars do not.

> **Where did that come from?**
> The word "stellar" is a Latin-based adjective meaning "of stars." However, it has been in general use only since the seventeenth century. Early on it did not necessarily refer literally to a star; it might be used for any star-like or point-like object in the sky.

5.4.2 Brightness *versus* luminosity

Stars differ in brightness. This is due to their varying distances from you *and* their different luminosities. That stars may be dim due to their distance was not thought about until the classical Greek Heraclitus floated the idea in 540 BC/BCE.

What is the difference between the two words, "brightness" and "luminosity"? A floor lamp appears to differ in *brightness*, depending upon how far away you stand from it. However, the bulb in the lamp has a constant *luminosity* – the intrinsic amount of light emanating from it. (The luminosity of an incandescent light bulb usually is advertised by a permanent stamp on it, *e.g.*, "60 watts" or "800 lumens.")

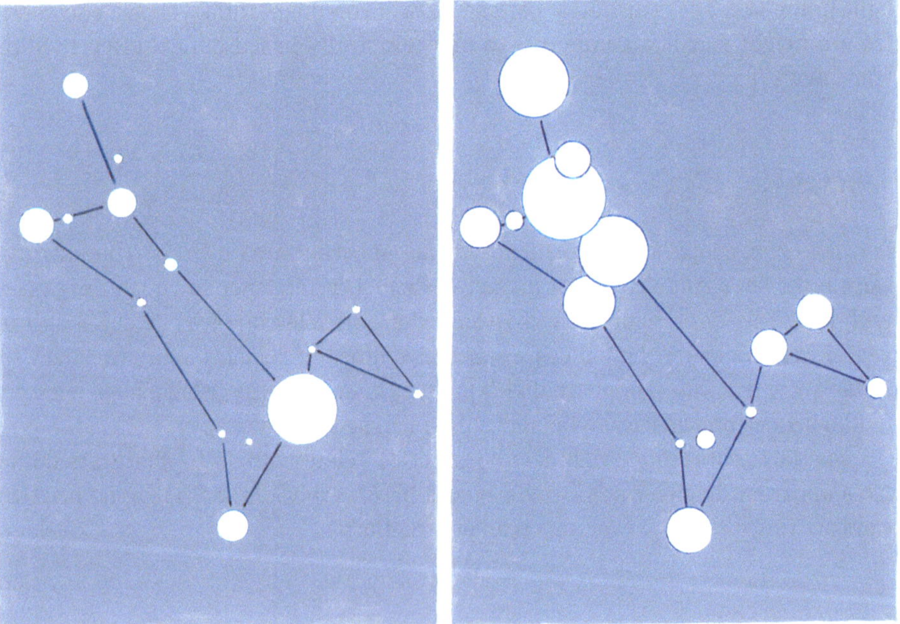

Figure 5.12: Left panel: Canis Major, where the size of the circle indicates brightness; right panel: the same constellation where the size of the circle indicates luminosity. Artwork by Christine Tarte.

The Sun is a good illustration. The Sun is in fact a star, no different from any other in the sky. It is so incredibly bright, not because it has anything other than an average stellar luminosity, but because it is *so* much closer to you than even the next-closest star. As there is no way to separate the effect of distance and luminosity in typical stars just by looking at them in the sky, hereafter this book uses "brightness" almost exclusively.

5.4.3 Stellar record holders

There is only one Sun. The closest among the thousands of nighttime stars visible to the unaided eye is the one named Rigel Kentaurus[4] [Alpha Centauri]. Though setting a record for proximity to you, Rigel Kentaurus still is terrifically far away. There is one star a bit closer, but it can only be seen through a telescope. Again, stellar distance is, alas, unknowable, to the naked-eye observer.

An abstract record-holder for another reason is the star Deneb [Alpha Cygni]. Among naked-eye-visible stars, it is the most luminous. Incredibly, it is somewhere between 100,000 and 200,000 times more luminous than our Sun. Yet it is not the brightest star in your sky. It is far away.

Recall that the brightest star in your sky is Sirius [Alpha Canis Majoris]. It wins this title by being a combination of *fairly* close (as stars go) and *fairly* luminous (again, as stars go).

Pollux is only the seventeenth brightest star in the sky, but still is easy to find. It is one of Bayer's exceptions: While it is the brightest star in the constellation Gemini, it goes by the catalog designation Beta Geminorum. Looking at this star, it is fun to know that it is the brightest among the set of those that astronomers have detected (in 2006) an orbiting planet – like our planet, the Earth, orbits the Sun.

5.4.4 Stellar magnitudes

Truth be told, the human eye/brain is not very good at quantifying differences in brightness. However, we still try.

Starting with the Hellenistic Greek astronomer Hipparchus, stars have been placed into brightness categories; each is called a **stellar magnitude**.[5] According to Hipparchus, first-magnitude stars are the brightest; sixth-magnitude stars are only just visible in very favorable skies like Hipparchus's.

---

4 as opposed to the star named simply Rigel
5 from the Greek for "greatness"

Today, using a formal definition of stellar magnitude, reference star Spica [Alpha Virginis] is assigned a magnitude of exactly 1.0. (There are a few stars that actually are brighter than this.) The so-called naked-eye range, of magnitudes 1–6 (five changes in magnitude), is arranged to be a span in brightness of exactly 100.

Not counting the Sun, there are . . .

- 20 stars of magnitude 1.0 (or brighter)
- 48 stars of magnitude 2.0 (or brighter)
- 171 stars of magnitude 3.0 (or brighter)
- 513 stars of magnitude 4.0 (or brighter)
- 1,602 stars of magnitude 5.0 (or brighter)
- 4,8 stars of magnitude 6.0 (or brighter)

Not everybody can see a star of sixth stellar magnitude. On the other hand, there are people with visual acuity such that they sometimes can make out thousands of additional, fainter stars than those listed above. In the next chapter, we investigate how not all of these stars are visible at the same time or from every place.

The stellar magnitude system remains useful because it is not an additive scale such as, for instance, the Celsius or Fahrenheit scale for temperature. Instead, it is multiplicative. Such a scale allows a single-place numeral to represent a quantity (star brightness) that varies over many orders of magnitude. But be careful: The idea that something decreases as its representative number increases is counterintuitive.

5.5 Stellar color

Differences in stellar color are real and have to do with the different temperatures of the stars. It is the brighter stars that more easily trigger our color vision. Plus, astronomers speak of color differences among the stars with an acknowledged exaggeration. The variation in color between the reddest of stars and the bluest is less than that between the different varieties of street lighting, all of which you may consider to be white! Star colors are not saturated – they can be thought of as containing a great deal of white mixed in with pure color. Moreover, they are point sources, which our brain tends to register as whiter than extended sources of the same color. In truth, only about 15 stars may be said to be "colorful" as seen by the naked eye.

Some prominent examples of stellar color: The bright stars named Gacrux [Gamma Crucis], Betelgeuse – another inversion, Betelgeuse is the second brightest star in the constellation Orion but designated Alpha Orionis – and Antares [Alpha Scorpii] normally are seen as red. Aldebaran [Alpha Tauri] is kind of orange. Most describe Vega [Alpha Lyrae] as white. Capella [Alpha Aurigae] and Procyon [Alpha Canis Minoris] usually are considered to be yellow. Rigel [Beta Orionis], Mimosa [Beta Crucis], and Bellatrix [Gamma Orionis] approach blue.

The reddest star likely visible to the naked eye is either Mu Cephei (the Garnet Star) or 119 Tauri (the Ruby Star). These stars are faint, but not so faint that you will be unable to perceive their color.

Alnitak [Zeta Orionis] is the bluest star that meets the same conditions. It is easy to find inasmuch as it is the easternmost (and southernmost) of the three Orion "belt stars."

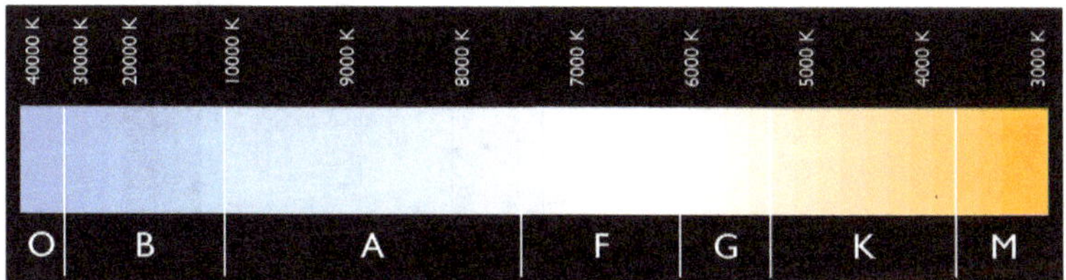

Figure 5.13: Star colors in order of surface temperature (degrees Kelvin). The letters refer to a spectral classification used by astronomers. Courtesy of Edgardo Garcia.

If you see a random, bright and colored star, it likely is red. This is because there are more particularly bright red stars in the sky than there are blue. This is not a coincidence. Astronomers have learned that red stars can be physically enormous. Because of their large surface area, they are correspondingly luminous. So, assuming stars are scattered at various distances from you (a crude but useful approximation), it is reasonable to assume that some of these big "high wattage" stars are near to you and appear quite bright. They are and do.

5.6 Unresolvable groups of stars

Besides individual stars strewn about the sky, there is another, distinct population of stars visible on a clear night. Nonetheless, you may not have recognized it as such. In reality, the Sun and its planets are located among a galaxy of stars that is disk-shaped. (See below.) You are located in the plane of that disk. So, in a particular set of directions, there are more stars than in any other. These are those pointing within the plane of our galaxy.

Imagine standing in a crowd of people roughly your own height. As you turn your head every which way, most of the time you see no faces. You see sky. You see shoes. But not faces. Yet in one plane, the plane 2 meters above the floor, you see *lots* of faces.

5.6.1 The Milky Way

Many of the stars in the galactic plane are extremely far away. Still, there are so many stars in this set of directions across the sky that their combined light blends together and is visible. From a dark vantage point, you see a seemingly continuous band of light produced by all these unresolved stars. It is named the Milky Way.

Figure 5.14: The Milky Way. Courtesy of Eberhard Grossgasteiger.

Notes from long ago
The Latin Via Lactea may be translated as a "road of milk" – literally, a Milky Way. The ancient Chinese called it the Yellow Road, and in John Milton's *Paradise Lost* (1667) it is the "broad and ample road, whose dust is gold." These latter names suggest a color that most of us do not detect with the naked eye.

The Milky Way is oriented at an angle to most nonstellar items of interest in your sky. (It is 60.2° with respect to the ecliptic; see, for example, Chapter 9.) Nonetheless, it always is brighter in the direction of the constellation Sagittarius. This is the direction toward the center of a huge lens-shaped system of astronomical objects that you sit in the midst of, the Milky Way Galaxy. The middle of the Galaxy bulges. That is why the Milky Way appears wider when looking in this direction, too.

There are several times more stars visible (per unit area of sky) in this direction than in the opposite direction (that of the constellation Auriga). You would see even *more* light from the direction of the galactic center; however, dark clouds of interstellar material block your view before you can see as far as the middle of the Milky Way Galaxy. (See Chapter 20.)

Notes from cultures around the world

It turns out that people who live in the southern hemisphere of the Earth get a better view of the Milky Way than do those who live in the northern hemisphere. Ethnoastronomer Duane Hamacher learned that the Torres Strait Islanders of the South Pacific Ocean use the Galactic bulge's position in the sky as an indicator of the changing seasons.

5.6.2 Example: Australia

Many aboriginal Australians make their home in the Great Outback, a place of breathtakingly dark skies. They notice dark patches *within* the Milky Way, which you now know to be caused by opaque, interstellar, dust clouds blocking our view of more distant Milky Way Galaxy stars. The most obvious of these apparent "gaps" in the Milky Way is the Great Rift, between the constellations Cygnus and the Scorpius-Sagittarius constellation border.

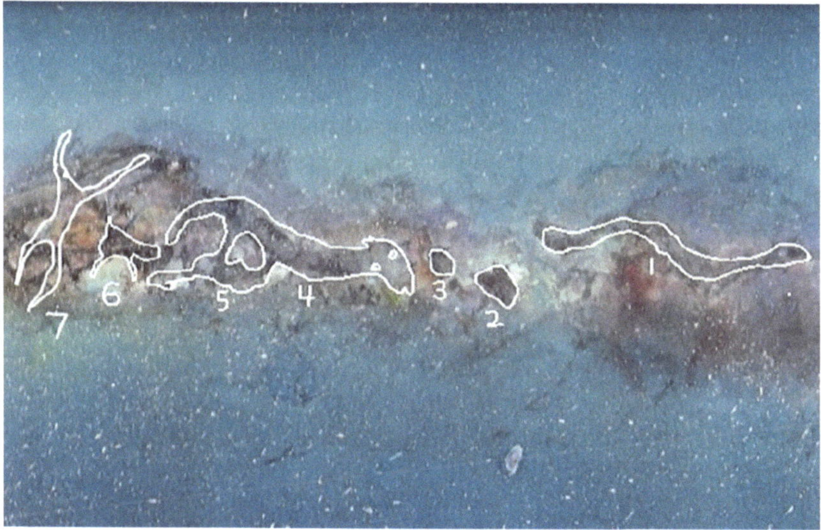

Figure 5.15: From South America, Incan dark constellations in the Milky Way. Courtesy of Steven Gullberg.

People all over the world have made up constellations from star patterns in the sky. The aborigines, among other indigenous people living at similar southern latitudes, invented constellations out of the patterns formed by these absences of stars. Examples include constellations that are said to represent the Sky Emu and Kangaroo.

5.7 Chapter summary

Following the age-old human habit of finding patterns, the sky is divided up into constellations of stars, into asterisms of stars, and even by the absence of stars. We experience a similar urge to name things. Many stars have proper names as well as catalogue designations. The most well-known of these is the Bayer System.

In addition to where they are in the sky, stars are described by their brightness and color. A useful means of doing the former is using the Bayer apparent-magnitude scale. Brightness, though, is not to be confused with luminosity, which cannot be determined by using only naked-eye observations.

Stars often come in groups, which may or may not be resolvable. The vastest example is the Milky Way. All other stars visible in the naked-eye sky belong to the Milky Way Galaxy.

5.8 Chapter review

1. What are some different ways humans have organized the sky?
2. What are constellations? What are two ways constellations have had cultural significance?
3. Why do constellations not have much physical significance?
4. Why cannot you see all of the constellations all yearlong?
5. What is the difference between an asterism and a constellation? What is an example of each?
6. Provide your own Earth-based example of the difference between brightness and luminosity.
7. What is stellar magnitude? What magnitude numbers are the brightest?
8. How can there be a dark constellation?
9. If you see a bright, colored star, why is it likely to be red?
10. Match the star on the left with its unique characteristic on the right.

Rigel Kent	brightest star in sky
Spica	closest (nighttime) star to the Earth
Mu Cephei, the "Garnet Star"	its brightness is the basis for the stellar magnitude system
Sirius	extremely red star

6 Navigating your sky

For a long time, people believed that the Earth stands still and that all sky motion is relative to the static Earth. This is incorrect. However, while just the opposite is true, it is also counterintuitive: For instance, if we are moving so fast, why does our atmosphere not form a continuous easterly gale? (Answer: The atmosphere is moving, too!)

6.1 Chapter learning outcomes

After reading this chapter, you will be able to:
- anticipate the apparent motion of celestial objects over the course of a night.
- assign a likelihood of seeing a particular celestial object (from your location).

6.2 Rotation of the Earth

You know – or, at least, have read – that the circular motion of the stars about a north-south axis is an illusion. It is caused by the fact that we stand on a rotating, spherical Earth. A star rises in the East because we are traveling up to seventeen thousand kilometers per hour toward the East. That is, the Earth spins counterclockwise as viewed from the direction we call north. Nevertheless, our brains are not well programed to handle relative motion.

6.2.1 What is moving?

Relative motion is a funny thing. Have you ever driven a car to a stop between two tall lorry?* When the light turns green, there is that odd moment when – just for an instant – it is unclear whether it is the view-blocking vehicles moving forward or you backward!

An exercise: If you turn counterclockwise in a complete circle while observing your horizon, you see familiar trees, hills, buildings, and wind-powered generators. Each enters the periphery of your vision, and then exits it, in turn.

But what if you were not intentionally moving your body? You could imagine seeing exactly the same thing if, however improbably, everything on the horizon began to turn around you (clockwise) in unison. (There used to be an attraction at Disneyland based on exactly this illusion.)

---

* truck

https://doi.org/10.1515/9783111441245-006

That it is the Earth that is rotating (clockwise as if seen from north) is readily demonstrable though. A **Foucault pendulum** is one hung such that its bob is at the same level as a circular set of pins surrounding its neutral position. The pendulum is set into motion. It swings in an invariant plane, while the Earth turns beneath it. The result is each adjacent pin is knocked over by the pendulum bob in turn.

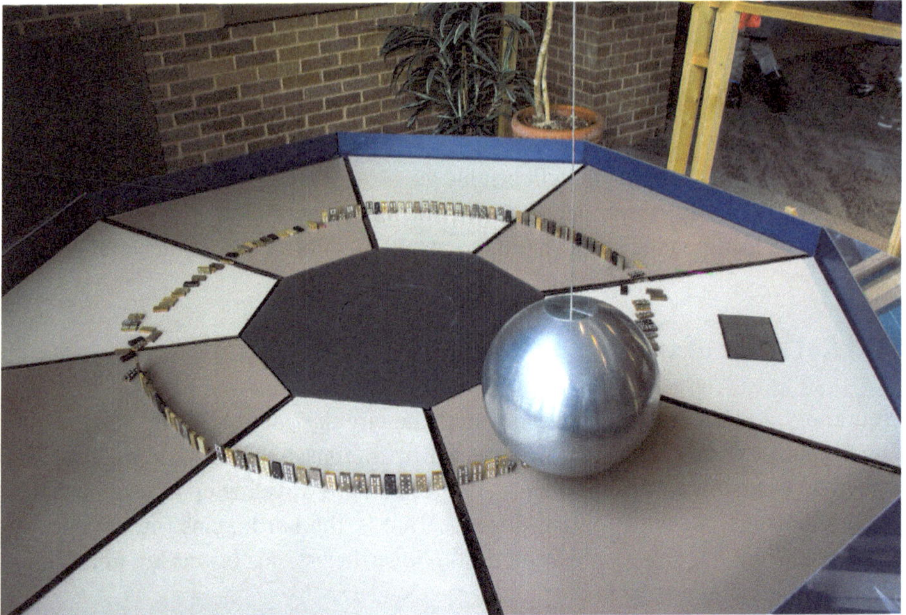

Figure 6.1: This pendulum is on display at the Cleveland Museum of Natural History (USA). All of the dominoes are knocked over in turn by the bob, as the dominoes rotate beneath *it*. Courtesy of Tim Evanson.

Then there is the **Coriolis Effect**. It is a pseudo force that seems to make objects traveling above the ground to change direction. It is caused by the fact that the speed of the rotating Earth beneath everything, from a moving air parcel to a projectile, changes with latitude. The apparent deflection creates spinning storms and forces artillery officers and snipers to take it into account when targeting.

6.2.2 Our language & relative motion

That the sky rotates is a pretty reasonable conclusion; it just happens to be wrong. Still, our language continues to acknowledge the misconception, since it was corrected, yet centuries later.

For example, in English, we routinely say words like "rising" and "setting" when referring to the sky. Such familiar vocabulary is used in this book. However, taken literally, such words continue to imply that the Earth is still and that the sky moves.

There is no such set of simple English words for the reality of the situation. Instead of "rose," try to insert a single word into the following sentence that does not imply a stationary Earth: "Just then, a bright star ____ above the horizon." There is none!

You are stuck with the language available. One way around this conundrum of relativity is to use a verb or adjective based upon "apparent." An example is, "Just then, a bright star appeared to rise above the horizon." However, such usage becomes cumbersome after a while.

This text inserts "apparent" occasionally, just as a reminder. Most of the time, though, the age-old relative words like "rise" and "set" are used without implicit literalism.

6.3 Rising & setting

How long does it take to notice that the sky changes due to the Earth's rotation? Consider a hypothetical star. For simplicity, it is rising due east and setting due west as it completes a circle centered upon you. The star will appear to travel 360° in 1,440 minutes. At most, a change in angular relationship (with respect to, say, the horizon) of double-digit degrees happens in only 45 minutes. This should be noticeable.

Forty-five minutes. That is not much time, but often, it is more than modern humans are willing to give the sky as they exit an enclosed workplace, in an enclosed car, to drive to an enclosed house.

If you are outside, it may well be daytime. This further limits the objects visible in the sky, and often attention is directed to something *down*: a walkway, lawn mower, golf ball, or mobile phone[1] held in hand. When you are indoors, you live in a virtual world without sky.

For these reasons, especially city dwellers, often do not know that the stars appear to move each night. Or, if willing to admit motion, that the manner in which the stars move is not haphazard. This is no fault of the urbanites: In the city, where the sky is defined as a narrow slit between high-rise buildings, it is even understandable.

The stars do look like they change place, and in an orderly way. Simplifying things significantly is the fact that the stars all appear to move together, in one unit. It is as if they were attached to a solid, moving background. They are not. Still, that is what it looks like.

The seemingly constant stars (in location with respect to each other, at least) form a sort of canvas on which any astronomical spectacle plays out. An axiom about astronomical objects is that, in reality, they all drift through space. However, it is the

---

1 cell phone

much closer planets, satellites, and minor bodies of the Solar System that change their relative positions in the sky on a timescale appreciable to us as they and the Earth orbit about the Sun. Thus, we are left principally with the stars. They appear fixed and permanently arranged in their constellations, serving as a ready reference for specifying the current address of other celestial objects.

6.4 One night

Consider a series of imaginary stars watched on a random night (although, they need not be imaginary). The first star that attracts interest appears at the horizon far to the southeast, increases in altitude as it follows a circular arc, and sets far to the southwest. Its altitude at star rise and star set is, by definition, zero. It reached its maximum altitude – it is said to **culminate** – halfway through the trip, when it is due south. Even this maximum altitude is not great, and the star spends comparatively little time at all above the horizon.

Figure 6.2: Looking south. Because this is a long-exposure photograph, the stars form trails of light as the Earth rotates. Courtesy of James Lowenthal.

Meanwhile, another star rises further north, though still in the southeast. This star also sets more northerly. It reaches a greater maximum altitude south, and it stays above the horizon longer, than did our first star.

What about a star that happens to rise due east? There is no physical reason why there might be such a star; the stars do not arrange themselves based on Earth's geography. Still, there are a lot of stars. That, one happens to rise due east and set due west is not farfetched. This star achieves a greater peak altitude (though still south) than either of the previous stars. It stays twelve hours above the horizon and twelve hours below.

A star rising in the northeast sets in the northwest. It may even pass overhead. Moreover, it spends greater than twelve hours in view.

Figure 6.3: Looking east. Because this is a long-exposure photograph, the stars form trails of light as the Earth rotates. Courtesy of Bob King.

6.5 Your horizon & zenith

Specialized vocabulary provides the convenience of compact sentences, which in turn can aid understanding. For example, instead of continuing to use the phrase "point directly overhead," the word **zenith** can be used. Your zenith is defined to be at an altitude of ninety degrees. It is the one place in the sky that is the same angular distance from every horizon. It is the only point in the sky with an altitude, but no azimuth. The horizon to zenith (90°) represents one-quarter of a whole circle, imaginarily drawn all the way around you (360°); twice the zenith's maximum value takes you from horizon to horizon (180°).

Figure 6.4: Looking west. Because this is a long-exposure photograph, the stars form trails of light as the Earth rotates. Courtesy of Eliot Herman.

Opposite to the zenith, 180° away, is your **nadir**. This is the point directly below you. As you might guess, this word has less utility in astronomy than "zenith," but not zero.

Figure 6.5: Perpendicular to the astronomical horizon, the direction above the figure is toward the sky observer's zenith; perpendicular to astronomical horizon, the direction below the figure is toward the sky observer's nadir. Artwork by Christine Tarte.

> **Where did that come from?**
>
> Zenith and nadir, along with almucantar, are gifts of the Arabic language. During the Middle Ages, every mosque would have had a muezzin. He was an astronomer who kept track of time for religious purposes. Christian monks similarly monitored the sky so as to accurately observe feast days and prayer times.
>
> There is an unbroken timeline of astronomy in the service of religion, beginning in pre-history and extending to the present.

It previously was noted that the horizon delimits exactly half of all you can see beyond the Earth, that it is a perfect circle. It consists of the set of points ninety degrees from your zenith.

Would not a mountaintop astronomer see more than 180°, from horizon to horizon? This certainly would be true if your pinnacle is a significant fraction of the radius of the Earth. In reality, the difference in elevation between the summit of Mount Everest and sea level is minuscule compared to the mean radius of the Earth. Therefore, the trigonometry of your sky is affected negligibly by mountain climbing.

Nor does a peak bring one closer to the stars. Stars are so far away that this difference in proximity is *extremely* close to zero. Anywhere on the Earth, the stars are effectively the exact same distance from you. Elevation may be advantageous to the naked-eye sky watcher, but not for this reason. It is because the length of 1 air mass is reduced.

6.6 Your celestial meridian

Back to one night under the sky. Stars always reach their highest point above the horizon –that is, greatest altitude – halfway through their nightly circular arc across your sky dome. This happens when they are on an imaginary vertical circle that intersects the horizon exactly north-south. It is called the **celestial meridian**.[2] The celestial meridian divides your sky into an eastern and western half. Stars increase in altitude east of the celestial meridian; they decrease in altitude west of the celestial meridian.

> **Notes from long ago**
>
> We go back in time to the Renaissance. The celestial meridian, as it would project onto the ground, was sometimes physically recreated. It was laid upon, or set into, church floors. American historian John Heilbron visited these meridian churches all over Europe.
>
> A few examples that he found are: the Basilica of San Petronio in Bologna, Basilica of Santa Maria del Fiore, and Duomo di Milano (each a cathedral). The celestial meridian appears in the Torre dei Venti (Tower of the Winds or Gregorian Tower) within the Vatican itself. Some architectural meridians even have named stars symbolically inserted along the meridian line, at their relative positions of culmination, north-south.

---

2 A generic meridian is a portion of a vertical circle with a positive altitude.

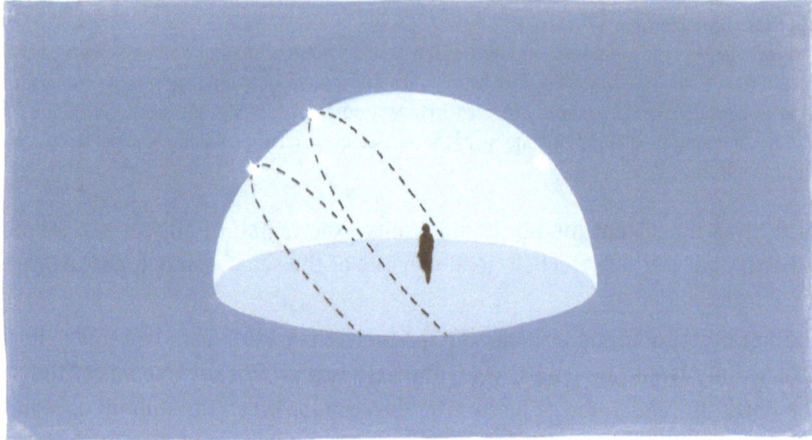

Figure 6.6: Stars culminate on the celestial meridian. Artwork by Christine Tarte.

The zenith helps better define the celestial meridian. Any number of circles around you could pass through north and south. Think of the segments of an orange. Only one passes through the zenith as well. This is the celestial meridian. (If you were to extend it below the horizon, it would eventually pass through the nadir, too.)

A star spends half of its time in your sky east of the celestial meridian and half its time in your sky west of the celestial meridian. As a star crosses the celestial meridian, it culminates. A culminating star is either due south or due north.

6.7 Example: Egypt

Do people pay attention to culminations in the sky?

The Egyptian pyramids hold a special fascination for people. At one time or another, there have been those who tried to see in them vast stores of lost knowledge, or all the known laws of physics. There is a tiny piece of this modern folklore that rings true.

6.7.1 The Great Pyramid of Khufu

The Great Pyramid of Khufu[3] (circa 2600 BC/BCE) is the only one of the Seven Great Wonders of the Ancient World still in existence. Not only is it the earliest of the Old Kingdom pyramids at Giza, but it is also the biggest.

The Great Pyramid is an early example of cardinal alignment: Its sides are exactly north-south and east-west. Here, "exactly" means within one-fifteenth of a degree!

---

3 or Cheops

Figure 6.7: This vintage 1867 photograph showing the Pyramid of Khufu was made by the famous French company, Maison Bonfils. It likely is staged with just a few individuals, in order to demonstrate scale.

What about the pyramid's interior? There are multiple rooms inside, but the most significant is the King's Chamber, which is thought to have been the pharaoh's tomb.

From the King's Chamber, two narrow shafts slant upward to the pyramid exterior. One is at a 31° angle to the horizontal; the other is at a 44.5° angle. Because the King's Chamber is not directly under the apex of the pyramid, the two shafts end up emerging from the pyramid at about the same height.

6.7.2 Egyptian connection with the sky

The mysterious shafts point into the sky. Were they passageways to the heavens? If so, where was the pharaoh heading?

Much later, the New Kingdom Egyptians marked time through the night by certain stars, or groups of stars. These culminated in succession. (See Chapter 8.) The description of one such group is unmistakable. It is a wide-shouldered, standing man. Today, it is easily recognizable as the constellation of the hunter, Orion: three stars (his belt) in the middle of a rectangle comprised of bright stars (his body). The three belt stars of similar brightness make up an asterism in their own right.

A central tenet of Egyptian religion in later days, which we know more about, was the idea that the soul of the deceased monarch would rise to the sky to meet up

Figure 6.8: A simplified schematic of the Great Pyramid's interior. Artwork by Christine Tarte.

Figure 6.9: Orion's shoulders are the stars Betelgeuse (left) and Bellatrix (right). His feet are the stars Saiph (left) and (Rigel (right). Courtesy of NASA.

with the god Osiris for judgment. This idea probably was important in Khufu's time, too. There are even references to the positions of the stars that make up this god, which include the three belt stars. Osiris was Orion.

6.7.3 A solution to the "Mystery" of Khufu

American Astronomer Virginia Trimble and Cairo-born Egyptologist Alexandre Badawy calculated where the belt stars were during pyramid-building times. At the latitude of Giza, their altitude, as they culminated, was 44.5° – the angle of the south shaft. The south pyramid shaft pointed, at the time of construction, toward the belt stars as they crossed the celestial meridian.

Maybe the Old Kingdom Egyptians meant some other stars, stars that culminate over the shaft at another time (those east or west of Orion)? Looking around in all directions, there simply is no other logical candidate. There is nothing else, nearly as bright and distinctive, which the tunnel might have been aimed at, during the time of its construction. The belt stars probably were the target.

6.8 Circumpolar stars

How familiar you are with the stars and constellations is likely a function of how often you see them in the sky. For most, some stars and constellations are more likely to be above the horizon at any random time, than others.

Our one-night tour of interestingly placed stars continues. A yet-more-northward-rising star will arc across your northern sky, not the southern.

The star under examination has crossed the **prime vertical.** It is a vertical circle, from due east to due west on the horizon, through the zenith. It is perpendicular to the horizon, at the horizon. If you were to continue the prime vertical below the horizon, it would pass through the nadir. The prime vertical is perpendicular to the celestial meridian too. It divides your sky into north and south halves. Stars that spend most of their time north of the prime vertical are probably those with which you are most acquainted.

Close to the point about which the stars seem to turn, a given star may never set at all! Rather, it rotates 360° above the horizon. It makes a complete **diurnal**[4] circle, while all the time in your sky.

Stars like this never dip below your horizon. For them, rising and setting is meaningless.

---

4 Latin for "happening once per day"

Such a star and sets of such stars (*i.e.,* constellations) are easy to keep track of and may become quite familiar. This is because, if it is night, you are guaranteed to see them. Any celestial object that behaves this way is said to be **circumpolar** from your location.

A circumpolar star culminates *twice.* It does so when it crosses the celestial meridian (counterclockwise, east to west) at its greatest altitude; this is upper culmination. It does so again when it crosses the celestial meridian (clockwise; west to east) at its lowest altitude; this is lower culmination.[5]

Figure 6.10: Looking north. Because this is a long-exposure photograph, the stars form trails of light as the Earth rotates. Courtesy of Martin Snow.

From the vantage point on the Earth that we have been using, there are six circumpolar constellations. These are the well-known Ursa Major and Ursa Minor; Draco, Perseus, Cassiopeia, and Cepheus; and the generally fainter Camelopardalis, Lynx, and Lacerta.

The most famous circumpolar star is the **North Star**,[6] which may travel in such a small diurnal circle that we are unlikely to notice its motion at all. It appears to be the one celestial object that is stationary in our sky.

---

5 For non-circumpolar stars, the adjective is dropped. Culmination assumes upper culmination. Lower culmination for these stars occurs below the horizon and is of little interest.
6 or Pole Star

Figure 6.11: A fanciful map of circumpolar constellations. It was once common to include imagined figures superimposed over the stars that were supposed to represent them. Courtesy of the Wellcome Trust.

Misconception alert

The current North Star is named Polaris. (See Chapter 8.) Due to its importance in navigation, there is an assumption that it must be a particularly bright star. It is not. Polaris's rank in brightness, among all the stars on the Celestial Sphere, is number 48. It is so well known because of its location, not its brilliance.

Logically, there must have been stars, when we focused our attention upon the South, which never *rose* above the horizon. These are **anti-circumpolar stars**.

6.9 Example: Egypt (Continued)

Recall the Great Pyramid of Khufu?. After the pharaoh met up with Osiris in the sky, presuming he was judged worthy, where did he head next?

The later Egyptians spoke of stars called the "Imperishable Ones." What kind of stars is "imperishable"? Circumpolar stars! They do not set/"die." Now, which ones?

Recall that the south-facing pyramid shaft points toward the stars of Orion's belt. But there is still a north-facing shaft that Trimble and Badawy examined. At Giza, the north-facing pyramid shaft pointed to the North Star – the ultimate circumpolar star. Khufu was on his way to join the "Imperishable Stars."

Tools of the naked-eye observer

The asterism of the Pointer Stars (the bright, outer-bowl stars of the Big Dipper, itself an asterism and, as you will remember, part of the constellation Ursa Major) happens to define a northward ray that points in the direction of the star Polaris [Alpha Ursa Minoris], part of the Little Dipper (an asterism found within the constellation Ursa Minor). As the segment between the two stars turns through 360°, it can be used to tell the time of night, like a clock hand. Any such "star clock," or representation of one, may be called a *nocturnal*.

Was the Great Pyramid an observatory? No. The shaft angles presented are averages; the passages are not perfectly straight. In fact, there is no clear line of sight through them at all. They are symbolic. Nevertheless, the incredible effort necessary to construct something as huge and long lasting as the pyramids nearly five millennia ago demonstrates the incredible power of the sky in the Egyptian religion and picture of the afterlife.

Notes from cultures around the world

Constellations or asterisms of stars often are easier to identify than single stars. This is especially true if the pattern is prominent, memorable, and made up of bright stars. In the case of a constellation or asterism that is circumpolar and lies far to the North, it may be just as useful an indicator of approximate north as the traditional way finder, the North Star.

A dramatic example of the Big Dipper's practical use in this regard might be found encoded in the folksong, *Follow the Drinkin' Gourd*, said to have been sung by enslaved people in the antebellum United States. Legend has it that the song is a secret set of instructions for escaping slaves, pointing them to freedom in the northern part of the country.

An explicit reference to the North Star, or even the Big Dipper, would be too obvious to those from whom the encoded directions were to be kept from. Indeed, it was best to maintain the fiction, in the presence of those who supported and maintained slavery, that African Americans did not know practical astronomy.

Thus, the "drinking gourd" was inserted into the words of the song as a substitute for the appropriate celestial marker. By following these stars northward, and other directions hidden in the lyrics, an escaped slave could connect up with the famous, so-called Underground Railroad – and freedom.

6.10 Chapter summary

Contrary to what is implied by our language, the Earth rotates. It is possible to demonstrate the rotation of the Earth. Our planet's diurnal motion results in stars rising, culminating, and setting. They do so in different paths, determined by how far away they are from the point about which they seem to turn: the projection of the axis mundi into the sky.

Very near the pivot point of your sky, there happens to be a star called the North Star. It is so named because, by definition, it is very near due north. It is also significant because it is the one star in the sky that changes location imperceptibly.

Stars that cross your celestial meridian twice per rotation of the Earth are circumpolar stars. Those that never appear on your celestial meridian are anti-circumpolar. Circumpolar stars had special significance to civilizations, both in the distant past and more recently.

6.11 Chapter review

1. Why does it appear that your sky is moving, even though it is the Earth that is moving?
2. What is the issue with the phrasing "rising" and "setting" when referring to your sky? Why do we still use this language if it is incorrect?
3. How do stars appear to move throughout the night?
4. What is your zenith? What is your nadir? How would your zenith and nadir be different if you were standing on the north pole?
5. What is the celestial meridian? What are some examples of how cultures have used the celestial meridian?
6. How can the zenith help you better define the celestial meridian?
7. What are circumpolar stars? Why are some stars circumpolar and others not?
8. What is the significance of the North Star? How is the asterism of the Pointer Stars (in the "Big Dipper") within the constellation of Ursa Major?
9. Describe an example of how past cultures have used the positions of stars in the sky.
10. Match the current behavior of the star on the left with the future behavior of that star the right.

rising south of east	will not rise and set; will spend 24 hours above your horizon

rising due east	may pass through your zenith; will spend more than twelve hours above your horizon
rising north of east	will culminate at a low altitude; will spends less than twelve hours above your horizon
circumpolar	will spend twelve hours above your horizon

7 The celestial sphere

Your sky consists of everything that you can see above the horizon. But what about below the horizon? Stars and constellations exist in those directions, too.

7.1 Chapter learning outcomes

After reading this chapter, you will be able to:
- label a globe of the Celestial Sphere.
- notice changes in your sky as you travel on the Earth.
- visualize changes in your sky over the interval of the solar year.

7.2 Two ways to describe things celestial

So far, your new vocabulary does a reasonable job of describing what takes place every night in your sky. You can call this set of terms and measures the horizon coordinate system.[1] It is made up of words such as "horizon," "zenith," "altitude," "azimuth," and "meridian" and based upon an imaginary dome of stars moving overhead.

7.2.1 Your personal reference system

Such a reference frame is *personal.* It is centered on the observer – you. Your horizon is slightly different than everybody else's horizon, if only for the fact that two people cannot stand on exactly the same place at the same time. Your zenith is slightly different from that of the person just beside you.

You carry your personal sky dome around with you wherever you go. As you walk about, your zenith moves with you, forever hovering above your head. You walk toward your horizon. But, like the proverbial "end of the rainbow," you never reach it. Your horizon forever stretches out ahead of you.

The dome of your sky is similar to an umbrella, as you carry it deployed on a rainy afternoon. Its rim "horizon" always surrounds you, while its point "zenith" remains fixed overhead.

As useful as the sky dome concept is, it is not universal. It can become confusing when two people at very different locations try to talk about the sky, because the specific references each uses are not the same.

---

[1] more technically, an alti-azimuth coordinate system

https://doi.org/10.1515/9783111441245-007

An absolute way of talking about the sky is called for, too. This would be one that everybody can imagine in exactly the same way.

7.2.2 A universal reference system

In addition to the horizon coordinate system, there is another, parallel vocabulary set for describing what is seen as taking place in all the skies of the world. It is the **Celestial Sphere**.

The Celestial Sphere takes into account something so far ignored: the fact that the stars continue to exist when they take their leave from your personal sky dome! They are still present, merely below your horizon.

Figure 7.1: Celestial globe: a physical model of the Celestial Sphere. Photograph by Aaron Spurr.

For the stars' apparent motion to be consistent, you must imagine a continuation of the hemisphere of sky that you see, namely a second hemisphere below the horizon. There is – for want of a better term – an "anti-sky" that you cannot see.

It is logical to imagine a single, complete sphere of stars entirely surrounding you. Such a sphere includes every possible direction, including those that you can never see due to the presence of the ground.

Like our sky dome, the Celestial Sphere is admittedly not a real thing. There is no big, finite shell of stars encapsulating the Earth. Instead, space effectively extends infinitely in all directions.

Were you to transport yourself away from the Earth far enough, you would be able to see the entire Celestial Sphere, with the Earth occupying but one tiny dot on it. However, our discussion of naked-eye astronomy does not include space travel!

Still, as we look up into space while situated on the Earth, it *looks* as if there is (or ought to be) such an encasement of stars.

Another thing: You can leverage the Celestial Sphere concept into a convenient tool for better describing the overall apparent motions seen in your sky. While there are other naked-eye objects that behave as if they are attached to a Celestial Sphere, for now, consider this book's use of phrases such as "stars and constellations" to cover everything that the fictional Celestial Sphere is well-suited to display.

Like the horizon-zenith reference, the Celestial Sphere is "you-centric." Only this time, "you" is everybody on the Earth. As the Celestial Sphere is really infinite in extent, the Earth is really infinitely small in comparison. It is as if the whole Earth were reduced to a mere point at the center of the sphere. This is why you see exactly 50% of the Celestial Sphere all the time, even though you are not at the center of the Earth.

Because you are geometrically situated in the middle, the biggest circle that can be drawn on the Celestial Sphere is one centered upon you. It is a great circle. Your horizon or the full circumference of any circle drawn around you are examples of great circles.

Where did that come from?

At one time, the Celestial Sphere was thought to be real. It was the invention of the classical Greek Eudoxus. The Sphere was popularized among Hellenistic Greeks and later throughout the Roman world, as well as that of the post-Romans during the Middle Ages. It was done so by none other than Aristotle (via his writings).

"Cracks" in the theory of a physical Celestial Sphere only became unignorable in the sixteenth-century. Its demolition was the work of the last, great, naked-eye astronomer, the Danish nobleman Tycho Brahe. Brahe plotted the location of stars he observed on a celestial globe of his own making. He demonstrated that there must be objects (*e.g.*, comets) that pass through the solid "sphere." (See Chapter 19.)

7.3 Celestial equator

On the Celestial Sphere, two Celestial Hemispheres, the North Celestial Hemisphere and the South Celestial Hemisphere, are divided by the Celestial **Equator**. The points on the Celestial Sphere farthest from the Celestial Equator are the North Celestial Pole and the South Celestial Pole. (See below.)

In your sky, the Celestial Equator is a great circle about you, just like other great circles such as the horizon or celestial meridian. You can easily see how any great circle must intersect any other, when both are on the Celestial Sphere, at two points.

Figure 7.2: North Celestial Hemisphere is blue; the South Celestial Hemisphere is pink.
Photograph by Aaron Spurr. Edited by the author.

In the Earth's northern hemisphere, the arc of the Celestial Equator above your horizon is an east-west semicircle in your southern sky. In the southern hemisphere, it is in your northern sky. Of course, it is worthwhile reminding that, just as is true for any feature of the Celestial Sphere, the Celestial Equator is not a physical construct.

7.4 Celestial poles

Every day, the Celestial Sphere appears to rotate once completely. The direction of progress is such that most stars appear to rise in the east and set in the west, just like the Sun. Circumpolar and anti-circumpolar stars are the exceptions to this rule.

Any rotating sphere must do so about, not one but, two points. These points define the Celestial **Poles**. The Celestial Poles are opposite each other on the Celestial Sphere (meaning 180° apart on any great circle that includes them). The Celestial Equator is between them, as far as possible away from either. You could say that it is the Celestial Poles that define the Celestial Equator.

While the Celestial Equator is always in your sky, from a given location, only one of the two Celestial Poles is at an altitude above your horizon. In the northern hemisphere, this is the North Celestial Pole. (The Pole Star is so named because of its proximity to this Celestial Pole.) In the southern hemisphere, it is the South Celestial Pole.

7.5 Where you live affects your sky

When referring to the opposite Earth hemisphere, the proper adjective is antipodean. Most of us are **antipodean** to people living in the southern hemisphere.

Notes from yesterday and today

Until about 400 years ago, few northern-hemisphere people had any conception of what either the celestial or terrestrial view would be like in the southern hemisphere. And even then it conjured mental pictures of strange-looking persons somehow walking head-over-heels or even with faces below their feet!

Thus, in regard to location, the word "antipodean" came to represent a theoretical extreme. In *Much Ado About Nothing*, William Shakespeare has Benedick so intent on avoiding Beatrice that the character says,

Will your Grace command me any service to the world's end? I will go on the slightest errand now to the Antipodes that you can devise to send me on . . . (Act II, Scene 1)

7.5.1 North celestial hemisphere & south celestial hemisphere

In the northern hemisphere, most (in one special location, all) of your sky is, at one time or another, made up of all the stars in the North Celestial Hemisphere and some (in one special location, none) of the South Celestial Hemisphere. The reverse is true if you are antipodean. Of course, for those who live in the southern hemisphere, you are antipodean; most of their sky similarly is made up of the stars belonging to the South Celestial Hemisphere.

Those familiar with maps will recognize that the geography of the Celestial Sphere simply borrows Earthly geographical terms and adds the adjective "celestial." This is

Figure 7.3: Celestial Equator emphasized. The North Celestial Pole is represented by the top of the celestial globe, and the South Celestial Pole is represented by the bottom of the celestial globe. Photograph by Aaron Spurr.

appropriate, as the Celestial Poles are just the extension of the axis mundi reaching infinitely into the sky. The North or South Celestial Pole is not a place. It is a direction.

Imagine a giant search light erected on the Earth's North (or South) Pole and aimed upward. Its beam travels infinitely, marking the way toward the North (or South) Celestial Pole. It never "reaches" it, because there is nothing to reach. Similarly, the Celestial Equator is the projection of the Earth's equator into the sky.

7.5.2 Rising & setting azimuth

Thinking of the sky as the intersection of the horizon with a Celestial Sphere makes clear why where on the horizon a star rises and sets is dependent upon how close or far it is from the Celestial Poles. Only a star equidistant from the Celestial Poles (*i.e.*, on the Celestial Equator) can ever rise due east and set due west. The rest of the stars, except circumpolar stars, rise either northeast or southeast and set either northwest or southwest.

Recall that all but the very brightest stars are subject to **extinction** from naked-eye view before reaching the horizon. Most of the time, as a star changes altitude, it also changes azimuth. Thus, the azimuth at which a star disappears is not exactly the same as the azimuth at which it technically sets. The dimmer the star, the longer is the interval between extinction and setting, and the greater the difference in azimuth between the two.

This azimuthal effect is more appreciable the closer you are to the Earth's poles and the further away you are from the equator. Notice that, in the northern hemisphere, extinction produces a systematic error in apparent setting direction toward the South and vice versa. Of course, the exact extinction-caused rising or setting azimuth depends upon local conditions of the atmosphere night to night.

7.6 Tour of skies at different latitudes

The purpose of looking at selected imaginary – or not so imaginary – stars in the last chapter was this: to examine how stars rising at specific azimuths on the sky dome (or, if you now prefer, the visible part of the Celestial Sphere) will subsequently behave through the course of a night, meaning until daybreak. This was all pictured from a specific spot on the Earth. Now you will explore what happens if you change your location. How do other skies look?

7.6.1 Traveling northward

You start your tour, expectedly, in the mid-latitudes of the northern hemisphere. You "walk" quite fast. Your entire itinerary takes place over the course of one night!

First you perambulate northward. You see the altitudes of the stars in your northern sky increase, as each culminates (or upper culminates). These North Celestial Hemisphere stars, not counting circumpolar stars, spend more and more time above your horizon. Many that rise and set at your starting place become additional circumpolar stars.

As you head further north, the North Star, the one star that you trusted to maintain its position in your home sky, now appears to be moving! It is increasing in altitude. Ninety degrees away, behind you as you face your direction of travel, the Celestial Equator is decreasing in altitude.

Fewer and fewer of the stars in your sky belong to the South Celestial Hemisphere. Those that you still glimpse (have not become anti-circumpolar) reach a lower and lower altitude, even as they culminate. They spend less and less time above your horizon.

You continue northward. Eventually you reach a place where the North Celestial Pole is at your zenith (altitude = 90°). The Celestial Equator lies along your horizon (altitude = 0°). All the stars that you see are stars of the North Celestial Hemisphere. They are all circumpolar.

Here the stars do not change in altitude during the night; instead they turn about you as if you were at the center of a carousel. You are standing on the Earth's North Pole.

Figure 7.4: Star trails as seen looking toward the zenith (upper right) at one of the Earth's poles. Courtesy of Ludovic Lubeigt.

Why is nothing rising nor setting anymore? It is because, once you pause in your hike at the pole, *you* are no longer moving. You are at the pivot point. The rotation of the Earth spins you, but no longer physically transports you west to east.

7.6.2 Traveling southward

Returning to your journey's start, you now explore what happens if you head due south. Everything takes place in reverse. The Celestial Equator in front of you increases in altitude. The North Celestial Pole behind you slips horizonward. Unfamiliar stars and constellations appear in the South: These are the forever anti-circumpolar at your home latitude. You get a better view of the South Celestial Hemisphere. Behind you, fewer stars are circumpolar. Eventually you reach latitude zero.

Standing on the equator, you have the advantage of seeing the entire Celestial Sphere as it revolves over you. The South Celestial Hemisphere takes up the southern half of your sky, the North Celestial Hemisphere takes up the northern half. The Celestial Equator runs through your zenith; the Celestial Poles are on your horizon. Nothing is circumpolar. You find yourself in a giant drum, rolling east to west.

The view from the North Pole looked very foreign to you. Here, too, the behavior of the sky is different from what you came to expect each night back home in the northern temperate latitudes.

If you linger, and were it not for the periodic rising and setting of the Sun, you would see all the stars on the Celestial Sphere. Of course, you would not be able to do so all at once.

Traveling even farther south results in a mirror image of what you saw as you headed north. The North Celestial Pole is below your horizon from now on. Other stars in the far celestial north begin to disappear (turn anti-circumpolar). Southern stars become still easier to see. Some become circumpolar, turning about the *South Celestial Pole.*

At this point in your trip, the sky likely looks peculiar to your "northern eyes." The Celestial Equator is in your northern sky, and most of your sky dome is taken up by the South Celestial Hemisphere. Even familiar star patterns, those that you still can see, probably look odd to you, compared to how you remember them at home, once you are south of the equator: They are "upside down."

7.7 Terrestrial/nautical navigation

On your imaginary journey, when you were at the North Pole, the altitude of the North Celestial Pole was 90°. When you were at the equator, the altitude of the North Celestial Pole was 0°. It also is true that at the South Pole, the altitude of the South Celestial Pole is 90°. The altitude of a Celestial Pole always equals the absolute value of your latitude!

Figure 7.5: Star trails as seen looking toward the horizon at the Earth's equator. Courtesy of ESO.

For instance, if you live in Nepal, your latitude is 30° N; the North Celestial Pole is at an altitude of 30° above your northern horizon. For those who live in Auckland, New Zealand, their latitude is almost 40° S. There, the altitude of the South Celestial Pole is almost 40° above the south horizon.

7.7.1 Finding your latitude

This spherical geometry is very useful for navigation. Measuring the altitude of the Celestial Pole (or, perhaps its nearby surrogate the North Star) might be difficult on a rocking ship's deck. Nonetheless, the principle is, at least, straightforward. Of course, this particular technique works only at night.

Misconception alert
The idea that stars thought of as invisible during the daytime can be made visible by looking from the bottom of a vertical tunnel or through a long, empty tube is a myth. In fact, doing so interferes with your contrast vision, making the attempt to discern such stars even harder.

You must know what hemisphere you are in to navigate in this way, but even a cursory familiarity with stars and constellations will inform you of that. Otherwise, the direction of circumpolar stars, counterclockwise in the North or clockwise in the South, provides you with the necessary information. After that determination, a measurement of a single angle will tell you where you are, north and south. No arithmetic is involved.

Tools of the naked-eye observer 
Until the end of the seventeenth century, a navigator might have used a *quadrant* to measure the altitude of objects in the sky. It consisted of a single plate in the shape of a quarter-circle disk marked off with increments of 90° along its partial circumference. An attached plumb bob pointing toward the nadir (under gravity) helped orient the quadrant. Altitude was read from the plate where the plumb line crossed it.

The quadrant evolved into the *sextant*. Its more convenient, smaller arc accomplished the same thing as the quadrant did by using mirrors.

7.7.2 Finding your longitude

Unfortunately, determining your longitude is more difficult. More than one ship has foundered because it sailed too far east or west and ended up on the rocks. The westward progression of stars due to diurnal motion introduces a time element to longitude determination by the skies.

To calculate longitude, you must be able to measure the time difference between local time and time at a location of known longitude. One attempted technique was to use the changing location of the Moon on the Celestial Sphere as a time indicator. Its position on the Celestial Sphere compared to what shipboard tables said that its location ought to be, at (say) a home port, would yield the longitude difference between ship and shore.

In order to be able to describe the Moon's location on the Celestial Sphere, star locations on the Sphere were necessary to use as a reference. A moment's thought will satisfy you that a featureless, "empty" Celestial Sphere is useless for this technique, inasmuch as any place on such a Sphere looks like any other place.

Navigation was a major motivation for carefully mapping star locations. Unfortunately, the longitude-determination technique described above proved to be easier described than practically accomplished.

A better solution turned out to be a technological one: A gimbaled clock was invented that was not affected by a bobbing or rolling superstructure. Set to home time, the difference between clock time and local time yielded longitude.

7.8 Revolution of the Earth

Be careful of the two words, "rotation" and "revolution." In common speech, often they are used interchangeably. Yet they mean different things.

Rotation refers to something spinning on an axis. The turning body may stay in one place.

Revolution refers to traveling around another body, in a closed pathway. The path is a circle or some other continuous curve. In the case of a planet and the Sun, that path is elliptical and called an orbit. The Earth both rotates and revolves in the same angular direction, but otherwise the two motions are unrelated.

Figure 7.6: The people in the revolving door are traveling on a circular path from one place to another. They are an example of revolution. The dancer is spinning rapidly but staying in one place. He is an example of rotation. Courtesy of Deuallangpoa [left image] and Hilmi Işılak [right image]. Edited by the author.

Because the Earth revolves about the Sun, from your point of view the Sun appears to *move* across the sky with respect to the stars. It does so slowly, a little less than 1° per day, out of 360° all the way around a great-circular route. The result is that the Sun reaches the same point on the Celestial Sphere once in approximately 366.256 days (specifically, **sidereal**[2] days), an interval of time, the sidereal year.

Once again, relative motion yields an incorrect impression. And once again we resort to a useful fiction: the Sun's motion about the Celestial Sphere.

Did you notice the appearance of the number 366? You may have expected the more familiar 365. The days in a 365.256-day year are *solar* days. (See below.)

Superimposed on the apparent motion of the whole Celestial Sphere turning once about you every sidereal day, is a diurnal change due to the fact that the Sun's rise and set locations on the Celestial Sphere slowly shift. The result is this: As an Earth inhabitant orbiting the Sun along with your planet, different parts of the Celestial Sphere are blocked from your view by the Sun, when it is above your horizon. At the same time, different parts of the Celestial Sphere are exposed when the Sun is below your horizon. That extra "day" gives, to your nighttime tour of the sky, a view of the entire Celestial Sphere visible at your latitude over the course of (by one definition) a year.

---

2 from the Latin, "of the stars"

7.9 Combined effect of the Earth's rotation & revolution

If you go outside at the same time every night, thereby taking the Earth's rotation "out of the equation," you see each star set 3 minutes, 56 seconds earlier than it did the night before. Similarly, another star will rise approximately four minutes earlier each night.

This is because clocks are based on the *Sun*. The commonly used 24-hour day[3] is not the sidereal day. It is the solar day. This is the average length time between the *Sun's* culminations.

Four minutes may not sound like much. Indeed, tonight's 11:00 PM sky looks a lot like last night's 11:00 PM sky. So will tomorrow's. But 3 minutes, 56 seconds summed every day eventually adds up to an angular change of 360°. Over ~365 ¼ days, it is as if there is an extra rotation of the Celestial Sphere in addition to the diurnal ones you expect. This is how you get from the sidereal year to the canonical solar year.

The effect would not happen if our clocks were set to coincide with the apparent motion of the stars only. In that case, you could choose any bright star. Capella, let us say. Define the day (as always, the rotation period of the Earth) as the time from Capella's culmination to when it does so again – in other words, the sidereal day. What you would notice, though, is that the Sun, as it appears to move westward on the Celestial Sphere each day, is then *not* at the same place in the sky at the same time. (Perhaps this is the celestial meridian, at noon.)

The Sun is much more important to affairs day in and day out than are the stars. You wish the Sun to set in the evening, rise in the morning, and culminate near noon. You want it to do this every single day.

This expectation fails if your habit is to measure time with respect to the stars. After a while, the Sun rises at midnight and sets at noon. This is confusing, to say the least.

So you measure your clock day with respect to the moving target that is the Sun. This guarantees that the Sun is at its highest altitude come midday, and nowhere to be found in your sky at midnight – every solar day of the solar year. There are a few places on the Earth where this common-sense-sounding rule does not apply; we will discuss them later.

7.10 Night skies of spring, summer, autumn, & winter

Because the Sun is the basis of your commonly used clock time, it is the Celestial Sphere that seems to be placed differently in your sky each solar day and not the Sun. (What immediately follows does not involve circumpolar stars and constellations.)

---

3 Any 24-hour period is called a nychthemeron! English allows us to use numerals in words, and "24-hours" is easier to say than this alphabet-only tongue twister. This book will not use "nychthemeron."

Imagine that you are looking at the night sky at 9:00 PM, an arbitrary but convenient time. It is February. You see rotating above your horizon a familiar set of stars and constellations, which in the North is called the wintertime evening sky.

However, by May, the Earth has traveled one quarter of its way around the Sun. There are stars and constellations that you used to see on February evenings that have been intruded upon by the Sun. They are lost in daylight.

On the other hand, there are stars and constellations that you could not see in February because of the Sun, which are now visible in May's 9:00 PM sky. You are presented with the stars and constellations rising, setting, and culminating in a springtime evening sky.

By August, summertime evening sky is the portion of the Celestial Sphere (visible at your latitude) that was in the same direction as the Sun last February. It was then in the daytime sky. It is your current nighttime sky. In other words, at 9:00 PM in July, you see the half Celestial Sphere (visible at your latitude) that is opposite of the one you saw at 9:00 PM in January. Daytime and nighttime skies are flipped.

The Earth moves another quarter of the way around the Sun. By November, more stars have been "lost" and "found" to the 9:00 PM sky. The stars and constellations above your horizon are those that you associate with the autumn evening sky.

With patience, over the course of the solar year, you see in your nighttime sky all the stars and constellations that ever appear above your horizon. While the words associated with seasons may change for the southern-hemispheric observer, the result is the same. (See Chapter 9.)

Certainly, assigning a star or constellation to your winter, spring, summer, or fall evening sky does not mean that you can see it all night during those seasons. It might rise while it is still daylight and once night arrives already be at appreciable altitude. Such a star is said to be **dock-pathed**. (A star that you can see from rise to set is said to be **night-pathed**.)

If you work first shift, you might be more interested in what the sky looks like in the early morning. We could just as well talk about the winter, spring, summer, and fall morning skies. Morning skies are out of sync with evening skies. The sky always looks different early at night than it does late at night, because of diurnal motion. But morning skies, too, are the same each time of the solar year.

So it does not matter. If you observe the, say, 4:00 AM sky all solar year round, you will not miss anything available there above the horizon given your latitude. Once again, "available" includes marginal stars (ones dallying with extinction). It also includes dock-pathed constellations (the other way): Daytime comes, and the star or constellation is still at an appreciable altitude.

Notes from long ago

A table of star or constellation positions designed to, in some way, indicate the time of year is called a *parapegmum*. A parapegmum is a physical object. Perhaps it is made of stone, clay, or wood.

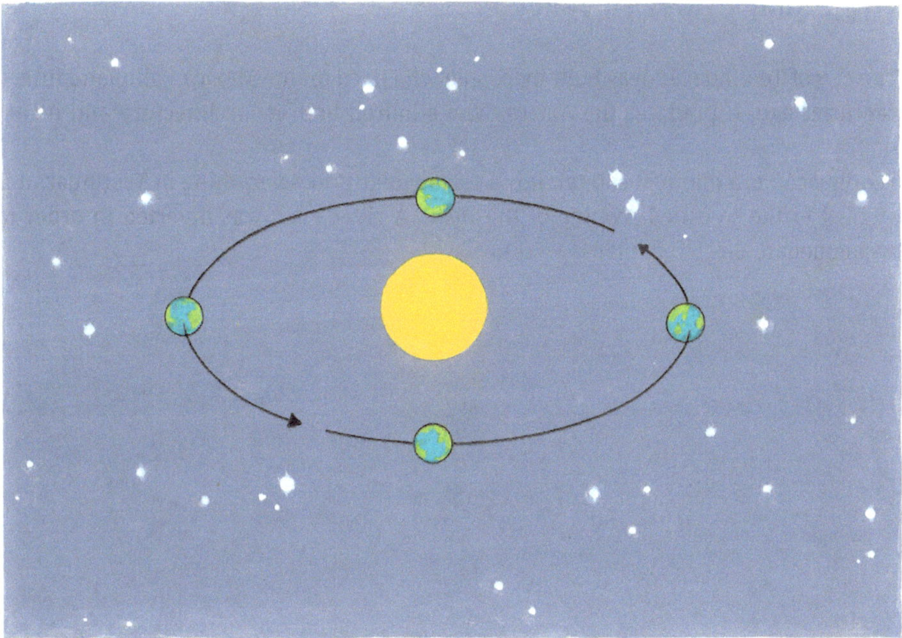

Figure 7.7: Picture yourself in this illustration on the nightside of the Earth (away from the Sun). At a given time and on a given date of the year, the stars and constellations, which you see at night, are those on the left of the illustration. Those on the right are below your horizon and will be in your daytime sky. (They are sunward on the Celestial Sphere.)

Three months later, you do the same thing. However, now the stars and constellations, which you see in your nighttime sky, are at the bottom of the illustration. While some are new, some of those that you saw three months previously are below your horizon; they now are in your daytime sky.

Three months later, you once again do the same thing. The stars and constellations, which you see in your nighttime sky, are those that were invisible to you six months ago (on the opposite, anti-sunward, side of the Celestial Sphere).

Your sky changes noticeably in yet another three months. You are back to your original nighttime view of the sky (and side of the Celestial Sphere) after one year. Not to scale. Artwork by Christine Tarte.

Holes may exist in the parapegmum so that a marker peg can be moved day after day. The pegged hole serves as a cursor next to text that indicates the appearance of the current sky. For instance, a heliacally rising star on that date might be listed.

Parapegma[4] often were placed in Roman public spaces such as baths. Sometimes they included typical weather during a given season.

---

4 plural form

7.11 Example: Mexico Valley

The city of Teotihuacán was built by people who lived on the Mexican altiplano fifteen centuries ago. It predates the Aztecs, who admired both its architecture and its antiquity.

Teotihuacán is a planned capital. It is a rectangular grid. Everything in Teotihuacán is aligned to the so-called Avenue of the Dead. A river even was diverted in order to accommodate this street plan.

Figure 7.8: Pyramid of the Moon and Plaza at Teotihuacán. Courtesy of Adrian Zwegers.

Cardinal alignment is so common among planned cities that we notice its absence more than its presence. The surprise at Teotihuacán is that the central avenue does *not* run north-south. It is skewed 15 ½° to the east of north/west of south. Why?

American New World archaeoastronomer Anthony Aveni concluded that the angle suggests something celestial. He decided to find out what it is.

Because it is so close to north, the "almost north-south" axis of Teotihuacán encircles little celestial area during the course of the sidereal day. In other words, the circumference of a 15 ½°-radius circle centered on the North Celestial Pole is small. However, at a right angle to the Avenue of the Dead, the city's "almost east-west" axis does encircle a lot of area on the Celestial Sphere. This axis traces a circle that has nearly as large a circumference as the Celestial Equator, a great circle.

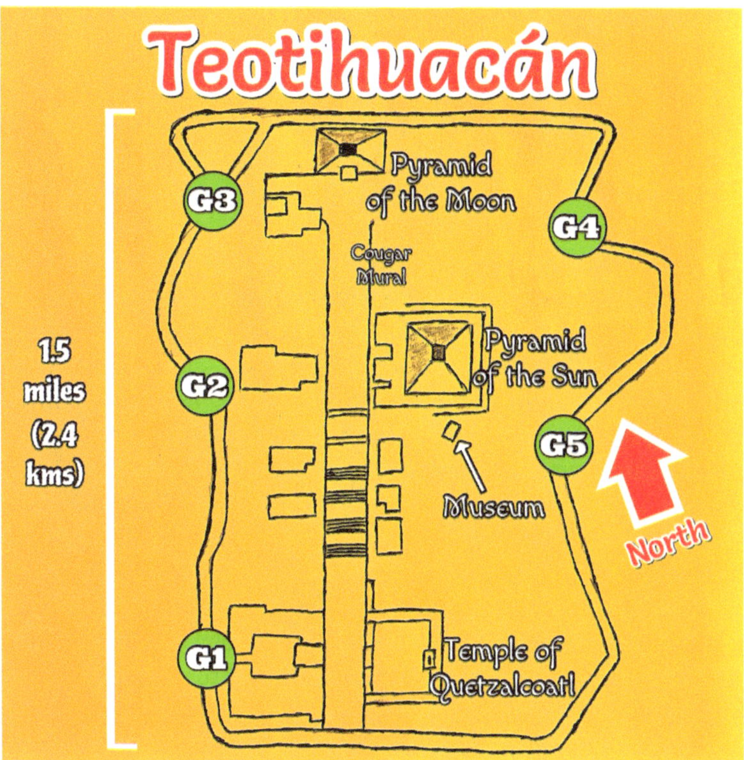

Figure 7.9: Stylized Map of Teotihuacán. Courtesy of Andrew Smith.

Even so, there is only one significant celestial object to which Teotihuacán points. Aveni found that, when the city was built, the Pleiades rose heliacally within 1° of the "almost east-west" line.

Why complicate the layout of a city choosing a stellar alignment instead of the more obvious cardinal alignment? While the Pleiades have no tangible effect on humans, obviously the Sun does. And there was a strong connection between the Pleiades and the Sun for the Teotihuacános.

The heliacal rise of the Pleiades at Teotihuacán took place on the same date that the Sun reached the zenith. The date on which the Sun shines overhead often is special to those who live close to the equator. (See Chapter 11.) While in no way proven, this alignment theory is compelling.

7.12 Chapter summary

The horizon coordinate system of altitude and azimuth is local. The Celestial Sphere is literally universal. How one appears with respect to the other is a function of your location on the Earth.

You are small compared to the infinite Celestial Sphere. For this reason, your sky includes half of the Celestial Sphere at all times.

The Celestial Sphere uses as markers the North and South Celestial Pole and Celestial Equator. The relationship between these and equivalent referents on the Earth is useful for navigation.

The combined rotation of the Earth and its revolution about the Sun result in two different definitions of the day, the sidereal and the solar. It also makes possible seeing all rising and setting stars at your latitude over the course of a year.

A most impressive case of celestial alignment may exist at Teotlhuacán.

7.13 Chapter review

1. How is your personal reference system – the horizon coordinate system – different than the universal reference system of the Celestial Sphere?
2. How does where you live affect the sky that you see?
3. Describe the changing sky on an imaginary vacation from the south temperate latitudes to the South Pole, then to the equator, and finally then to the north temperate latitudes.
4. How is a Celestial Pole useful for nautical navigation?
5. Why is it harder to determine your longitude than your latitude?
6. What is the difference between the Earth's rotation and revolution? What unit of time do we call the Earth's rotation period? What unit of time do we call the Earth's revolution period?
7. How does the combined effect of Earth's rotation and revolution cause a difference in the 24-hour day *versus* the sidereal day?
8. If we measured our clock day to the stars instead of the Sun, what will happen to the apparent position of the Sun at noon over time?
9. If you step outside at the same time each evening, why are the stars you see in your sky during the summer different from those you see in the winter? Draw a picture with the Earth, the Sun, and stars to help you explain.
10. Why do archeoastronomers believe that the ancient city of Teotihuacán in Mexico does not run north-south, but rather is skewed 15 ½° to the East of a north-south line?

8 Celestial coordinates and the precession of the Equinoxes

The entire Celestial Sphere is the closest thing there is to an unchanging reference in the sky. Of course, in reality, the stars are not glued to the Sphere; they do move through space. Yet, stars are so far away that, even though they possess great velocities, it takes an incredibly long interval of time for them to appear to change positions in your sky.

8.1 Chapter learning outcomes

After reading this chapter, you will be able to:
- distinguish between apparent and proper motion.
- read coordinates on the Celestial Sphere.
- take into account the effect of precession.

8.2 Proper motion

How long does it take for real motion on the Celestial Sphere, called **proper motion**, to (for instance) distort a constellation pattern? The interval of time involved is almost always longer than that over which we have any practical concern.

An exception is when three stars happen to lie in a "straight" line on the Celestial Sphere. The human eye is well trained to detect deviations from a line.

For instance, Greek astronomers in the early era of the Roman Empire noted that the easily identified stars Zubenelgenubi [Alpha Librae], Arcturus [Alpha Boötis], and Mizar [Zeta Ursae Majoris] seemed to form a perfect line. Today, due to proper motion, Arcturus is three-quarters of a degree out of line to the West. This is a noticeable defect, even to the naked eye.

If you know the direction and speed at which a star travels on the Celestial Sphere, you can predict its future location. You also can figure out where it has been.

Australian astronomer Ray Norris speculates about the Pleiades. While people with good eyes and good observing conditions claim to see more than seven of its stars, few come up with exactly seven. And yet the world over, independent cultures reference *seven* Pleiades.

Norris suggests that myths about the personification of the "lost" Pleiad – most amazingly similar to each other – began when proper motion had just started to make resolution of two of the Pleiads difficult. It was the time when people started to count them as one, for a total of six.

https://doi.org/10.1515/9783111441245-008

But this was tens of thousands of years[1] ago! Norris wonders if the myth about how the seventh Pleiad was "lost," in its various versions, represent humanity's oldest story.

Tens of thousands of years ago is beyond our purview. Normally, shorter-term proper motion is detected in (at least) nearby stars only with the resolution provided by a telescope. Even then, doing so takes a patient astronomer with an excellent star map. Proper motion is largely irrelevant to this book.

Figure 8.1: The star with the greatest proper motion is Barnard's Star. Its change in position on the Celestial Sphere, between these two images, is 3½ arc minutes. Courtesy of Steve Quirk.

8.3 Coordinate system on the Celestial Sphere

Because the Celestial Sphere acts as a frame of reference for the stars and constellations, a coordinate system is built using it, just as another one (altitude and azimuth)

---

1 When very long periods of time are considered, there is little value in making a distinction between sidereal and solar years.

is built "onto" your sky. On a globe of the Celestial Sphere, it serves the same purpose as latitude and longitude does on an Earth globe. (See Chapter 20.) Yet, later in this chapter, you will see that even this coordinate system, seemingly affixed to the Celestial Sphere, must be modified over great lengths of time.

8.3.1 Declination

On the Celestial Sphere, angular distance north and south is measured in degrees of arc just like altitude and azimuth. This dimension on the Celestial Sphere is called **declination.**[2]

Everything on the Celestial Sphere has a declination. Declination ranges from 0° at the Celestial Equator to +90° at the North Celestial Pole and −90° at the South Celestial Pole. The result is something like a number line. Using positive numbers in the North Celestial Hemisphere and negative numbers in the South Celestial Hemisphere is admittedly "hemisphere-centric." As you might guess, the system was developed in the North. Declination is very much like terrestrial latitude, except with an explicit "+" or "−" that avoids the somewhat awkward "N." or "S." suffix.

Declination serves the history of astronomy. Suppose an ancient astronomer composed, or is purported to have composed, a catalog of stars. But otherwise, we do not know much about this inventory nor its compiler.

The stars listed are unlikely to be those of the entire Celestial Sphere. The catalog's cutoff in declination indicates those stars not well seen from the astronomer's latitude and thus suggests from where they were observed.

Alternately, if the roster includes stellar magnitudes, there is a declination, at which atmospheric extinction seriously undervalues a star's brightness. If this effect is noticeable in the catalog, it is further evidence of provenance.

Astronomical detective work such as this may involve a case in which we already know, from the astronomer's biography, from where he should have observed. If the catalog's latitude cutoff matches with the declination appropriate for that latitude, great. If not, he may have obtained the data for his register elsewhere, or from somebody else – perhaps an earlier observer.

8.3.2 Right ascension

By induction from declination, you might think that east and west are marked on the Celestial Sphere by a similar coordinate, one that looks very much like terrestrial lon-

---

2 from the Latin for "bending away"

gitude. However, the east-west coordinate on the Celestial Sphere is measured in angular hours of **right ascension**.

Right ascension starts with 0 h and continues eastward in equal-sized increments to 24 h. Because right ascension marks out position on a circle, 0 h and 24 h are the same. (A shortened superscript notation is used in the same way that ° stands for "degrees.") This start/finish point for right ascension is the First Point of Aries; see below.

The reason that right ascension borrows a unit that sounds like one for timekeeping is that, while terrestrial longitude does not change in any other reference frame, in the celestial reference frame of your sky, the location of a given right ascension does appear to change with the Earth's diurnal motion. And that rotation is recorded by the passage of time.

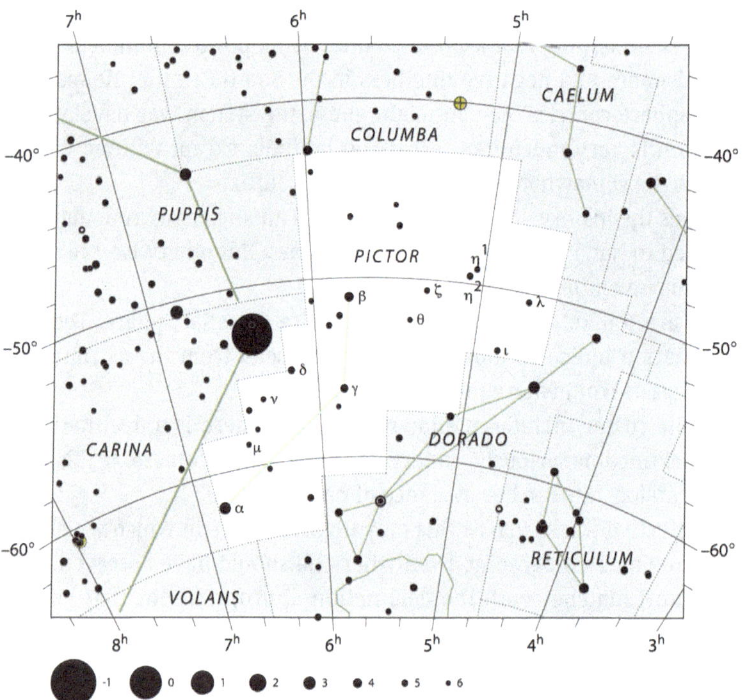

Figure 8.2: Curved lines on this constellation map are of equal declination; radial lines are of equal right ascension. Their coordinate values are recorded in the margins. Courtesy of the IAU.

Using a 24-hour scale makes the change from the Celestial Sphere reference to the sky dome reference easier. For instance, you can refer to an object being a certain **hour angle** away from your celestial meridian.

As an example, consider a star. It is at an hour angle of −3 h. Therefore, it is 45° east of your celestial meridian. Or a star at an hour angle of +3 h. It is 45° west of your celestial meridian.

To know when a particular star will be on your celestial meridian, subtract the right ascension currently on the meridian from the right ascension of the object. If the answer is a negative number, it corresponds to the number of hours of time you will have to wait. If the answer is a positive number, it means that you have missed the event and will have to wait approximately 24 h.

Declination often is abbreviated DEC; right ascension, RA. Declination, like any angle, can be sub-divided into sixtieths of a degree (minute of arc, with the symbol '). Right ascension, like time, can be further sub-divided into sixtieths (minutes, with the symbol ^{m}). The latter is just like a clock. The notation is especially important in right ascension so as not to confuse the measurement of angular position with the measurement of time.

When the location of a celestial object is given in celestial coordinates, the **epoch** of those coordinates often is quoted as well. The epoch is the date in the neighborhood of which the coordinates are meant to be used. It might look like this: "(2000)." How important is this information?

Tools of the naked-eye observer

Technically, they are a kind of celestial *planisphere*. These devises often are marketed as "Star Wheels" or "Star Finders." They are simply constructed, and usually inexpensive. Physically, a celestial planisphere is a volvelle, a flat construction with moveable parts, typically made out of paper or plastic. A different one is required for different brackets of latitude.

The bottom layer of this two-part mechanism is a map projection of a partial Celestial Sphere – more than a Celestial Hemisphere – onto a (usually round) plane segment. The top layer is opaque, except for an aperture, through which the bottom is seen. This top layer rotates about a point. As it is caused to do so, it exposes different parts of the map on the bottom layer.

The axis of rotation is the Celestial Pole; curves representing constant RA and DEC may be drawn surrounding it. Normally, motion is produced by simply turning the edge of the movable upper layer. The rotating layer is an independent part, jacketed within a frame that holds the two pieces together. Alternately, given a window, it is mounted on a pivot.

To manipulate the celestial planisphere, a scale on the perimeter of the top rotating layer (most often, hours of the day) is set to correspond to another scale (most often, hours of the day) on the circumference of the fixed bottom layer. Matching the current date and time, your sky is displayed, with the edge of the aperture forming the horizon and its center, the zenith.

The celestial planisphere is the modern version of an ancient instrument called the *astrolabe*. It was invented by the classical Greeks and perfected much later by Islamic astronomers.

8.4 Wobble of the Earth's axis

Yes, the stars are "fixed" on the Celestial Sphere over a human lifetime. Many human lifetimes. However, the Celestial Sphere, *as a whole*, changes over historical times.

Although the stars retain their relationships (ignoring proper motion), the entire Sphere changes its pivot point over the millennia. The result is that declinations and right ascension change extremely slowly. Hence, it is necessary to display the epoch of these coordinates in/on a celestial table, map, or globe.

8.4.1 Precession

Right now, the North Celestial Pole points to a particular place in the sky that happens to be near the moderately bright star Polaris [Alpha Ursa Minoris]. This has been true for a long, long time. It has been throughout all of our lifetimes and those of our parents and their parents, *etc*. Nevertheless, it was not always the case, and eventually will not be so anymore.

The axis mundi moves at a constant rate on a closed path; it revolves about a center. The phenomenon is called (formally) the **Precession of the Equinoxes**.

This name differentiates it from other types of precession encountered in astronomy. Notwithstanding, it is safe in the context of this book to refer simply to precession.

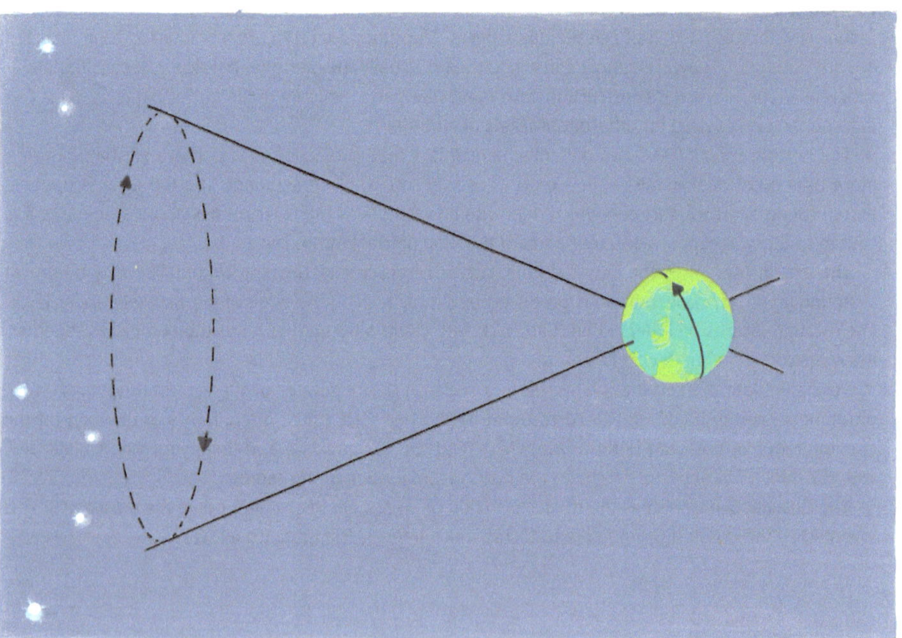

Figure 8.3: Axis mundi pointing in the direction it does today and also in the direction it will point in the far future (or past). The changing direction of the Earth's axis follows the circle of precession, with one cycle taking place every 26,000 years. Artwork by Christine Tarte.

> **Misconception alert**
>
> Unfortunately, two terms have similar meanings but in fact refer to different things. It is easy to confuse the word "precession," the word used in this book, with the more common word "procession." "Precession" shares an etymology with "precede."

Precession is caused by the axis of the Earth wobbling. The North and South Celestial Poles both trace out a circle on the Celestial Sphere. Clockwise, as viewed from the North, is the direction of advance.

The Earth is not perfectly circular; it is of a slightly greater diameter, as measured in the plane of the equator than it is as measured through the poles. This inequality creates a moment arm tugged upon by the gravity of the Sun and Moon. The result is precession.

8.4.2 The Great Year

The reason why you probably have not heard of precessional motion, unlike diurnal or annual motion, is the time involved: Precession changes the positions of fiducial marks on the Celestial Sphere (the Celestial Equator, the Celestial Poles, *etc.*) at a rate of about one degree every 72 solar years. It takes 26,000 such years to make one complete turn through the precession circle! This period is called the **Great Year.**[3]

> **Where did that come from?**
>
> This Great Year is sometimes inappropriately called the Platonic Year. The term is a misnomer inasmuch as it was discovered well after the age of Plato.
>
> Plato hypothesized that there would be a time in the far future at which all celestial objects returned to the same relative location. He and his later followers called the period of such a coincidence the "Perfect Year" or Great Year. The Great Year eventually was given a much more restricted meaning as the period of precession. With time, Plato's definition of "Great Year" became that for the eponymous Platonic Year.

Think about watching a top. You easily can observe the wobble that all tops have. It usually is slow enough so that you can count how many wobbles occur in, say, one minute.

At the same time, the top may be spinning so fast that its motion is a blur: You cannot count the hundreds of rotations per minute the top makes. In this example, the ratio of the period of the top to the period of its wobble is hundreds-to-one.

Ponder how many rotations the Earth makes in 26,000 years. Obviously, it is an even more extreme ratio.

The day and year, by any definition, mark comparatively short periods of time relative to the human lifetime. The Great Year does not. From year-to-year and – par-

---

3 having nothing to do with the more common use of the word "year," the period of the Earth's revolution

ticularly – day-to-day, we cannot tell that the position (for instance) of the Celestial Pole is slowly moving.

It is not untruthful to tell your children that the "pole star" Polaris marks the direction of north. This will be true for all practical purposes in their lifetimes and their children's and *their* children's.

Notes from long ago

So subtle is the precessional change in a human lifetime that it is a triumph of naked-eye astronomy that it was discovered so early by a Greek astronomer named Hipparchus *circa* 150 BC/BCE. Unfortunately, later astronomers confused themselves by thinking that the rate of precession changes or even that precession reverses direction. For a time, such hypotheses added yet another (nonexistent) motion, affecting the sky, which is unnecessary.

You may have heard of the **tropical year.** The tropical year takes into account precession.[4] The difference between the solar year and the tropical year is so small that, in the context of this book, the two terms can be used interchangeably.

8.5 Different pole stars

Because of precession, eventually the North Celestial Pole will noticeably move away from Polaris, many hundreds of years from now. Some other star nearer to it may serve as the North Star. The point in the sky about which everything seems to turn will be nearer this new Pole Star, than to Polaris. In twelve-thousand years, for instance, the bright star Vega [Alpha Lyrae] may serve to indicate true north. However, it will not be nearly as close to the Celestial Pole as Polaris is now.

Currently, there is effectively no "South Star." Most of the Celestial Sphere is not occupied by a star! The closest naked-eye star to the South Celestial Pole is Sigma Octantis, but it is magnitude 5.5 – only just visible.

7,200 years from now, Alsephina [Delta Velorum] will become a reasonable "South Star." It is of second magnitude, the same as Polaris.

Precession does not affect your sky in a practical sense – unless you are dealing with the lengths of time familiar to archaeologists. Remember how the Khufu pyramid's airshaft pointed (past tense) toward the North Star? *That* star was not Polaris, it was Thuban. This star is the brightest in the modern northern constellation Draco [Alpha Draconis]. Thuban was the North Star in the time of the Old Kingdom Egyptians. In fact, it was about as close to the North Celestial Pole as Polaris is now.

An interesting thing about North Stars is that there does not have to be one. There will not be another star as bright as Polaris near the North Celestial Pole for

---

4 However, there is inconsistency in the definitions used for the tropical year

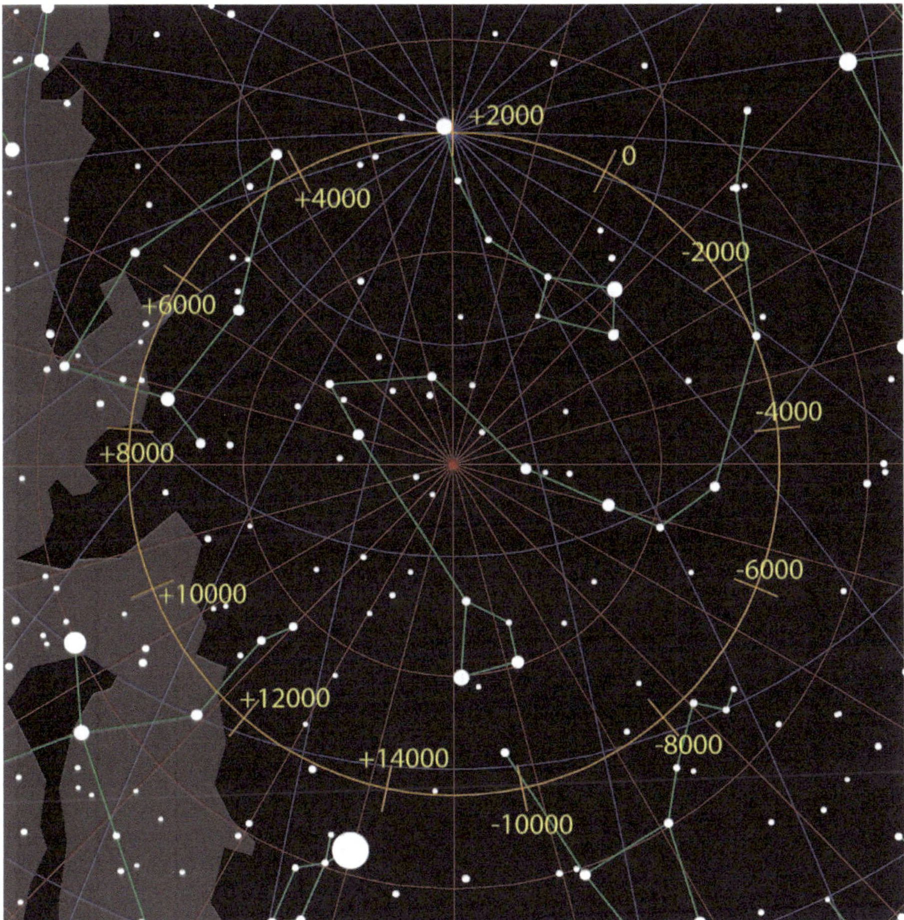

Figure 8.4: A computer program maps potential Pole Stars of the future over the length of the precessional cycle. Courtesy of Tauʻolunga.

five thousand years. Then, it will be Alderamin [Alpha Cephei], and that star still will be a few degrees away from the North Celestial Pole at its closest.

Indeed, there is no close-to-the-Pole star, other than Thuban and Polaris. The fact that there happened to be one in 2700 BC/BCE underscores why the ancient Egyptians might have been interested in the North Celestial Pole. The fact that there is today a North Star, so as to make discussion of the Egyptians relatable, smacks of great coincidence.

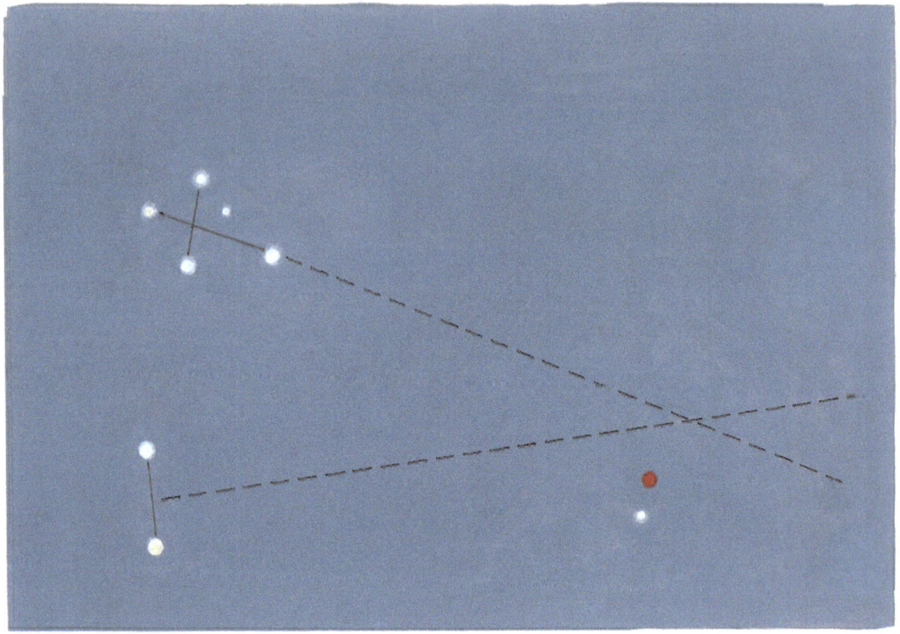

Figure 8.5: A traditional way of finding the South Celestial Pole, using two familiar southern constellations. Artwork by Christine Tarte.

8.6 No pole star at all

How could one find north at night if there is no North Star? One way might be to pick out two circumpolar stars, one above the North Celestial Pole and one below it. Wait for the two stars to have the same azimuth. A line[5] connecting the two stars, then, is perpendicular to the horizon; a plumb bob will attest to this. Mark this azimuth (by placing a foresight post on the horizon, for example). Then, wait for the stars to apparently rotate 180°. The first star now is below the North Celestial Pole, and the second star is above the North Celestial Pole. Once again, they share an azimuth. Mark this azimuth (with another post or some such tool). North is at an azimuth halfway between the two posts. South can be found by this method, too.

Of course, you must choose your two stars wisely. If, one of the two posted events occurs in the daytime, you have to wait a while until it occurs at night. This is *a* method. It is hard to say that it is a very practical one.

---

5 Notice that a "line" in spherical geometry is really a circle!

8.6.1 Finding the South Celestial Pole

There are other tricks. Remember that right now, there is no well-placed star near the South Celestial Pole to mark it. It is a lot to ask that there be a satisfactory North Star and South Star at the same time. More often, it is one or the other.

8.6.2 Example: Brazil

How do you easily find the South Celestial Pole today? As it happens, the stars that define the long segment of the constellation Crux, the so-called Southern Cross, are by themselves an asterism. They point in the *direction* of the South Celestial Pole.

But how far away from Crux is the Celestial Pole? Indigenous Brazilians are credited with determining that the distance is four lengths of the "cross."

If you are willing to wait, think of the "cross" (again, "top to bottom") as forming a ray. Repeat this exercise later. The two rays, the current one and the memorized one, intersect at the South Celestial Pole.

8.7 Precession & chronology

Precession theoretically may help us date events. Other archaeological techniques yield the construction date of the Great Pyramid. However, what if we did not have these methods at our disposal? Instead, what if we only had the suspicion that the pyramid is aligned to a North Star?

Suppose that it is you. You simply could calculate when Thuban, the likely candidate, was in the right place to be in the direction of the pyramid "airshaft" (within a century or two). You could then hypothesize that this approximate date was the date of construction – or, at least, design – for the pyramid.

8.7.1 An age when the North Star was not Polaris

Precession does not affect circumpolar stars only (and which stars will *be* circumpolar stars); it affects the entire Celestial Sphere. This includes the rising and setting azimuths, and the culminating altitudes, for stars that are not circumpolar. Another way to very tentatively date the Khufu pyramid would be to figure out when Orion's belt stars were in the direction of the south shaft, assuming that we already strongly suspected a connection between the pyramid and Orion.

Luckily, these Orion stars are far from the poles of precession. Stars near one of the poles of precession do not change their location much at all on the Celestial Sphere, even over lengths of time comparable to the Great Year.

What good fortune! If the precession period was much longer than it is, it would be of no use at all to archaeologists who are interested only in time intervals comparable to the age of human civilization. If the precession period was much shorter than it is, there might be an insurmountable ambiguity as to how many revolutions about the pole of precession had been made between the event they wish to try to date and the present.

For example, if the precession period were a mere thousand years, Thuban aligns with the Great Pyramid airshaft every thousand years. There would be several possible construction dates occurring during the long timespan of human habitation in Egypt.

In 26,000 years, humans have gone from pre-civilization, a time from which we have no monuments to date, to modernity. For establishing construction times by celestial alignments they incorporate, 26 millennia are just about right.

Historians and archaeologists do not use dates determined from precessional arguments alone. Still, they serve as useful tools.

8.7.2 Precession affects rising and setting dates

And then there is Teotihuacán. Precession establishes where on the horizon, the Pleiades set at that site. By the tenth century, the city axis no longer pointed toward the Pleiades, due to precession. The civilization of the Teotihuacános was dying out by then, anyway.

The much later Aztecs called the ruins of Teotihuacán, the birthplace of their gods. Anthony Aveni suspects that subsequent Mexican Plateau inhabitants were so impressed with the empty city of Teotihuacán that they aligned their cities the same way. This was true, even though the significance of the orientation was lost to the newcomers.

Some of these "copy" cities are found a hundred kilometers from Teotihuacán. Their locations and later construction dates, in the precession cycle, cause their alignments to deviate noticeably from 15 ½° of azimuth.

8.7.3 First point of Aries

When the Sun first moved "into" the constellation Aries, the event used to be a reasonable marker for the beginning of spring in appropriate climates. (See Chapter 9.) The solar year was once considered by some to start in temperate latitudes when the Sun was at this **First Point of Aries**. Any starting point on a circle is arbitrary; the First point of Aries also is the starting/ending point for right ascension – 0 h equals 24 h.

Yet, eventually, precession destroys coincidences between star phenomena and Sun phenomena. For instance, if you mark the date to sow seeds by the rising of certain stars at dawn, there will come an era when you (or more likely, your descendants) will plant at the wrong time of solar year.

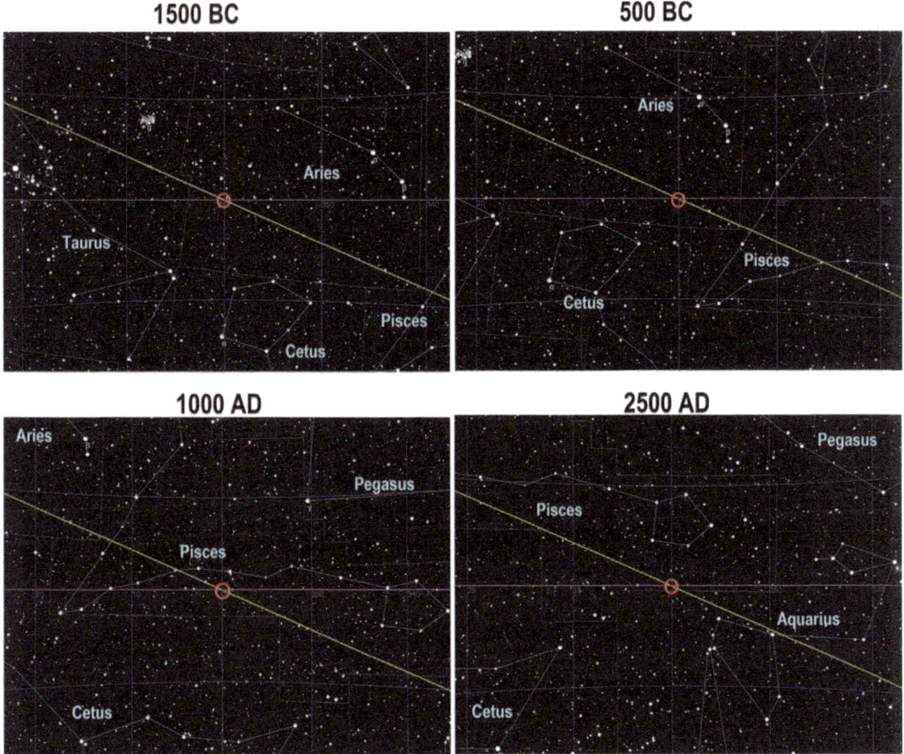

Figure 8.6: Star maps showing how the First Point of Aries changes in location. It arrives in Aquarius in (about) the year 2500. Images courtesy of Tom Ruen.

Modify the rule – replace "dawn" and "rise" with "dusk" and "set," for example – and extend the tradition for centuries. Still, the time comes when the calendar stars must be replaced, or your people will suffer the agricultural consequences. All such folk practices inevitably must die out.

Notes from yesterday and today

As you can see, precession not only effects the location of the Celestial Poles but the point at which the ecliptic and Celestial Equator meet. (See Chapter 9.) Curiously, astrologers still use the "Sun signs" from two thousand years ago. These now are more than one constellation off, due to precession!

While it was in Aries during the Hellenic age of Greece, the First Point of Aries actually has been in Pisces since 498. The transition is called a *Change of Ages* and happens about every 2,140 years. (The Christian symbol of the fish and the spread of Christianity during the Age of Pisces has been reflected upon.) The Sun's arrival at the more-and-more anachronistically named First Point of Aries eventually will be still one more constellation over. This will be when the so-called "Age of Aquarius" dawns, some six centuries hence.

8.8 Adding up what affects the appearance of your sky

It is now possible to summarize the four, and only four, motions that determine the stars you see in your sky. In order of extent to which they most likely affect your view, they are:

i. the rotation of the Earth (diurnal)
ii. your location on the Earth (latitudinal)
iii. the revolution of the Earth about the Sun (annual)
iv. when you happen to be born in the cycle, caused by the wobble of the Earth's axis (precessional)

Nevertheless, all this apparent change in the sky may seem dizzying. But remember that the apparent motions caused by the Earth's rotation, revolution, and even precession are readily predictable and always circular. Now that you understand the motions demonstrable by using the Celestial Sphere, it is time to discuss in more detail the motion of the single-most important object on the Celestial Sphere.

8.9 Chapter summary

Declination and right ascension are coordinates used to specify a direction on the Celestial Sphere. A rare star that changes its declination and right ascension, with respect to the stars around it, exhibits proper motion.

Hour angle is the number of hours it will take for a celestial object to reach your celestial meridian. Alternately, it is the number of hours ago a celestial object reached your celestial meridian.

The Celestial Sphere undergoes one cycle of precession (due to the Earth's axial wobble) in a Great Year. This Precession of the Equinoxes changes the First Point of Aries and causes Polaris to be your North Star only temporarily during one cycle. There have been, and will be, different North Stars in the past and future. Often, there is no North Star or South Star at all.

Precession has the potential to be useful for dating architecture that is thought to be aligned with a known celestial object.

8.10 Chapter review

1. What is the difference between apparent motion and proper motion? Why is proper motion largely irrelevant for naked-eye astronomy?
2. What are the right ascension and declination of the following described point on the Celestial Sphere: 1) It is halfway between the Celestial Equator and the Celestial Pole. 2) It is due South of the First Point of Aries.
3. What are Pole Stars? The South Celestial Pole does not have a Pole Star. How do people in the southern hemisphere determine where the South Celestial Pole is?
4. How might precession be useful to help us figure out when the pyramids were built?
5. Three stars used to appear to form a line segment in the sky. Now they do not. What is the cause of the change?
6. There are four motions that determine the sky you see at night.
   A. Describe each motion.
   B. Draw a picture of these motions.
   C. Explain the period of each motion.
7. You know that an (imaginary) building is aligned to the North Celestial Pole. Yet, it is not pointing in the direction of Polaris. It is pointing toward a place on the Celestial Sphere one-quarter of the way backward around the circumference of the circle of precession. How old is the building?
8. Long ago, a uranographer never left home in the mid temperate latitudes. He claimed to have produced a star catalog, all on his own, which includes stars near the South Celestial Pole. How do you know that he was lying?

9 The Sun

The stars you see in the sky are numerous, but there are only two luminaries. In the *daytime* sky, the most obvious (perhaps the only) astronomical object visible is the Sun. Its light overwhelms that of everything else combined.

The Sun's influence has been obvious from pre-history: Crops respond to it. Animals – including people – change their behavior, depending upon its presence. You only have to extend your palm to feel its life-giving warmth.

What is more, the Sun is our most obvious timepiece, especially for keeping track of the time of day. Plus, its position on the Celestial Sphere slowly changes, whereas those of the stars do not; it makes one circuit in a solar year.

9.1 Chapter learning outcomes

After reading this chapter, you will be able to:
– set a clock by the Sun.
– argue against a commonly misstated cause of seasons and for the correct cause of the seasons.

9.2 Morning, noon & night

When the Sun is still rising (east of the celestial meridian), we write that it is **ante-meridiem**.[1] When the Sun is setting (west of the celestial meridian), we say that it is post-meridiem. While these Latin terms remain in use, they are invariably abbreviated as something like "a. m." and "p. m." (Here, AM and PM are used.) Thus, the Sun defines the morning and afternoon.

The Sun appears on the celestial meridian at noon. In the case of the Sun, this is not necessarily its culmination, but the difference is only a matter of seconds.

Until recent times, astronomers started their day at noon, not midnight. At noon, there is a specific occurrence at a specific time (the Sun on the celestial meridian) that can be noted and used to calibrate their very exact clocks. There is no such comparable event twelve hours later, though a midnight transition from one day to the next is much more convenient for everybody else.

---

1 Ante is Latin for "before"; post is "after"

https://doi.org/10.1515/9783111441245-009

9.2.1 How we set our clocks

While we supposedly still set our 24-hour clocks with respect to the Sun, sometimes the real Sun appears to arrive a bit late to the celestial meridian, according to our watch. Sometimes, it appears a bit early. This would not happen if the Sun moved through exactly the same angle, in the same direction, on the Celestial Sphere, each day.

> **Notes from yesterday and today**
>
> What if it is cloudy? Our simplest means of both telling time and direction ("the Sun rises in the East and sets in the West") becomes unavailable!
>
> But unlike that of any other celestial object, the Sun's light is bright enough that it illuminates you even if its disk itself is not visible. This opens the possibility of finding time or direction based on sunlight alone.
>
> Many sources of light – the Sun is no exception – produce light that is *polarized*. Physicists find it useful to treat light as a wave. In the case of polarized light, this wave undulates in a preferred plane.
>
> There exists a translucent mineral called *Icelandic Spar*.[2] It breaks along a uniform cleavage. If you look at the sky through a piece of flat Icelandic spar, only sunlight polarized in a certain plane passes through. The light that you see brightens and dims as you look in different directions and you change the orientation of the rock specimen in your hand.
>
> Specifically, you see sky brightness change, depending on whether you look left, right, above, or below the Sun. A little practice with this technique allows you to find the approximate position of the Sun *behind* the clouds.
>
> Supposedly, the mineral was the Vikings' "sunstone." Old Norse sagas suggest that they used this polarization-based method to navigate their Long Boats across the often clouded-over North Sea. You may feel that observing through anything violates the spirit of strictly "naked-eye astronomy."
>
> This would not have mattered to a Norseman trying to decide how much sailing time they had until nightfall, or the way to Greenland, though.

However, the Earth as well as all the other planets, travel in their elliptical orbits faster than average when they are closest to the Sun. This point is called **perihelion**. Planets travel more slowly than average when farthest from the Sun. This point is called **aphelion**.[3] This is a natural outcome of Isaac Newton's Theory of Gravitation. The practical result is that the Sun appears to accelerate and decelerate in our sky, with respect to our clocks. It is called the **solar anomaly**.

The solar anomaly is at its greatest in November, when the Sun's arrival on your celestial meridian can occur as much as sixteen minutes ahead of clock time. In February, it can be as much as fourteen minutes late.

---

2 formerly, Iceland crystal; also, optical calcite, crystalized calcium carbonite, and clear calcite

3 These terms are by way of Greek; Peri- means "near," apo- means "away from"

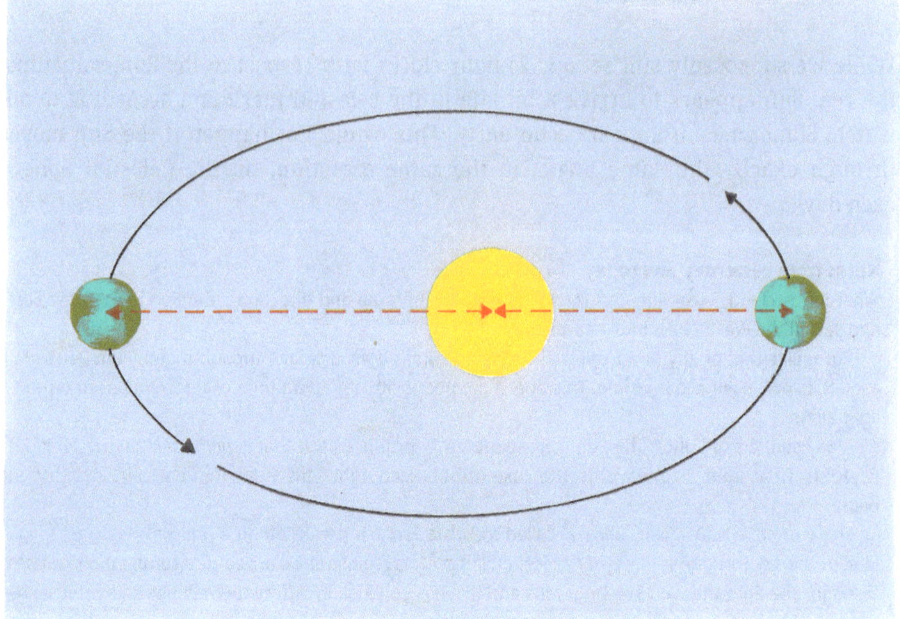

Figure 9.1: Aphelion (left) and perihelion (right). (The difference in the distance between the Earth and Sun is exaggerated.) Artwork by Christine Tarte.

A further (admittedly, more minor) complication: The Sun's motion on the Celestial Sphere is not always directly east to west. Ideally, you would wish your "master" clock's "face" to occupy a single plane; you do not want your clock's "hands" to undertake motion in and out of the dial. You want them to stay parallel to it.

Ultimately, you want all of your clocks to reflect a constant rate of time passing (constant rate of angular motion). So, you would prefer that the luminary on which they are based do so too.

There is not much that you can do about the arc across the sky, as a function of time, which the real Sun makes on a given day. However, you can make an easy adjustment to your modern timepiece to account for the real Sun "running" fast or slow. The correction is based on the well-calculated Equation of Time. For a given location and time of solar year, the **Equation of Time** tells you to add or subtract some number of minutes to Sun time in order to get the clock time.

A graphical representation of the Equation of Time is called the **analemma**. It is usually an asymmetrical figure of eight that illustrates the position of the Sun at a given time, from a given latitude, in your sky, throughout the solar year. Typically, that time is noon. The crossover point that makes the "eight" occurs in April and September.

The orientation of the "eight" is latitude-dependent. It is perpendicular to the horizon at one of the Earth's poles, but parallel to the horizon at the Equator. (See below.)

The width of the analemma is the number of minutes (+ or −) your clock time deviates from the real time of the Sun's arrival on your celestial meridian. The reason why one lobe of the figure eight is "fatter" in width than the other has to do with the timing of aphelion and perihelion. Indeed, were the Earth's orbit to be circular, the analemma at the equator would be a straight-line segment.

Figure 9.2: Analemma constructed by photographing the Sun at the same time of day throughout the year (2011) and then superimposing the photographs. Courtesy of Giuseppe Donatiello.

Tools of the naked-eye observer
The analemma traditionally appears on globes. Used this way, it is situated so as to indicate the latitude at which the Sun is directly overhead on every day of the solar year. The analemma was invented by eighteenth-century French nobleman/astronomer Jean-Paul de Fuchy.
  An analemma for the Moon can be constructed, too. However, it is rarely seen.

These days, few people literally set their clocks by the real Sun. They do so based upon some other clock. **Horologists** (clockmakers) still give lip service to the Sun by saying that we keep track of time based upon this luminary. In reality, most of us do so through intermediary chronometers based upon a fictitious, constant-rate Sun.

9.2.2 Hours

Similarly, you want the hours (along with minutes and seconds), into which a day is divided, to be of constant duration. For better or worse, you prefer waiting an hour in the doctor's office, on a given day and at a given place, to take just as long as waiting an hour in the dentist's office, at a different place on a different day.

Notes from long ago

Why are there 24 hours in a day? Why not, say, ten? Or one hundred?

The year of the ancient Egyptians began with the first morning appearance of the brightest star in the sky, Sirius. They noticed something though. After about ten days, Sirius no longer rose to mark the end of night. So, they chose another star that would fit this role. And ten days after that, it was another star. The result was a set of 36 stars, more-or-less evenly spaced in a circle around the Celestial Sphere. Each rose heliacally, in turn, for about 1/36th of the year. These special stars were called *decans*.

The utility of decans was that they could be used to tell time at night. For example, as Sirius rose, if it was still dark, you would know that the next decan behind it would soon rise. Moreover, you would know that the next decan after that would rise after about the same interval of time.

Even over the shortest night in Egypt, a minimum of twelve decans would appear sequentially at the eastern horizon during the night. Counting decans permitted the counting of *time*.

Meanwhile, the Egyptians were marking time between sunrise and sunset in a different manner. They used a stake and a scale, upon which the stake's shadow fell, divided into ten intervals. In truth, this shadow-method did not work well for defining the first or tenth interval, when the shadow was very long.

But what about the time of day during which the Sun does not cast a shadow, because the luminary is below the horizon, yet when it is still too light to see decans? The Egyptians solved this problem by adding an hour of twilight to the morning and evening of each day.

Tallying the Egyptian day up from dawn to dusk to dawn results in: 1 h of twilight, plus 10 h of daytime, plus one hour of twilight, plus 12 h of night. This sum equals 24 h. So, close to the equator, this worked reasonably well. (But see the main text below.)

Hours of constant length were not always the case. Before traditional clocks, Sun *events* were used to break up the day. By this, sunrise, noon, and sunset are meant.

Noon to noon is a (nearly) constant duration. However, the time between sunrise and sunset is affected greatly by the time of year and your location on the Earth. The time between sunset and sunrise changes similarly. The length of nighttime, of course, is half a cycle out-of-phase with the length of daytime.

People have a choice: Put up with a varying length of hours, assign more hours to daytime (or nighttime), as required, or skip keeping track of such subdivisions of time altogether. Today, the second choice is universally agreed upon: hours of constant length, independent of whether they are hours of daytime or hours of nighttime.

Notes from cultures around the world

The daily prayer times for traditional Muslims are not governed by the clock. The Islamic and Jewish days begin at sunset; an Islamic prayer is to be made right afterward.

Muslims also pray after nightfall. By this, they mean near astronomical twilight. This equals, variously, a solar altitude of −15°, −17°, or −18°, depending upon the textual source. Another definition is ninety minutes after sunset.

A third prayer is to be performed before sunrise: The Sun's altitude equals −15°, or −18° or −19°, once more, depending upon the source. A fourth prayer is offered just after the Sun culminates.

It is not necessary to watch the brilliant Sun and estimate the location of the celestial meridian to find this midday Islamic-prayer time. There is a less awkward method. At (local) noon, the Sun is at its highest place in the sky for that day, so the shadow cast by a stake is at its shortest. In north temperate latitudes, this shadow always points north. In south temperate latitudes, it is south.

Still, it might be difficult to decide upon the moment of the shortest shadow. More conveniently, mark the shadow length any time before noon. Do so again when it reaches this same shadow length after noon. As the Sun was at the same altitude at both times, noon occurred at a time halfway between. Of course, knowing when noon *was* is not the same thing as knowing when noon *is*.

The last Islamic prayer is the most interesting from an astronomical point of view. It is to be made after a stake's shadow is twice the length of – or, alternately, is as long as – the stake, plus its noon length. Notice that it does not matter what the overall length of this shadow is on any particular day; that depends upon the time of the year and your latitude – as well as, of course, the length of the stake.

Others skip the measurement and say about ten minutes before sunset.

The times of day at which shadow lengths occur vary, as a function of season and latitude. However, they always happen (except for, perhaps, in the Arctic and Antarctic; See Chapter 11).

9.2.3 Time zones

The Sun cannot be at the same place in everybody's sky at once. It reaches its noon position in your sky at a different time, depending upon at which longitude you live. The farther west you reside, the later noon arrives.

This may be of little concern to you: The time intervals between your local sunrise, noon, sunset, and any other such similar sky event, are unaffected. It gets complicated though when people from different longitudes start communicating with each other. Or start traveling routinely from place to place in an easterly or westerly direction. Whose watch is right?

To make things easier, and remove ambiguity, humans have divided the world into 24 **time zones**. The Sun is on the celestial meridian at noon in the middle of each time zone. It is a bit early in the eastern part of the time zone, and a bit late in the western part of the time zone. However, despite this, everybody living in a given time zone (at least, in principle) observes the same time.

The first time zone is arbitrarily that at which the center longitude runs though Greenwich, England. For a long time, the Greenwich Observatory was the preeminent institution of astronomical timekeeping.

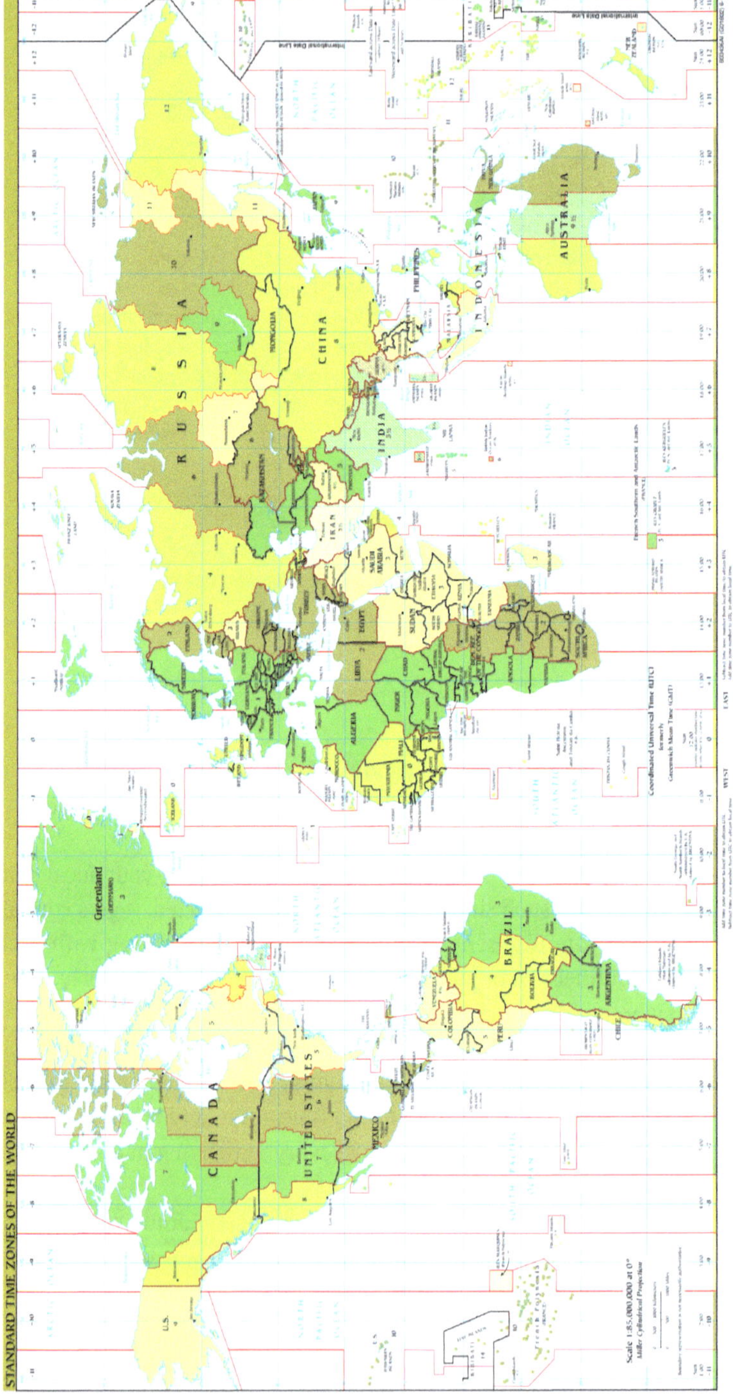

Figure 9.3: Modern time zones and the International Date Line. Complicating this system is the decision made in some politically defined regions to introduce Daylight Savings Time through part of the year. This arbitrary clock reset begins and ends on different dates in different places and may change from year to year. It has no basis in astronomy.

In theory, the borders of time zones should be semi-great circles of constant longitude. In practice, they have been adjusted here and there, for geopolitical reasons. A few places such as Afghanistan, India, and Iran use time zones that are not an integral number of hours different from Greenwich. A few more ignore time zones altogether: The Peoples' Republic of China uses a *single*, national, time stretching across sixty degrees of longitude.

Nonetheless, most of the world's population uses one of 24 different times, the one that best fits their longitude. This is certainly better than everybody observing their own local time. Confusion is lessened, and commerce benefits.

The one catch to time zones occurs if you continuously travel westward (perhaps on a jet airplane). Eventually, you will cross into a time zone that puts you at clock midnight. Soon thereafter, you are one hour earlier – into the day before!

Such a metaphorical time machine is ridiculous. Therefore, one time-zone border is called the International Date Line. It is agreed that crossing it, either east or west, requires that you to reset your date as well as the hour. Westward travel gains a day, eastward travel loses a day.

The International Date Line longitude is chosen so that it does not cross many land masses and is further finagled so that no country has to operate in two different "days" at once. The establishment of time zones is a great example of international cooperation.

9.3 The ecliptic

In time spans greater than a day, it is more obvious that the Sun does not stay put like the stars. The annual apparent path of the Sun through the sky and stellar background is called the **ecliptic**. The ecliptic runs through the Celestial Sphere; the path of this great circle is centered upon you.

9.3.1 The zodiac

In a Western tradition, the Sun enters (in turn) each of twelve constellations along the ecliptic and then repeats the progression. These are the **zodiac** constellations made famous by the belief system, previously introduced as astrology. Because of the way modern constellation boundaries are drawn by astronomers, the Sun spends a brief part of the year passing through the thirteenth constellation of Ophiuchus; astrologers use their own equally spaced constellation boundaries and do not recognize Ophiuchus.

Figure 9.4: Astronomical clock in the fIfteenth-century Torre dell' Orologio, Venice, Italy. Its dial is decorated with twelve golden, zodiac constellations. The representations are stylized.

Where did that come from?

The Babylonians may have invented the idea of a zodiac, but the word is Greek. "Zodiac" comes from the fact that many of these constellations are supposed to represent animal figures. Zo- means animal, as in "zoo." Dia- means circle, as in "dial." The zodiac is the Circle of Animals. In Mesopotamia, these special constellations appeared in the evening sky as a "signal" to farmers during the region's rainy season.

The figures of the zodiac constellations sometimes defy our modern attempts to match them with existing Greek myths and legends. The zodiac may be a heritor of actual, earlier, Babylonian constellations. Yet, there are some patterns to be seen within the set.

For instance, some of the zodiac constellations that are always to be found at low altitude in the South have water themes, *e.g.* Pisces (the Fish), Aquarius (the Water Bearer), and Capricorn (the Sea Goat; mythically half-goat, half-fish). Greece is surrounded by the Mediterranean in the South.

9.3.2 (Mis)use of the zodiac in astrology

Natal astrology is the belief that your future (indeed, your whole life) is determined by the zodiac constellation that the Sun was projected onto at the time of your birth (as well as other sky phenomena). That 8,000,000,000 human fates are categorized into a mere twelve groups appears simplistic; yet that is how the system is arranged.

However, astrologers still use the "Sun signs" from more than a thousand years ago. These now are more than one constellation off, due to precession!

> **Misconception alert**
> You might think that your "sign," the constellation that the Sun is "in" during your birth month, would be displayed prominently in the sky at the time of your birthday. Yet, even if you ignore the effect of precession, this is still not the case. Because the Sun is "covering up" (in front of) the relevant constellation on your natal date, "your" zodiac constellation is the *daytime* constellation that is then *hardest* to see.

9.3.3 Ecliptical coordinates

The ecliptic suggests another way of specifying location on the Celestial Sphere, besides declination and right ascension. These are **ecliptical coordinates**. They are useful, at least for objects that always stay on or near the ecliptic, such as planets. (See Chapter 18.)

The ecliptical coordinates of the Sun require only one parameter. The ecliptical latitude of the Sun is always zero. To specify the remaining coordinate, ecliptical longitude, you start at 0° (chosen to be equal to the First Point of Aries) and simply continue through a 360° counterclockwise circle (as viewed from north), until you return. The Sun appears to travel through all ecliptical longitudes in one solar year.

Ecliptical coordinates, not right ascension and declination, were favored in Europe until the about the time of the invention of the telescope. The Chinese though, always have used coordinates based upon the Celestial Equator and Poles.

9.4 Why do we have seasons?

The ecliptic is not the same thing as the Celestial Equator. The former is defined by the Earth's revolution about the Sun; the latter is defined by the Earth's rotation axis (the axis mundi). The difference lies in the fact that the axis of the Earth is tilted 23 ½° from a line perpendicular to the plane of the Earth's solar orbit.

An equivalent way of saying the above is that the ecliptic is tilted 23 ½° (more accurately, 23.4°) with respect to the Celestial Equator. This tilt is formally called the **obliquity** of the ecliptic.

9.4.1 Angular momentum

Why is it necessary at all to introduce a new imaginary circle on the Celestial Sphere, the ecliptic, when discussing the Sun? The rotation and revolution of the Earth are

Figure 9.5: Why are table globes mounted in a "tipped" frame? Would it not be easier to place the globe on a simple, single axis? The "tip" is set to the obliquity angle of the Earth. Thus, the globe is displayed as it would appear in the plane of the ecliptic. Photograph by the Author.

independent of each other. If the Celestial Equator turned in the plane of the ecliptic, it would be a tremendous coincidence.

All the other planets rotate in planes that are tilted with respect to the planes of their orbits about the Sun, too. Even the Sun rotates in a plane slightly different than the ecliptic.

A given obliquity does not change significantly. You might have thought that if the Earth moves with respect to the Sun, the orientation of its rotation axis should move as well. This is not the case.

Over intervals of time with which we are practically concerned (that is, ignoring precession), the Earth's axis mundi always points in the same direction. The North Celestial Pole is near the star Polaris, throughout the year. Indeed, a changing axis orientation would require some unknown force – perhaps, an interplanetary giant who hits the axis mundi with a hammer! Save this impossible feat, the axis remains at a fixed orientation in space. This axial stability is due to the same physics that is put to use in a stabilizing gyroscope.

If you ever ride a bicycle, you notice that – somewhat counterintuitively – it is harder to tilt over when traveling *fast*. Your rapidly rotating tires resist change to their plane of rotation.

Angular velocity is both speed and rotation-axis direction. A linear-moving object has momentum, and will continue to move in a straight line (at a constant speed) unless acted upon by an outside force. So too, does a rotating object have **angular momentum,** and will continue to rotate (at a constant angular velocity) unless acted upon by an outside torque.

Assuming that your bike-tire axes are perpendicular to your path, they stay that way. It is when you must slow, say at a stop sign, that – unless you are an acrobat – you likely have to keep the bike from tipping (by putting your feet out).

9.4.2 Obliquity

Returning to a familiar question, "Why do we have seasons?" The answer is that the Earth's obliquity is responsible for the seasons.

The answer is *not* because the Earth is closer to the Sun at one time of the solar year than another, and thus closer to the source of heat at one time and farther at another. If that were the correct explanation of the seasons, it would conflict with something you learned from your geography teacher: that the seasons experienced by our planet's two hemispheres are different, even though the date is the same.

Imagine that you are suffering through winter. At the same time, antipodeans are enjoying summer and vice versa.

This is a situation in which a little knowledge is a "dangerous" thing. You know that the Earth's orbit around the Sun is elliptical, with the Sun at one of the ellipse's two foci. Therefore, the Earth actually is a bit closer to the Sun at one place in its path (perihelion) than at another place in its path (aphelion).

But figures often exaggerate the Earth's orbital eccentricity to illustrate a point. In truth, the Earth's elliptical orbit it is only slightly out-of-round. The difference between the Earth's closest and farthest distances from the Sun is only a little over three percent – not enough to cause the dramatic changing of the seasons that you experience. The degree of this difference changes over hundreds of thousands of years, but never amounts to very much. In an ironic twist for those of us who trudge through northern winters, the Earth in our era is slightly *closer* to the Sun, early in the chilliest month of our calendar, January.

No, the following fact is much more important to seasonal change. Think about the summer. Your hemisphere is tipped *toward* the light and heat coming from the Sun. During the following winter, the same hemisphere is tipped away from the direction of the Sun's light and heat.

9.4.3 The changing altitudes of the sun throughout the year

Returning to our view of the sky: Consider a summer's noon. Remember that noon is the time at which the Sun culminates.

To glance up at the Sun at this time, you must tilt your head far back. The Sun is at a significant altitude. It is not at your zenith, but it is high in the southern sky.

Now consider the same circumstances in your winter. This time you do not need to tilt your head so far back to glimpse the Sun. *The Sun is at a lesser altitude than it was six "months" ago.* In fact, at any hour of a winter's day, the Sun is lower in altitude than it is at the corresponding hour of a summer's day.

Figure 9.6: It is northern winter, southern summer. Your friend lives at the same longitude that you do, but at the same latitude south as you do north. You are both looking at the same object (e.g., the Sun) in the sky.You do not have to bend much; the object is close to your horizon. However, your friend much contorts to see what you see, because the object is at such a high altitude. Photograph by the author.

Figure 9.7: In this AI-generated scene, it is midday and sunny. Nonetheless, by programing mid-latitude winter, long shadows indicate that a low-altitude Sun has been assumed.

At the same time and on the same date on which you notice how high the Sun is in your sky, compared to a solar year's average noon, your twin at the same latitude in the opposite hemisphere comments on how low the Sun is in their sky, (again) compared to a solar year's average noon. Except for the fact that he or she is looking north, the Sun is in the same place in the sky (at the same altitude) that it was for you half-a-year before or will be half-a-year after.

Your summer's day is a winter's day for your twin. And vice versa. The two of you see the same thing; the noon Sun at a certain altitude, but at opposite times of the solar year.

Figure 9.8: In this scene, conditions are similar to those in the previous figure, except that it is summer. The shadow is short, indicating a high-altitude Sun.

You may know all of this, but not consciously. Picture in your mind that last holiday card you received. It is one showing an artistic rendering of a sunny, outdoors, scene in winter. Regardless of the artist, a faithful one always includes long *shadows* in the painting. This is indirect evidence of the Sun's low altitude.

Next, imagine a (now rare!) vacation, picture postcard that you receive, one depicting a sunny summer's day at the seaside. Do long shadows seem appropriate to your recollection? No. Summer is a time when shadows are scarce, even to the degree that you create artificial shade using a beach umbrella. This is indirect evidence of the Sun height in the sky.

None of this would happen if the Earth's obliquity was 0°. Full disclosure requires saying that the obliquity the Earth oscillates by 2.4 degrees, back and forth about a mean at which we are close to now, over a 41,000-year cycle. This rocking has implications for the Earth's climate. Yet, the period and amplitude of the variation is such that it is of no practical significance to today's Sun watcher. (The Earth's obliquity is shrinking right now.)

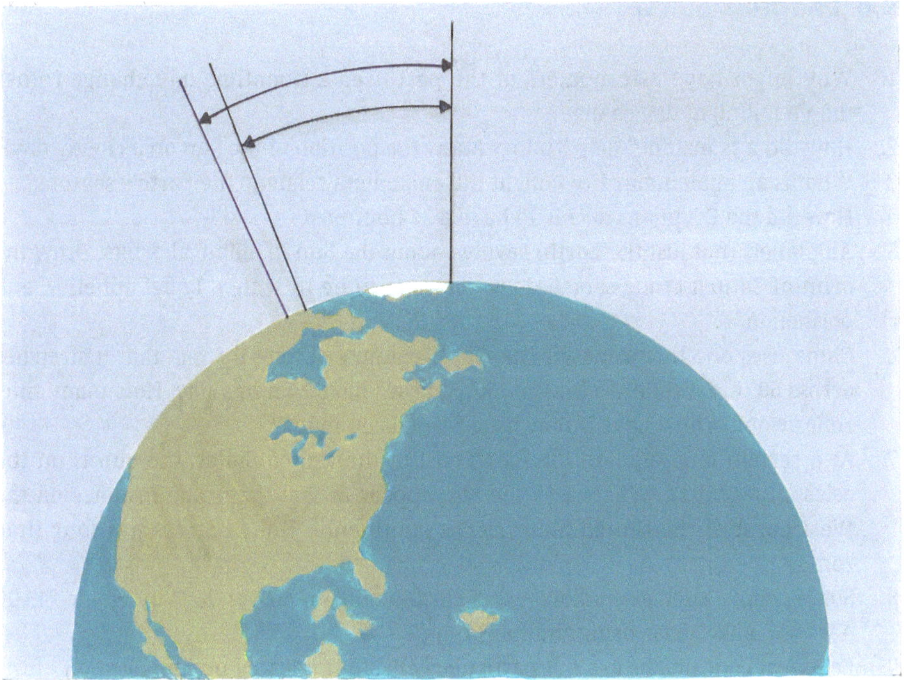

Figure 9.9: Earth's changing obliquity. Artwork by Chris Tarte.

9.5 Chapter summary

We use an imaginary Sun that moves on the Celestial Sphere at a steady rate for the purpose of timekeeping. We further divorce ourselves from the real Sun by introducing hours and time zones to simplify the timing of travel and communication.

The apparent annual path of the Sun is the ecliptic. The position on the Celestial Sphere of the Sun, or other celestial objects (such as planets) that take similar paths, may be recorded in ecliptical coordinates.

The ecliptic and Celestial Equator differ because of the obliquity of the Earth. This obliquity, and the fact that the axis mundi constantly points in the same direction (during human-scale intervals of the Great Year), results in our seasons. The principal cause of the seasons is *not* the eccentricity of the Earth's orbit.

You see the altitude of the Sun cycle through the solar year. Its position in your sky at a given time may be predicted using an analemma.

⚡ 9.6 Chapter review

1. Why might have astronomers of the past used a noontime day change rather than a midnight day change?
2. How did a "sunstone" help Vikings know the position of the Sun on a cloudy day?
3. What is an analemma? How might the analemma relate to the Earth's seasons?
4. How did the Egyptians decide to have a 24-hour day?
5. All planets (not just the Earth) revolve about the Sun in elliptical orbits. Draw the orbit of Saturn (Your sketch does not have to be to scale.). Label aphelion and perihelion.
6. China uses one time zone for the whole country despite the fact that it stretches across 60° of latitude. A circle is 360°, and we have 24 h in a day. How many time zones would China have if they used them as intended?
7. At a certain longitude on the East Coast of the United States, the Sun is on the celestial meridian. Where will the Sun appear at that that same instance on the West Coast of the United States? (The continental United States has four time zones.)
8. Some people, such as soldiers, use a clock with 24 hours on it. Why would "13:00 AM" not make sense using "military time"?
9. Ecliptical coordinates are not particularly useful in describing the location of the star Polaris. Why?
10. Match the season on the left with the circumstances on the right.

summer	Sun at lower-than-average culmination altitude
fall	culmination altitude of the Sun increasing
winter	Sun at higher-than-average culmination altitude
spring	culmination altitude of the Sun decreasing

11. Many people wrongly think that the seasons are caused by the Earth moving closer or further from the Sun. Why is this inaccurate? What really causes the seasons?
12. Consult a map or Earth globe. Is there a country that uses the exact same time as you but in which, the seasons are opposite. If so, what are the names of the two countries?

10 Your year through the seasons

The changing altitude of the Sun, as well as other seasonal differences in your sky, can be illustrated by using a map of the Celestial Sphere.

The proper shape of the Earth (as well as the fictitious Celestial Sphere) is a sphere. It is best modeled by a spherical **globe**. But globes can be impractical. So, you "peel" the skin off your Earth globe and flatten it out. It becomes a **cylindrical projection** map.

10.1 Chapter learning outcomes

After reading this chapter, you will be able to:
- draw a cylindrical projection map of the Celestial Sphere.
- distinguish between the Equinoxes, Solstices, and cross-quarter days based upon the Sun's location in your sky.

10.2 A celestial map

When you force the spherical surface of the Earth onto a plane, distortions occur. The North and South Poles are no longer points. They take up the entire top and bottom edge of the map, respectively. The equator is no longer a circle; it is a straight line segment. And, most oddly, instead of being continuous, the Earth's surface is interrupted – at the right and left margins of the map. (We remember that the left and right sides are in reality the same thing.)

Physical distortions include the overly large appearance of Greenland[1], and an Antarctica that looks very little like its true shape. You put up with all of these distortions for convenience.

What works for the Earth, works for the Celestial Sphere. It, too, can be turned into a planar map for convenience. Instead of oceans and continents, a celestial map shows stars.

You take a globe of the Celestial Sphere and "peel" *its* surface. This flattens out into a map that has all the same distortions as your Earth map.

The North and South Celestial Poles, and the Celestial Equator, are all now represented by line segments. The Celestial Hemispheres are rectangles, and constellations far north or south are stretched out. Most egregiously, the Celestial Sphere appears to terminate on its "left" and "right" sides.

The major difference between the Earth map and the Celestial Sphere map is that your map of the stars is made to be used looking up, whereas your terrestrial

---

1 Perhaps this explains a recent American President's interest in this sparsely populated country

https://doi.org/10.1515/9783111441245-010

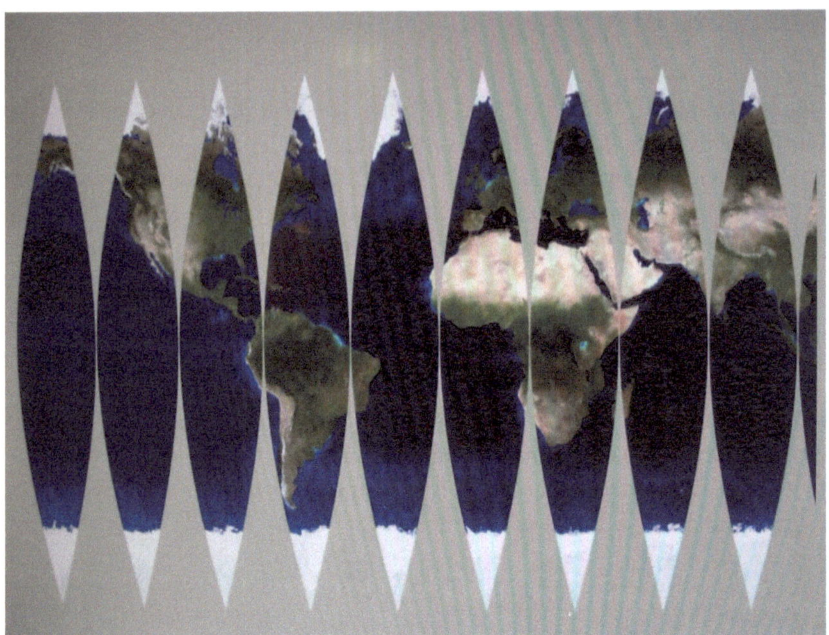

Figure 10.1: Segments from which a globe is made illustrate the first step in creating a cylindrical projection map. Courtesy of Geoffrey.

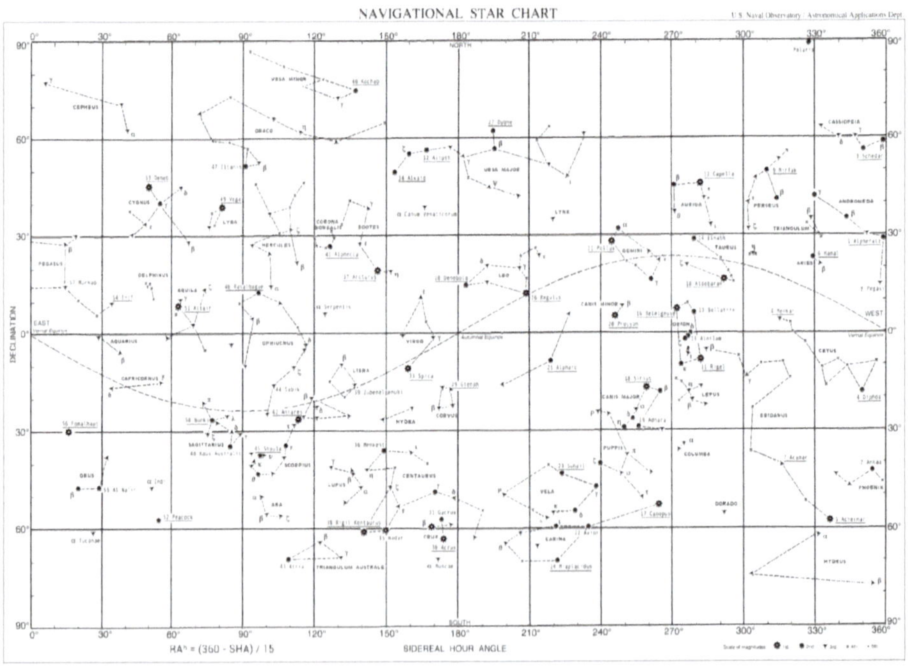

Figure 10.2: Cylindrical projection celestial map. USNO.

map – think of a road map – is meant to be used looking down. So, when north is "up," its traditional orientation, east is to the left and west is to the right on the celestial map – until you pick it up, flip it, and include it in your view upward. Again, you put up with it all for convenience.

10.2.1 Adding the Sun to your map

Now add the Sun to your map. Here you encounter a problem. The Sun never remains at one specific point on the Celestial Sphere as each star does. The best that you can do is to plot its path, the ecliptic, on your chart.

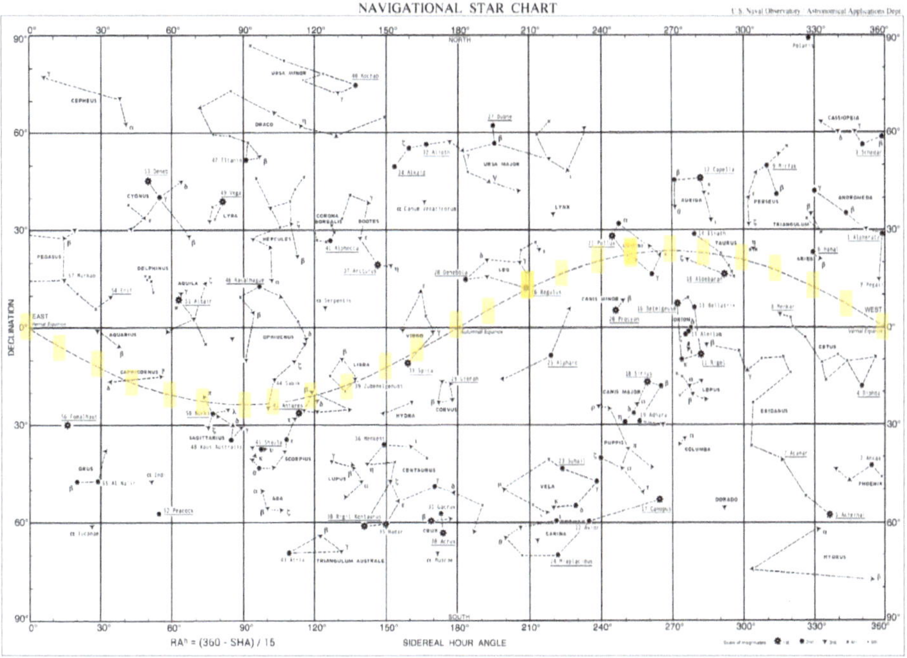

Figure 10.3: Same as Figure 10.2, but with the ecliptic highlighted. USNO. Edited by the author.

The Sun appears to circle the Celestial Sphere once per year, so the ecliptic on your map is a continuous curve all the way from the left-hand side to the right-hand side. The shape of the ecliptic is peculiar. It looks like a wave undulating across the map. But you are ready for this. You know that the ecliptic is really a great circle. Its wavy appearance on the diagram is just another map distortion. In order to have a "straight" Celestial Equator, you must abide a "wiggling" ecliptic.

All this is worth it, though. Because the Sun is at a given point on your map once each year, the map doubles as a sort of calendar. You can draw (say) 365 vertical lines of constant right ascension, right-to-left across the diagram, so as to correspond to the days of any particular calendar year. The current date is represented by the line of right ascension on which the Sun sits.

10.2.2 Implications of your map-as-calendar

Using your map-as-calendar, certain interesting phenomena pop out at you. For instance, the Sun spends half of the solar year in the North Celestial Hemisphere and half in the South Celestial Hemisphere. Equal amounts of the ecliptic are above and below the Celestial Equator on your Celestial Sphere map. It is the times when the Sun is in the North Celestial Hemisphere that the Sun looks particularly high in your daytime sky. For your antipodean acquaintances it is when the Sun is particularly high in the South Celestial Hemisphere.

At the risk of mixing up the geographical hemispheres with the uranographical Celestial Hemispheres, you might say that the Sun is in "your" Hemisphere. You call these times "summery."

It is the times when the Sun is in the opposite Celestial Hemisphere that the Sun looks particularly low in daytime sky. You might say that it is not in "your" Hemisphere. You call these times "wintery."

10.3 Solstices

Notice that there is a point, on your just-constructed celestial map-as-calendar, where the Sun gets as far north as it ever will. This corresponds to the day of the solar year when the Sun is at its maximum noon altitude. The date corresponds to the Summer **Solstice**[2] in the northern hemisphere. (Change the first sentence of this paragraph to read "south as it ever will" in the southern hemisphere.)

The Sun seems to pause at its Summer Solstice altitude for a day or so ("hang time"). It then reverses direction (north to south or south to north) on your Celestial Sphere map and "sinks" in your noontime sky.

This particular day is formally considered the first day of summer, but informally it is the beginning of midsummer in some cultures, based upon climate. In the northern hemisphere it happens on 20 or 21 June each year. In the southern hemisphere the date is 21, 22, or 23 December.

Back to the northern hemisphere, there is that turnaround time when the Sun gets as far south as it ever will. This corresponds to the day of the solar year when

---

2 In the Latin, it is solstitium, meaning "Sun stopped" or "Sun still"

the Sun is at its minimum noon altitude. The date corresponds to the Winter Solstice.

The Sun seems to pause at its Solstice altitude (again, "hang time"). It then reverses direction (south to north or north to south) on the Celestial Sphere map and "elevates" in your noontime sky.

This date is formally considered the first day of winter, but informally it is the beginning of midwinter in some cultures, based upon climate. It happens on 21, 22, or 23 December each year. In the southern hemisphere the date is 20 or 21 June.

> **Where did that come from?**
>
> Throughout history people kept track of the Solstices. Think how you would react if you saw the Sun getting lower and lower each day. A logical question would be: Eventually will there be a day when the Sun does not make it above the horizon at all? Perhaps never to return! This would be calamitous.
>
> Therefore, it is a great relief when the Sun does not seem to be any significantly lower in altitude than it was at the same time the day before. It is extremely noteworthy when it "turns the corner" and starts to noticeably get higher in the sky, day-by-day.
>
> That would signal a time for great rejoicing. Is it coincidental that Christmas, New Years, and other festivals are celebrated so near after the northern Winter Solstice?

10.3.1 Keeping track of the Solstices

Once you realize that a Solstice is annual, you may wish to cheerfully commemorate it. (See the box above.) Or at least solemnly acknowledge it. You will want to mark the event on the actual date – not merely after there is evidence in the sky that it has passed. Knowing the upcoming date becomes important. You need to be able to anticipate it, so that there is the necessary time to schedule, plan, and prepare.

10.3.2 Example: Ireland

Newgrange[3] is a neolithic tomb. Physically, it is a mound 14 m high with a diameter of 90 m. An entranceway leads into a tunnel 19 m long that ends in a 6 m tall cruciform chamber. Newgrange is an example of a passage grave.

A south-facing door automatically suggests some astronomical intent at Newgrange. The Sun, Moon, and planets all appear in the southern sky for an observer living this far north. But nothing is meant to shine through the door of Newgrange regularly.

It was archaeologist Michael O'Kelly, hired to restore the monument, who discovered a small transit *above* the door. It is 1 m wide by 90 cm tall. It had been plugged

---

3 Sí an Bhrú in Irish

Figure 10.4: Newgrange is surrounded by 97 kerb stones. It is estimated that the entire complex weighs 200,000 tons. Courtesy of Ron Cogswell.

by two removable blocks of quartz. Indeed, scratch marks indicate that the block was removed and reinstalled many times.

Around the time of Winter Solstice sunrise at Newgrange, the Sun shines through this transit, illuminating the inside of the tomb. It does so for 17 minutes. This has happened for over 5,000 years.

More than a week either side of the Solstice is not good enough to mark the Solstice itself on some kind of calendar. However, it is anticipatory. A week allows you to prepare for the forthcoming Solstice. While we do not know for certain that this information was used at Newgrange, we do know of many cultures that create alignments purposely – not to the Solstice – but to a date some number of days ahead of it.

Newgrange demonstrates a symbolic connection between human life and the "life" of the Sun: old and weak at the Winter Solstice, vigorous and strong ("at the top of its game") at the Summer Solstice. The Sun will be "reborn" and repeat the cycle. Was it thought that those interred at Newgrange might do so as well?

10.3.3 Azimuth of the Sun during different seasons

How might you determine the Solstice dates experimentally? Let us start with the obvious fact that observing sunrise and sunset is easier than "observing" noon.

Figure 10.5: Thirteen sunrises, one for each "month" of the year, plus one. A hill outside of Taipei appears in each. It serves as a reference, demonstrating the changing solar azimuth. Courtesy of Meiying Lee.

Inasmuch as the Sun is not always on the Celestial Equator, its sunrise azimuth and sunset azimuth change. Only on two days per solar year is the phrase "the Sun rises in the East and sets in the West" strictly true. At all other times, sunrise is somewhat north or somewhat south of east, and sunset is somewhat north or somewhat south of west.

How far north is the Sun when its (apparent) altitude is 0°, and once again when its (apparent) altitude at 0°, depends upon your latitude and how far the Sun is into North Celestial Hemisphere. That is, north, away from the Celestial Equator.

How far south the Sun is when its (apparent) altitude is 0° and once again when its (apparent) altitude at 0° depends upon your latitude and how far the Sun is into the South Celestial Hemisphere. That is, south, away from the Celestial Equator.

A northeastern Sun, as daytime begins, corresponds to a similarly northwestern Sun as it ends. The Sun's right ascension and, especially, its declination change little over this less than 1/365th of the solar year.

The gradual motion of the sunrise or sunset azimuth, along the horizon, is like that of a pendulum on a grandfather clock. The bob swings slowly back and forth, centered on due east or due west, from a northward point (at the northern hemisphere's Summer Solstice or the southern hemisphere's Winter Solstice) to a southward point (at the northern hemisphere's Winter Solstice or the southern hemisphere's Summer Solstice) and back again. Its period of oscillation is one solar year.

The farther away from the equator you live, north or south, the wider the pendulum swings. This makes finding a Solstice, based upon its rise/set azimuth, easiest at high latitude.

To do so, record the date on which the Sun is at a particular azimuth some days before the expected Solstice. Then record the date on which the Sun returns to this azimuth after the Solstice. The date of the Solstice will be halfway between the two recorded dates.

Architecture was frequently aligned to Solstice azimuths. For instance, the Aztecs of central Mexico (middle of the second millennium) routinely marked the Solstices in this way. Today it is almost unheard of to do so. How quickly traditions can die!

10.3.4 Twilight during different seasons

The length of twilight varies with the seasons, too. This is true regardless of which definition of twilight you use. The summertime Sun must cross from either the northern or southern half of your sky dome (depending upon in which hemisphere one lives) to the other half. This is not true in wintertime. The result is that, in the summer, the Sun spends more time just below the horizon and produces longer twilights, the time during which the Sun is below the horizon but still causes illumination above the horizon, than it does in the winter.

Notes from cultures around the world
Angkor Wat is a Hindu, later Buddhist, temple. It is located in Yaśodharapura, the capital of the one-time Khmer Empire. It was built by King Suryavarman II in the twelfth century. The site was restored in the twentieth century.

Angkor Wat is unusual in that, as a whole, it faces west. West is the direction associated with death, because the Sun "dies" here each day, albeit in a more mundane sense than it does at the Winter Solstice. This suggests that its purpose might have been as an eventual mausoleum for the King.

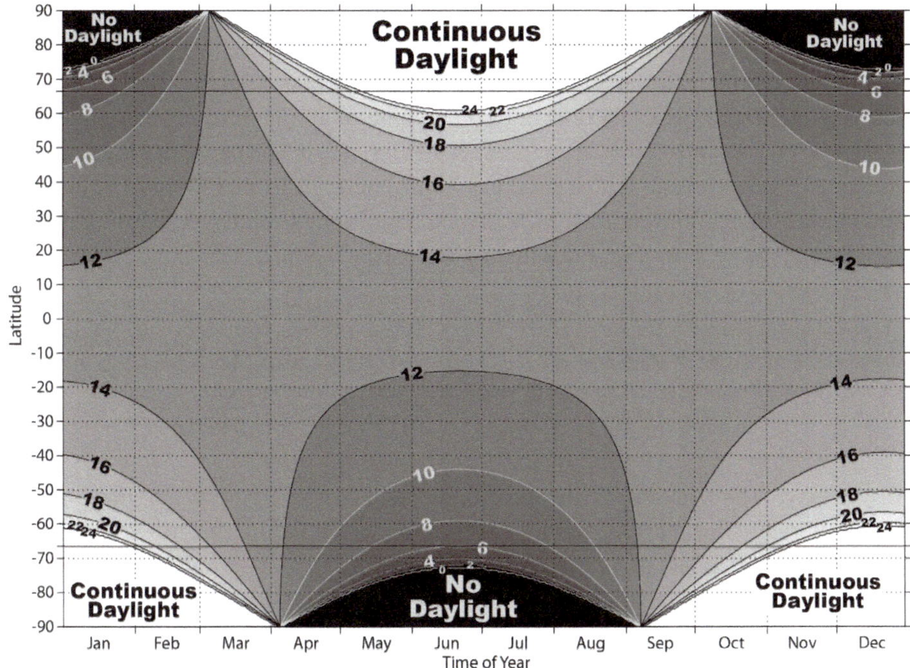

Figure 10.6: The word "daylight" is not rigorously defined. Here, the computer programmer chooses to use it to mean from civil dawn to civil dusk. The curves plotted on the graph are the length of daylight in hours, as a function of latitude and time of year. Choose a latitude. Perhaps it is 40° N. As the year goes by (at this latitude) notice the amount of daylight at one Solstice plus that at the other Solstice does not equal 24 h. This is because of the longer twilight in the winter. Courtesy of TheSevenSeas.

The complex consists of walls, colonnades, galleries, shrines, and a moat. It covers more than a square kilometer.

Astronomical inscriptions abound. Many alignments appear within Angkor Wat's architecture, most prominently those with respect to the two Solstices.

10.4 Equinoxes

So much for the Solstices. Now examine two other special locations/dates on our celestial map-as-calendar.

These are the two places, corresponding to two times of the solar year, when the Sun is not north or south of the Celestial Equator. It is *on* the Celestial Equator. These dates are about halfway between those of the Solstices.

When the Sun appears to be moving southward and crosses into the South Celestial Hemisphere from the North Celestial Hemisphere, you may well mark the occasion, too. This date is called the Autumnal **Equinox**[4] in the northern hemisphere. It is the first day of fall to some. The Equinox "falls" on 22 or 23 September.

The date on which the Sun moves back from the South Celestial Hemisphere to the North Celestial Hemisphere is the Vernal Equinox in the northern hemisphere. It is most often recognized by northerners as the first day of spring, 20 or 21 March. Flip "autumnal" and "vernal" in the southern hemisphere. (Our modern **civil calendar**'s and clock's affectations, known as leap year and the not-related-to-astronomy Daylight Savings Time, may alter the date of a Solstice or Equinox; see Chapter 14.)

> **Misconception alert**
>
> As opposed to "Solstice," where the adjectives following it sound familiar (Winter, Summer), the Latin words still often used for distinguishing the Equinoxes may not (Autumnal, Vernal). For this reason, it is easier to get the Equinoxes mixed up. Try the following mnemonic: "Spring Up, Fall Down."

Yes, we say, "The Sun rises in the East" and "The Sun sets in the West," even though this is literally false most of the time. However, on the Equinoxes, the Sun *does* rise due east and set due west.

For this reason, you might think that it is easier to establish the date of an Equinox than it is the date of a Solstice. Yes and no. That the Sun's position on the horizon at an Equinox is in a cardinal direction certainly helps. However, recall the pendulum. Like a real pendulum, the Sun appears to move across the horizon (change azimuth) the fastest at "mid-swing." This occurs at the Equinoxes, making it easier to miss the exact day.

10.4.1 Limb & vertex

On a weather broadcast, the meteorologist often includes sunrise and sunset times for the date. You may notice that, even on the Equinox dates, the sunrise time is not just the sunset time with the "AM" and "PM" switched. There are not exactly twelve hours between the two as there logically should be for an object on the Celestial Equator.

The Sun is not a point; the luminary looks like a disk. When discussing sunrise and sunset, are we talking about the geometrical center of the Sun above or below the horizon? It is likely that we are not, because when the middle of the Sun is at the horizon, there is plenty of light from the half of the disk above the horizon to still call the result daytime.

The edge of a sphere appearing as a disk is called its **limb**. The Sun's highest limb (furthest from the horizon) is its **vertex**. You are more likely to think of sunrise as the moment when any part of the solar disk emerges from below the horizon, when you

---

4 In Latin it is aequus, meaning "equal" and nox, meaning "night"

first see the vertex. This is first **gleam**. As you are likely to count as daytime any time when any part of the Sun is above the horizon, you probably do not call "sunset" until every bit of the Sun's disk, including the vertex, is below the horizon. This is last gleam. Our definitions of sunrise and sunset artificially lengthen daytime.

10.4.2 Sunrise, sunset, & atmospheric refraction

Yet the vertex is not the main factor in what seems to be overly long Equinox daytime.

The apparent positions of celestial objects are shifted by atmospheric refraction. The geometric direction to the object and the direction from which the light comes are different.

Except at the zenith, atmospheric refraction always moves the apparent position of a celestial object away from the horizon. Further complicating matters is your elevation. The definition of one air mass assumes sea level; on top of a mountain, it corresponds to a shorter distance. Thus, the greater your elevation, the smaller the magnitude of the apparent shift in celestial position caused by atmospheric refraction. Uranographers have to take atmospheric refraction into account, but most of the time its effect is too small to be noticeable to the rest of us.

The luminaries are an exception to that last statement. You easily can see the Sun all the way down to the horizon, where there is the most atmospheric refraction. The last/first rays of the Sunshine upon you after/before any of the real disk of the Sun (meaning the Sun's true geometric direction) already has/has not set/risen. Its light is bent up over the horizon.

Sunset is delayed, and sunrise comes early. The result is that the length of daytime is extended by up to five minutes, depending upon your latitude.

Do not confuse the dates of maximum daytime with the earliest sunrise and latest sunset. The latter two dates are independent of each other and need not correspond to a Solstice.

10.5 Divisions on your celestial map

Let us return once more to your celestial map-as-calendar. Just as it is divided in two by the traditionally horizontal Celestial Equator, it can be divided into four parts by vertical line segments. While the Celestial Equator is an obvious south-north division, our map-as-calendar is really a cylinder; there is no obvious place to put an east-west edge or middle.

Yet, because of the importance of the Solstices and Equinoxes, an uranographer traditionally places line segments on the celestial map-as-calendar. These run north-to-south through the points that represent the Sun's location at these auspicious times. Such demarcations are named the Winter Solstitial **Colure**, Vernal Equinoctial Colure,

Summer Solstitial Colure, and Autumnal Equinoctial Colure, respectively (perhaps with an adjustment to be made due to the hemisphere in which one lives). Each of the resulting four sections represents where the Sun "resides" in springtime, summertime, falltime,[5] and wintertime, respectively.

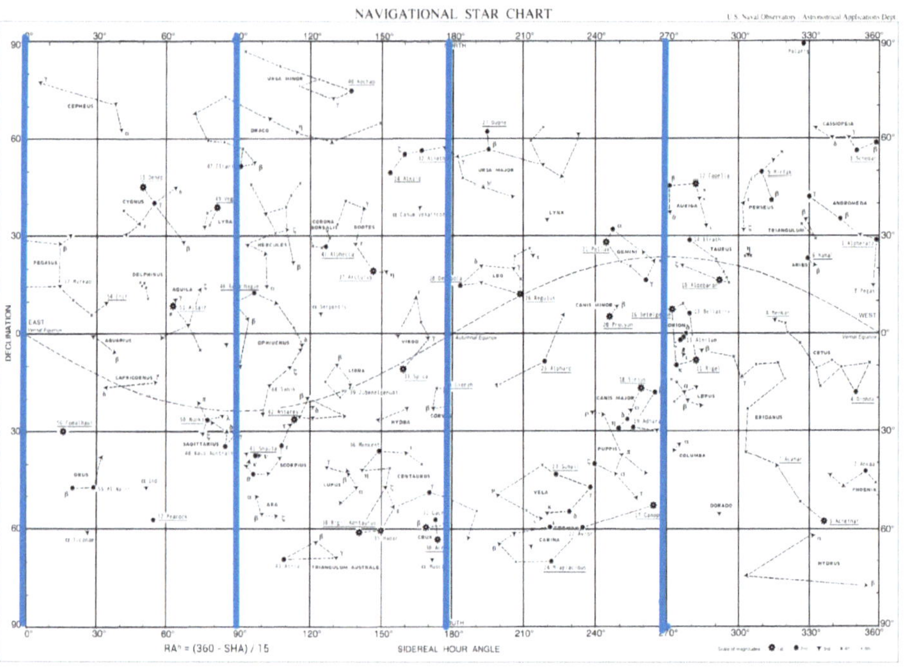

Figure 10.7: Same as Figure 10.2, with the colures highlighted. They appear applicable to the northern hemisphere, in the same order as they are listed in the text above. USNO. Edited by the author.

Tools of the naked-eye observer

Unlike the Earth, an actual globe of the Celestial Sphere need not be completely opaque. An example is the framework sphere called an *armillary*.

An armillary is a minimalistic set of intersecting rings that may include a physical Celestial Equator, ecliptic, Equinoctial Colure, and Solstitial Colure. Together, they create a stylized view of the Celestial Sphere seen from what must be (by definition) an imaginary point of view outside it.

Measurements in the sky can be made using an armillary with scales on its rings. Indeed, if your purpose is a simple task such as establishing the times of the Equinoxes, this instrument may be stripped down ever further. You only need one ring, the *equatorial ring*. Placed in the plane of the Celestial Equator, the Sun will shine all day on the top of this ring in the summer and the bottom of the ring all day in the winter. Only on an Equinox will this observation become ambiguous.

---

5 Use of this term reflects your author's background. The word is little spoken outside of the central United States.

The armillary, or equatorial ring, was more likely to be used as an astronomical teaching device, though. Today it has gone out of fashion.

Instead of referring to the *times* of the Solstices and Equinoxes collectively, you can shorten the expression by referring to the **quarter days**.

Notes from yesterday and today

Portugal has a long nautical tradition. It became a modern nation on the decks of its sailing ships. Because an armillary (preceding Box) can be used in support of navigation, it is part of the state seal. Portugal is the only country that uses an astronomical instrument for its symbol.

While on the topic, it is worthwhile mentioning that Australia and New Zealand are the only nations on the flags of which appear a real (not stylized) celestial object. It is the asterism of the Southern Cross.

10.6 Cross-quarter days

For those whom the Solstices mark midsummer and midwinter, when do summer and winter begin? Spring and autumn? These beginning and end dates may occur approximately halfway between the quarter days. They are called **cross-quarter days**.[6]

There is a cross-quarter day approximately six weeks after the Vernal Equinox (six weeks before the Summer Solstice). There is another cross-quarter day approximately six weeks after the Summer Solstice (six weeks before the Autumnal Equinox). This cross-quarter day is followed by one approximately six weeks after the Autumnal Equinox (six weeks before the Winter Solstice). A final cross-quarter day happens approximately six weeks after the Winter Solstice (six weeks before the Vernal Equinox).

Alexander Thom's survey of British stone circles and their multiple orientations was taken up and made more statistically rigorous by Clive Ruggles. Eventually he had an impressive number of orientations, which he plotted on a graph of frequency versus azimuth. If all the orientations were random, one would expect the plot to be more or less flat. It was not. There were peaks at azimuths that corresponded to special days, with troughs in between. These special days were the Solstices and Equinoxes, but also, with lower statistical confidence, the cross-quarter days. (Considered together, the eight quarter and cross-quarter days are colloquially called the Wheel of the Year.)

Unlike on the quarter days, the Sun undergoes no obvious change in its motion on cross-quarter days. At most we can say that they are inflection points on the ecliptic as it appears on our celestial map.

Thus, are cross-quarter-day alignments to be believed? The answer may well be "yes." There is a history of acknowledging cross-quarter days that survives through the present. Let us start with the fact that they have names.

---

6 or mid-quarter days

10.6.1 Names of cross-quarter days

Also unlike the Solstices and Equinoxes, the names of the cross-quarter days are not universally agreed upon. Those that come to us from ancient Europe are the best known. Yet, as cross-quarter days have a more minor role in modern cultures, even these sound obscure to many.

The names of the cross-quarter days are derived from the so-called "Celtic Calendar." However, there is debate over whether there ever was one such single calendar or, for that matter, one culture called the Celtic people.

After the Vernal Equinox is Beltane (about 1 May). Young couples were allowed to meet, before another six weeks led to marriage. They had the time between Beltane and the Summer Solstice to get to know each other! June still is a traditional month for nuptials.

This cross-quarter day is followed by the one named Lughnasad (about 1 August), when crops ripen. If some of the cross-quarter dates seem a little "off," later calendar reform may be to blame.

Continuing: On Samhain (about 1 November), the summer is clearly ended. The spirits of the dead, in the form of animals, looked for souls to steal. Dressing up as an animal or carving a human face confused these spirits. (In Mexico this date still is the "Dia de Loas Muertos," the Day of the Dead.) The next day after Samhain, the spirits were calmed.

Figure 10.8: 2007 celebration of Imbolc in England. Courtesy of Steven Earnshaw.

Finally, Imbolc (about 1 February), when lambs were born, was the time to estimate how much longer winter cold would last. How long was it to be until planting was possible?

These were each festival dates, just as were the quarter days. The Christian Church, eager to stamp out "pagan" celebrations around the world, came upon the idea of replacing them with its own feast days. These were on or near the same date.

The most obvious of such overlaps is Christmas. We still use a (supposedly) Celtic word for the period near the Winter Solstice: It is "Yule," and is a synonym for the more or less, nonreligious celebration of Christmas.

Less well known is the "Midsommar" (Summer Solstice) surrogate, the Feast of Saint John the Baptist. It occurs with the same precession-induced time lag of a couple of days past the Solstice date.

Easter is historically associated in time with the Jewish Passover. It is the beginning of this older holiday that is tied to the Vernal Equinox. The same linkage is true of Rosh Hashanah and the Autumnal Equinox.

The cultural transference of the cross-quarter days by the Christians was even more successful. Beltane became the Feast of Saint Joseph (1 May); unbaptized children who had died were symbolically christened. Lughnasad, the beginning of the *harvest*, became Lammas (1 August; Loaf Mass Day). Samhain became All Saints Day (1 November). Imbolc became Candlemas (2 February; the presentation of Jesus at the Temple). On this last day, people brought to the church important things – like candles – to be blessed.

10.6.2 Contemporary cross-quarter days

Today, cross-quarter days have been trivialized into more whimsical, unofficial holidays. Most famously, the cross-quarter day in the fall survives as Halloween (31 October).

That of winter is the even-more-curious Ground Hog's Day (unique to the USA and Canada). The spring cross-quarter day evolved into customs associated with May Day.

The summer cross-quarter day is largely forgotten. While the United Nations designates 30 July as the International Day of Friendship, and 9 August as the International Day of the World's Indigenous Peoples, it is difficult to connect these commemorations directly to the cross-quarter day. There are various August cross-quarter-day celebrations – here and there – in Europe, though.

An exception to the "back seat" that cross-quarter days have taken occurs in Japan. Here, the climate still lends itself to starting the four major seasons on the cross-quarter days.

10.7 Seasons are not of equal length

The Sun does not reach each of the Solstices, Equinoxes, and cross-quarter days in equal intervals of time. This may have been first noticed by the classical Greek Euctemon around 430 BC/BCE. The cause, unknown to Euctemon, is the solar anomaly.

Previously, it was noted that the elliptically solar-orbiting Earth arrives at perihelion (the point at which it travels fastest) in the month of January. It arrives at aphelion (the point at which it travels slowest), one half year later, in July.

If perihelion and aphelion occurred coincidentally on the Solstices, the lengths of spring and summer, and of autumn and winter, would be of equal duration. However, this is almost never true. The line connecting perihelion and aphelion, called the **Line of Apsides,** rotates at an extremely gradual rate around the Earth's orbital plane. It rounds to one day every 58 solar years! The direction is counterclockwise as viewed from above the North Pole.

This would not happen if the Sun and Earth were in isolation. However, our planet shares the Solar System with other planets. Their gravities – particularly that of massive Jupiter and Saturn – modify the Earth's orbit in this regular way.

Presently, the resulting asymmetry is increasing. In the year 1246, the season-lengths nearly *were* the same. In the year 6430, aphelion and perihelion will coincide with the Equinoxes. In the distant future, the pairs of seasonal lengths will once again be equal.

The sidereal and solar year are so close to each other in duration. It is not always been necessary to distinguish between them except in certain situations.

Here is yet another definition of the year, useful when discussing certain topics, that has a small-but-noticeable effect on what you see. It is the **anomalistic year**, measured as the time it takes the Earth to travel from peri-/aphelion to peri-/aphelion. In other words, it is the length of time from the Line of Apsides to the Line of Apsides to the Line of Apsides again. Its average length is 365.260 days.

Compare two **civil years** (1 January through 31 December; see Chapter 14): For one, an anomalistic year happens to begin just before your civil year does. Because it is the longer of the two, that anomalistic year will end just after the end of your civil year.

During this particular civil year, the Earth is near perihelion more than it is near aphelion. The solar anomaly results in a faster Earth speed for more time, this civil year, than a slower Earth speed. The length of the solar year simulated by this civil year will be a little short.

What if the anomalistic year starts mid-calendar year? For this particular calendar year, the anomalistic year begins such that the Earth will be near aphelion more than it will be near perihelion during the civil year.

The solar anomaly results in a slower Earth speed for more time that civil year than it does a higher Earth speed. The length of the solar year simulated by this civil year will be a little long.

Another way of saying the same thing: If in the first example you time a solar year starting at midnight on New Year's Day, it will end at a time later than in the second example.

The duration of these two sample solar years, begun and completed so as to correspond to a civil year, nonetheless are not the same. The first example is shorter than the second. This is a very restricted way of thinking about the concept of year. But the odd result is that, technically, your civil year should vary in length!

10.8 Not all seasons are based upon astronomy

It must be acknowledged that today people are increasingly removed from a direct relationship with the solely astronomical seasons. Your definitions of the seasons may have devolved into strictly calendrical sets of three months: They begin with 1 December, 1 March, 1 June, and 1 September, respectively. These are the start dates of the so-called **meteorological seasons**.

There also are location-specific climatological seasons such as the wet/rainy season, dry season, hurricane season, wildfire season, and the very technical monsoon season. In parts of the tropics, it is useful to divide the solar year into six seasons!

Seasons may be defined based on biological phenomena, for instance, deciduous trees:

Revernal – budding
Vernal – leaves appear
Estival – full leaf
Serotinal – leaves begin to change color
Autumnal – leaves in full color
Hibernal – bare

Or birds:

Revernal – migration from winter to summer habitat
Vernal – nest building
Estival – eggs hatch
Serotinal – fledglings reach maturity
Autumnal – migration from summer to winter habitat
Hibernal – least avian movement

Clearly, these definitions of seasons have nothing to do with astronomy.

Returning to those that do, there was a time when, in one familiar civilization, the seasons were more complicated. The solar year was divided into *eight* seasons.

For the Romans, according to the contemporary Marcus Varro, the first {1} occurred from the beginning of the westerly winds in February to the Vernal Equinox and lasted for more or less 45 days. Another {2} took place until the heliacal rise of the

Pleiades (around 7 May) and lasted for approximately 44 days. From then {3} to the Summer Solstice was about 48 days. There were only about 27 days {4} until the heliacal rising of Sirius (close to 21 July). Now came a long stretch {5} of 67 days until the Autumnal Equinox. Another season {6} was 32-or-so days long and covered the remaining period until the acronical set of the Pleiades (about 28 October). It was now 57 days {7} until the Winter Solstice. The remaining 45 days {8} lasted until the seasons repeated.

What was required to mark all these seasons? You needed to know about the stars, Sun, *and* weather.

The record for rote memorization, though, must go to the traditional Japanese calendar. In it, there are 72 seasons!

10.9 Chapter summary

A cylindrical projection map of the Celestial Sphere doubles as a calendar. On it, special days stand out: the Solstices and Equinoxes, which astronomers consider to be the first days of each of the seasons (though this definition of seasons is not the only one possible).

An archaeoastronomical monument such as the Newgrange passage grave illustrates how the Sun's rising or setting azimuth changes through the seasons. However, something as seemingly simple as sunrise or sunset is complicated by the fact that the Sun is a disk (for instance, its limb has a vertex) and by atmospheric refraction. Twilight also changes as a function of season.

Between the Solstice and Equinox dates, there are the less-often-referenced cross-quarter days. The time between Solstices, Equinoxes, or cross-quarter days is not constant. The Earth's perihelion and aphelion can occur during any season. Another way to state this is that the anomalistic year, defined by the Earth's apsides, has a unique period.

10.10 Chapter review

1. How is the position of the Sun in your sky different in the summer than the winter?
2. How do we know when a Solstice occurs?
3. Why might more automobile accidents be common on and near the Autumnal and Vernal Equinoxes?
4. Why might you see the Sun rising before it has reached altitude 0°?
5. Describe the significance of cross-quarter days for one culture.

6. Draw a line segment that represents a calendar year. Mark the Solstices, Equinoxes, and cross-quarter days. Label the Solstices and Equinoxes by name. What do you notice about the spacing of these events on your "calendar"?
7. Why are the seasons not of equal length? In other words, explain the solar anomaly.
8. In what sense are the seasons a part of the natural world? In what sense are seasons human constructions?
9. Why might an Australian celebrate Christmas at the beach with a barbeque?
10. Why is it said that a celestial map can also be a calendar?

11 Seasonal insolation and daytime/nighttime

11.1 Chapter learning outcomes

After reading this chapter, you will be able to:
- list three noticeable effects of the change in seasons.
- determine your location on the Earth based upon seasonal changes.
- point to significant circles of latitude on the terrestrial globe.

11.2 Insolation

Admittedly, it is not the altitude or azimuth of the Sun that you commonly think of as marking the difference between the seasons. For most of us, what the seasons are famous for is this: When it is winter, it gets cold! When it is summer, it gets hot! Average temperature probably is the major thing you associate with the seasons.

Why the change in warmth? The varying average temperature with the seasons has everything to do with the Earth's obliquity and the sliding altitude of the Sun.

11.2.1 Efficiency of heating by the Sun

It is the Sun that provides virtually all the heat that the surface of the Earth receives. When absorbed by something opaque, most of that light's energy is turned to heat. It is true that about 30% of sunlight reaching the Earth is reflected back by the atmosphere, clouds, and air pollution. However, we are interested in the heat provided by the remaining 70%.

Imagine a discrete bundle of the Sun's rays. You can do this because there is only a finite amount of solar ultraviolet through infrared intercepted by the sphere that is the Earth at any time.

For simplicity, consider midday. When the Sun is at a low altitude, the bundle of energy intercepts the Earth's surface obliquely. That amount of energy is spread over a large area. Any one square foot, meter, yard, or acre does not receive much of it, and not much warming takes place.

You understand this, at least subconsciously. When you wish to warm your hands at a campfire, without much thought you hold them out with flat palms closest to the source of heat; you do not point all your fingers toward the flames.

Think of solar energy as paint in a bucket. If you splash it out onto the floor at a narrow-to-the-horizontal angle, a lot of floor gets coated, but at no place is the paint very thick. If you dump the bucket nearly straight down upon the floor, it is deposited thickly in one (albeit small) place. You will not have to worry about giving this spot a second application!

https://doi.org/10.1515/9783111441245-011

Figure 11.1: Each yellow bar is of equal width and illustrates the same unit amount of heat delivered upon two different latitudes on the Earth. However, because of the sphericity of the Earth, these units intercept the Earth's surface at different angles. The top unit is spread over a greater area (represented by the longer arc of the Earth-bar intersection) than the bottom one; its heat is diluted, and insolation is less effective than in the bottom unit (See text below.). Artwork by Christine Tarte.

Back to the real Earth and Sun. Meanwhile, in the opposite hemisphere, the Sun is at a higher altitude in the sky, and an equal bundle of solar rays hitting the Earth here is not spread over such a great area. There is little dilution. The heating is more concentrated and more effective.

The heating efficiency of the Sun at a given place on the Earth is called **insolation.**[1] Our illustrations took place at noon, but all through the day the insolation will be more effective where the Sun's energy strikes the Earth closer to the vertical.

The fewer number of hours during which the Sun appears above the horizon on a winter's day further reduces diurnal insolation at that time of solar year, while the greater number of daytime hours in the summer adds to insolation during those days.

---

1 not "insulation"

The difference in insolation is most dramatic when comparing the Solstices. The difference is detectable, to a lesser and lesser extent, for some time before or after the Solstices, too.

11.2.2 Example: Chaco Canyon, Colorado Plateau, North America

You hear a lot about designing energy-efficient buildings. The concept is old news, though: Such architecture was practiced in the American Southwest a millennium ago by the Ancestral Puebloan people.

The culture that inhabited Chaco Canyon *circa* 1100 apparently had solar insolation in mind when they built their 800-room, four-story, apartment-style community. It is now named Pueblo Bonito. The building is carved into the Canyon wall facing south. The high north wall reflects heat during the winter. Terracing allows heat absorbed by the front rooms during cold days to reradiate to the back rooms at night. Moreover, the design and location provide plenty of shade to common spaces in the Summer. At all times of solar year, thick walls serve as heat sinks to equilibrate temperature extremes.

Figure 11.2: Pueblo Bonito, Chaco Culture National Historical Park. It covers 1.2 hectares. Courtesy of Arian Zwegers.

11.2.3 Thermal lag of the Earth

The particular weather you experience at a particular location on the Earth varies from date to date and year to year. Nevertheless, most of us associate one particular time of solar year as that during which it is hottest and another time of solar year as that during which it is coldest on average. These dates normally are *not* the Solstices. What at first seems counter to the theory of insolation here presented is, in fact, further proof of its validity. You should not *expect* those dates to be the same.

How do you make tea? Do you place a full kettle of tap water on the stove, immediately remove it from the burner, and then soak a tea bag? This is likely not your recipe. The result would be lukewarm tea at best. It takes *time* on the stove to heat up an entire kettle of water.

It takes even longer to warm up – or cool down – an entire Hemisphere of the Earth. There is a **seasonal lag** between maximum insolation and the date on which your hemisphere is hottest. This delay may be six weeks long. Similarly, it takes time to cool an entire Hemisphere of the Earth. Again, there is a time lag, of similar length, between minimum insolation and the date on which your hemisphere is coldest.

In a northern mid-latitude climate, you might wish that the chilliest day of winter is in late December. In fact, these days usually are mild compared to what you experience in later January. Likewise, your summer is most oppressive in July/August, not June. The exact seasonal lag depends upon how close you live to the sea (a good heat sink) and your latitude. Those in southern mid-latitudes have a similar experience half a year earlier/later.

The northern hemisphere is typically a bit hotter in summer (and colder in winter) than is the southern hemisphere. Again, this has virtually nothing to do with the aphelion and perihelion of the Earth.

There are several real reasons. One is simply that there is more land and less ocean in the northern hemisphere as opposed to the southern. The hottest day that the Earth ever experienced occurred during the northern summer: 21 July 2024 (as of this writing).

11.3 Winter versus summer daytime & nighttime

The length of daytime is shorter in the winter than in the summer. Conversely, the length of nighttime is longer in the winter than in the summer. Contrary to common belief, it is the altitude of the Sun and length of daytime that cause deciduous trees to change leaf color, not temperature.

Just as a star in the South Celestial Hemisphere spends little time above your horizon if you live in the northern hemisphere, so does the Sun when *it* is in the South Celestial Hemisphere. It rises late and sets early. The result is the short daytime hours of your fall and winter (and long nighttime hours). In the southern hemisphere, the reference is to the North Celestial Hemisphere.

In the northern spring and summer, when the Sun is in the North Celestial Hemisphere, it spends more than half a day above your horizon. You enjoy the long daytime for summer recreation, perhaps at the expense of a short nighttime conducive to sleep. Of course, it is the other way around in the autumn and winter. The reference is to the South Celestial Hemisphere if you live in the southern hemisphere.

11.3.1 From terminator to terminator in summer

Any sphere placed in a beam of light is 50% illuminated. The rest of the sphere is in its own shadow. (The Earth produces no light of its own.) The Earth in sunlight is no exception: Half of the world is always in daytime, half is not (ignoring the minor effect of atmospheric refraction).

As the Earth rotates, a given location on the Earth may move from daytime to nighttime. The "border" between the two is the **terminator**. It is the sunrise/sunset circle, all around the Earth.

Figure 11.3: The Earth's terminator. Courtesy of NASA.

Any spot between the poles of the Earth travels in a circle of constant latitude, due to diurnal motion. Because the Earth turns at a constant speed, the length of the arc in daytime determines the length of time that any place on that circle spends in daytime.

The more of the arc that is lit, the longer daytime you experience at that location. The less of that arc that is lit, the longer nighttime you experience at that location.

Think of the Earth with its North Pole tilted toward the Sun. It is the northern Summer Solstice. For any location you choose in the mid-latitudes of the northern hemisphere, most of your latitude circle is illuminated on this Solstitial day; you experience the longer daytime (and shorter nighttime) of summer. Looking at your clock, the Sun seems to come up early and stay up late. Durations are exactly the other way around at the comparable latitude in the southern hemisphere.

Figure 11.4: The duration of your daytime on the northern Summer Solstice is proportional to the length of your latitude's sunlit arc, terminator to terminator. Photograph by Aaron Spurr.

As the solar year goes by, the geometry of the Earth and Sun changes. Eventually the Earth's North Pole no longer points toward the Sun. It points at a 90° angle to that direction.

The Earth has traveled one quarter of its annual journey about the Sun since your Summer Solstice. It is the northern Autumnal Equinox and southern Vernal Equinox. The axis mundi is not pointing toward *or* away from the Sun. Half our mid-latitude circle is illuminated, half is not. Equivalently, the length of the sunlit arc through which you travel on that day is equal in length to that of the shadowed arc. On this day, half of your time is in daytime, half is in nighttime.

In fact, it does not matter where you are on the Earth at an Equinox. The terminator passes over both poles. No latitude is affected differently from any other latitude. Wherever you are, you spend about 12 hours in light and about 12 hours in the dark.

Figure 11.5: The duration of your nighttime on the northern Winter Solstice is proportional to the length of your latitude's shadowed arc, terminator to terminator. Photograph by Aaron Spurr.

11.3.2 From terminator to terminator in winter

Later in the solar year, the Earth has made it halfway around the Sun since you began this tour of our planet's annual revolution at the northern Summer Solstice. At the northern Winter Solstice, the illuminated arc lengths in the two hemispheres are reversed. Just as yours was longer than that in the southern hemisphere one Solstice ago (equivalent to more than 12 hours), it now is shorter (equivalent to more than 12 hours). Looking at your clock, the Sun seems to come up late and go down early. Durations are exactly the other way around at the comparable latitude in the southern hemisphere.

Many days later, the Earth is now three-quarters of the way through its annual trip about the Sun. The northern Vernal Equinox (and southern Autumnal Equinox) brings you back to the spherical geometry of equal portions of daytime and nighttime. The bottom line: Ignoring some minor effects, the length of daytime and nighttime in either hemisphere is the same as it was in its co-hemisphere half a solar year before, a symmetric reversal you associate with a mirror image.

Addressed here is the duration of daytime and nighttime. Recall that the actual date, hour, minute, and second of sunrise and of sunset is governed by the fact that the Earth's velocity, as it revolves about the Sun, is different at different times of solar year. Moreover, the dates of earliest/latest sunset and earliest/latest sunrise do not necessarily occur on a Solstice.

11.4 Seasons at different latitudes

If you lived at the equator, this discussion about the seasons might be met by you with a yawn. Because of the symmetry of the globe, the sunrise and sunset hours do not vary much at the equator. It is as if it were near an Equinox all the time. Indeed, the effects of the different seasons are all muted near the equator.

On the other hand, seasonal effects become more extreme as you travel closer to either of the Earth's poles. For instance, traveling north, you come to a latitude at which the length of northern winter solstitial nighttime equals 24 hours.

You have reached the latitude at which, on this day, the Sun does not rise at all: 66 ½° N. (Again, the finite size of the Sun's apparent disk and minor effect of atmospheric refraction are ignored.) You stand on the Arctic Circle.

Notice that this latitude is just 90° minus the obliquity of the Earth. Farther north, there are more of these extreme days on either side of the Solstice.

11.4.1 The Arctic & Antarctic circles

High-latitude residents get their missing daytime hours back: At the Arctic Circle, on the northern Summer Solstice, there is a day on which the Sun never sets. Twenty-four hours of sunlight is often an excuse for a big party in these lands. The tourist brochures call it the "Land of the Midnight Sun," skipping over the fact that, half a solar year before and after, it is also the "Land of the Midday Gloom." Again, if you go farther north, there are more of these extreme days, those of continuous sunlight. (Keep in mind that atmospheric refraction makes slight adjustments to the dates and latitudes of these 24-hour events necessary.)

The biggest cities north of the Arctic Circle are in Russia. Topping the list is Murmansk, at more than 309,000 people.

Misconception alert

Few people actually live at a latitude above the Arctic Circle, and those who do live fairly close to it. Therefore, twilight dictates that even at the Solstice beginning winter, they may not experience the black of night for 24 hours. Even at the Solstice beginning summer, they may not experience the Sun in their sky all night long ("White Nights").

Traveling poleward from latitude 54 ½° north or south, it gets no darker than nautical twilight on the day of the Summer Solstice. Traveling poleward from latitude 60 ½° north or south, on the same date it gets no darker than civil twilight. Similar statements can be made about the Winter Solstice.

Figure 11.6: The Arctic Circle is in orange, the equator is in white, and the Antarctic Circle is in red. Photograph by Aaron Spurr.

Of course, you might be an international jetsetter. Once the Sun is about to set at your far-northern latitude, you can fly off to the same latitude in the southern hemisphere and *never* have to watch the Sun go down. There is an Antarctic Circle, too, at 66 ½° S. (The atmospheric refraction caveat is necessary here as well.) In the far South, extreme latitude effects are annually reversed. There are no permanent residents south of the Antarctic Circle, though, unless you include some species of seal and penguin.

Where did that come from?
It is not surprising that the Arctic has something to do with astronomy. The root word comes from the Greek *arktikos* for "bear." So?

Two prominent arctic asterisms, so because they are always high and circumpolar in the sky there, are the Latin-named Ursa Major and Ursa Minor. These are the Great Bear and Lesser Bear, respectively.

Figure 11.7: At 78° 56′ N., the 100-or-so residents of Ny-Ålesund, Norway, experience a Winter Solstice on which sky-brightness is no greater than nautical twilight. Courtesy of Harvey Barrison.

Things seem to get really out of hand near the Earth's North and South Poles. Of course, because the altitude of the Sun never gets very high at these extreme latitudes, it is always cold, either daytime or (especially) nighttime.

On the pole, the Earth spins, but as you stand upon the intersection of its surface and the axis mundi, this rotation no longer transports you anywhere. The Earth's diurnal motion no longer affects the altitude of the Sun.

In summer, the luminary just circles the horizon over the course of the day, maintaining nearly constant altitude. The only change in solar altitude that occurs at all does so over the longer period of the Earth's revolution. That revolution carries pole sitters across the terminator twice per solar year. The cycle of daytime and nighttime, and of the solar year, have become one and the same: half a year of illumination and half a year of darkness.

Figure 11.8: Time-lapse demonstrating 24 hours of daytime. In the northern hemisphere, this phenomenon is experienced from the most northern parts of Norway, Finland, Sweden, Greenland, Iceland, Canada, the United States, and Russia. There are no countries south of the Antarctic Circle. Courtesy of Виктор Габышев.

11.4.2 The Tropics of Cancer & Capricorn

There are two other circles of latitude conventionally highlighted on the globe besides the equator, Arctic Circle, and Antarctic Circle. These are labeled the Tropic of Cancer (23 ½° N.) and Tropic of Capricorn (23 ½° S.). The fact that each is at an angle from the equator equal to the Earth's obliquity hints that they have something to do with the Sun.

North of the Tropic of Cancer, the Sun always appears in your southern sky, no matter what time of solar year. Sometimes it is higher than average at noon, sometimes it is lower. But it is always south.

South of the Tropic of Capricorn, the Sun always appears in your northern sky, no matter what time of solar year. Sometimes it is higher than average at noon, sometimes it is lower. But it is always north.

> **Notes from yesterday & today**
>
> Tiny Necker Island is the last islet in the Hawai'ian chain worthy of the name. It looks too small for permanent human habitation–less than a fifth of a square kilometer in area! Yet it is the site of 33 indigenous temple platforms.
>
> What is special about Necker, such that native Hawai'ians would voyage to it for ceremonial purposes? Could it be that it sits on the Tropic of Cancer?

For those of us at temperate latitudes or higher, the Sun *never* reaches an altitude of 90°. Not on any noon of the solar year. So, what about the phrase, "The Sun is overhead at noon"? For most people, most of the time, it is false. Between the Tropics are the latitudes where the Sun *can be* at the zenith. Exactly on a Tropic, this occurs just once, at a Solstice.

The Tropic of Cancer passes through the countries Algeria, the Bahamas, Bangladesh, China, Egypt, India, Libya, Mali, Mauritania, Mexico, Myanmar, Niger, Oman, Saudi Arabia, Taiwan, the United Arab Emirates, and Western Sahara. The Tropic of Capricorn passes through Argentina, Australia, Botswana, Brazil, Chile, French Polynesia, Madagascar, Mozambique, Namibia, Paraguay, and South Africa.

Closer to the equator, this zenith phenomenon occurs twice per solar year. It happens farther and farther apart in time as the equator is approached, on dates symmetrically straddling that of a Solstice. At the equator, the noon Sun arrives at the zenith on dates as far from the Solstices as possible, those of the Equinoxes.

What about navigation when the Sun is at your zenith? It is no longer possible to use the direction of the Sun to indicate north, south, east, and west!

Summarizing, as the Sun is always within 23 ½° of the Celestial Equator, it is only between the Tropics that the Sun can be directly overhead. Forever within 23 ½° of the zenith at noon, it is *always* high in the tropical sky, thus explaining the year-round warmth experienced at those latitudes.

In the tropics, most celestial action takes place high in the sky. This is convenient, because the jungle often found in tropical lands obscures the view of events near the horizon.

> **Notes from long ago**
>
> A Zapotec city (circa fourth century; in Spanish, Monte Alban) lies within the Tropics. German architect Horst Hartung found that, in a Monte Alban building, the Sun shines down a vertical tunnel in the ceiling on the two zenith-passage days. Such an opening is called a *zenith tube*. The tube is so narrow that sunshine reaches the floor only on these two days (plus maybe a day before or after). There is a much later zenith tube at Xochicalco.
>
> Another ruined city in Mexico (*circa* 600) has been given the Spanish name Alta Vista. It is located within 20 km of the Tropic of Cancer. This is so suspiciously close that we may speculate whether Alta Vista was established intentionally where it was, based upon knowledge of the Tropic.

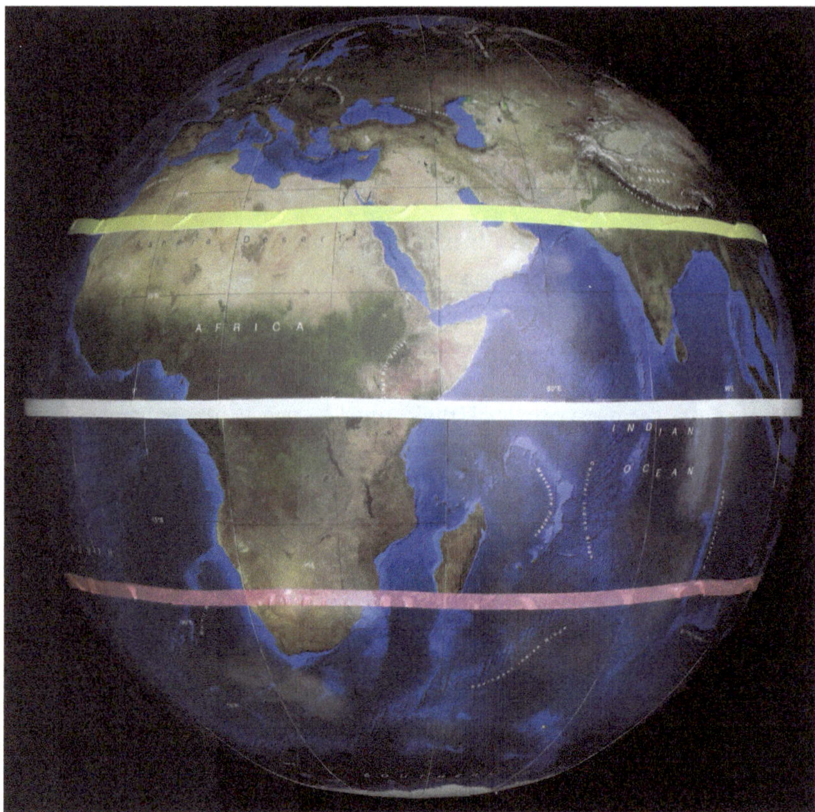

Figure 11.9: The Tropic of Cancer is in yellow, the equator is in white, and the Tropic of Capricorn is in pink. Photograph by Aaron Spurr.

Figure 11.10: At the right time on the right day for their location, these automobiles cast no shadow at all. The Sun is at the zenith. (We have discussed the problem of scale in astronomy; did you notice that these are *toy* cars?) Courtesy of Jason Thien.

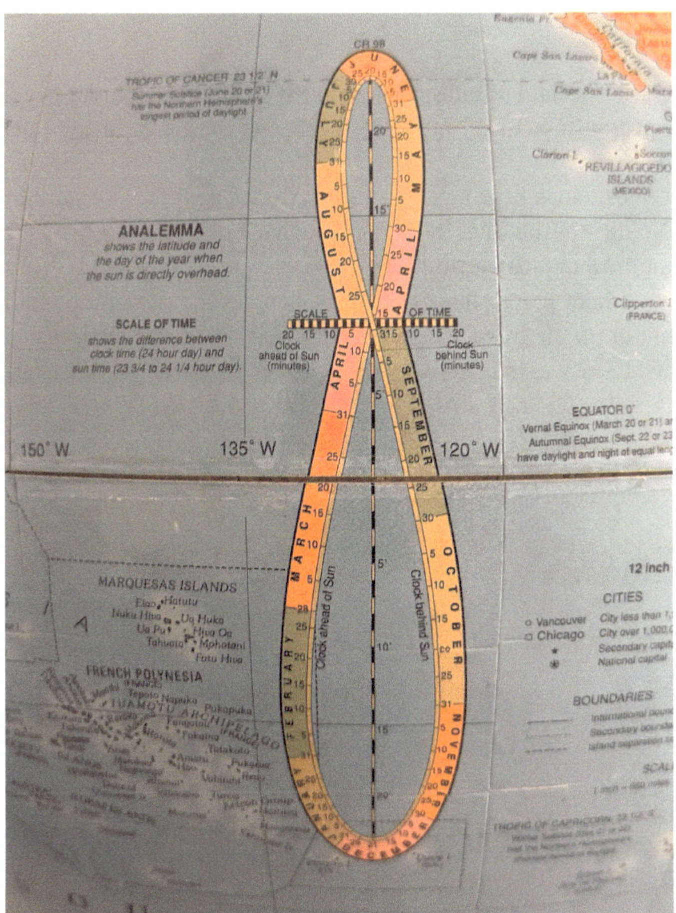

Figure 11.11: An analemma centered upon the equator traditionally appears on terrestrial globes. It is dated and used to indicate at what latitude the Sun reaches the zenith for each day of the year. Photograph by the author.

Incidentally, the Tropic of Cancer and Tropic of Capricorn technically are misnamed. Once upon a time, the Sun did reach its Solstitial destinations on the Celestial Sphere when it was in the constellations of Cancer and Capricorn. Precession since has changed that, but the names remain. It is for the same reason that the First Point of Aries is not situated in the constellation of Aries, anymore.

11.4.3 Erathosthenes

Who started painting these special latitudes on a globe? It was the Hellenistic Greek Eratosthenes, the first person to actually measure the size of the Earth in the third century BC/BCE.

The story is that Eratosthenes heard an interesting rumor about a city south of his: At noon on a certain date of the solar year, the Sun shone down a well in Syene. He knew that this meant the Sun must be at the zenith.

This was not the case where Erathosthenes lived; the Sun never reached the zenith in Alexandria. So, legend has it, he hired a man to visit Syene. His assignment was to check out the well, and while he was at it, to pace off the distance between the two cities!

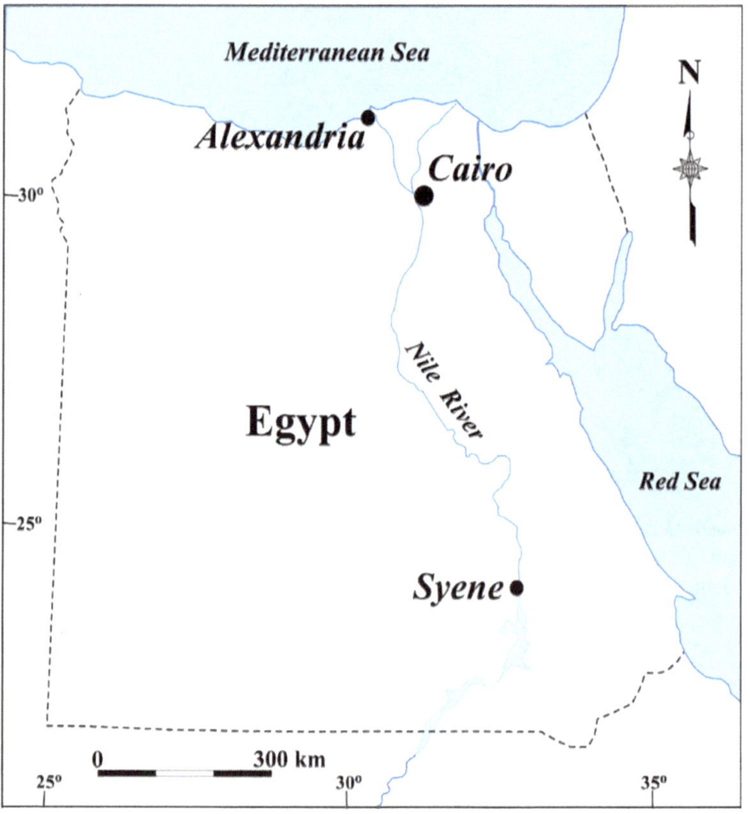

Figure 11.12: Alexandria to Syene, modern Aswan. Courtesy of the CIA.

The report on the well turned out to be true. Erathosthenes was then able to use {1} the altitude of his home Sun, at the same time it illuminated the bottom of the famous well; and {2} the physical distance between the northern and southern city on the circumference of the spherical Earth in order to calculate the planet's entire circumference. From that, simple geometry yielded the radius (or diameter) of the Earth. Eratosthenes was the first person to know how large the Earth is, even though he himself traveled over very little of it.

Figure 11.13: The Eratosthenes experiment works with vertical towers, too; these are easier to illustrate. Notice that the tower in Syene casts no shadow, but that in Alexandria does. The shadow makes an angle with the top of the tower.[2] This angle is equivalent to the angle, measured at the center of the Earth, between Alexandria and Syene.

Knowing the physical distance between these two places, a simple proportion shows what fraction of the Earth's circumference the Alexandria-Syene distance is. The result, of either the tower or well experiment, is the physical circumference of the Earth. Photograph by the author.

11.4.4 Duration of twilight

 is a drawn-out affair in the Arctic and Antarctic. The Sun makes a very shallow angle with respect to the horizon and spends a long time just below the horizon. If all three kinds of twilight occur at all, the transition can take much of a day.

---

2 For mathematics buffs: Tower, shadow edge, and length of shadow on the ground form a right triangle on a realistic scale where the Earth beneath the towers is effectively flat.

Twilight is brief within the tropics. The tropical Sun escapes the horizon quickly. It does so by having a much greater component of motion that is perpendicular to the horizon than in the Antarctic or Arctic. Astronomical dawn to full-on daytime, or civil dusk to full-on nighttime, may be an interval of less than an hour. For mid-latitudes it is two-to-three hours.

Twilight comes and goes quickly near the Earth's equator. There, daytime to nighttime may take as little as 20 minutes.

At whichever latitude you are, if your real horizon differs greatly from the astronomical horizon, your experience may be different. In a mountain valley, the late afternoon Sun will dip behind surrounding peaks while there is still the brightness (and color) you usually associate with midday.

A similar phenomenon occurs in the morning. Sunset/sunrise in such a place always is a sudden, even surprising event.

11.5 Chapter summary

A second seasonal effect is the local efficiency of insolation. The construction of buildings to take advantage of changing insolation is not new; it is seen at archaeological sites. A third seasonal effect is the changing length of daytime and nighttime.

Special latitudes on the Earth include the Arctic and Antarctic Circles. Here, at a Solstice, the Sun is below or above the horizon for 24 hours. Farther poleward, more-and-more days include total nighttime or total daytime. The most extreme cases are the Earth's poles, where the solar year is divided into two periods: that in which there is constant nighttime and that in which there is constant daytime.

Two other special latitudes on the Earth are the Tropics of Cancer and Capricorn. Here, on the latitude-appropriate Solstice, the Sun appears at your zenith. It does so on two days straddling that Solstice, further and further apart, as you travel equatorward. On the equator, the Sun is at your zenith on the two Equinoxes.

11.6 Chapter review

1. How does the Earth's obliquity account for why it is colder in the winter and warmer in the summer?
2. Why are the hottest days of the year typically in late July/early August for the northern hemisphere, and late January/early February for the southern hemisphere, and not on the Summer Solstice?
3. What makes Chaco Canyon a good example of construction for an efficient use of solar insolation?
4. If the North and South Poles receive 24 hours of daytime during certain times of the year, why are they still cold?

5. Why does the equator's daytime and nighttime not differ as much through the year as do the poles?

6. Even if you stay home, at what latitude are you traveling furthest during the course of the day?

7. Draw a picture to demonstrate the significance of the Tropics of Cancer and Capricorn. Why might "Cancer" and "Capricorn" no longer be the most appropriate names?

8. Match the Sun's behavior on the left with the date on the right.

Most direct insolation	southern December Solstice
Longest daytime hours	northern June Solstice
Latest sunrise	northern December Solstice
Sun in the constellation of	southern June Solstice
Capricorn during ancient times	

9. If the bottom of a well in an African city (not Syene) was illuminated by the Sun twice a year instead of once, what could you say about its latitude compared to that of Syene?

10. Consult a map or globe. Which is the one country in which both the equator and one of the Tropics passes through?

11. How was the Sun used by Eratosthenes to calculate the Earth's size?

12. Why is the length of twilight different at the poles than at the equator?

12 The Moon

The Moon is the brightest object on the nighttime Celestial Sphere, though second overall, by far, compared to the Sun. Besides the Sun, the Moon is the only other regularly appearing object in the sky, displaying a disk shape. It is one of the two luminaries. Those who would tell you that you must wait for nightfall to see the Moon likely have not *looked* for the Moon during the day.

12.1 Chapter learning outcomes

After reading this chapter, you will be able to:
- predict the tides.
- locate examples of the major albedo-defined regions visible on the Moon.
- model the different types of libration.

12.2 Differences between the two luminaries

However, there are three important differences between the Moon's appearance and that of the Sun. The Moon's shape, in outline, is inconstant (*i.e.*, exhibits phases). Moreover, it is not uniform; there are darker-than-the-average smudges on its disk. Plus, you cannot hold your hand up to the Moon to feel its warmth, as you can toward the Sun. It produces no light or heat of its own, and merely reflects them from the Sun. All these idiosyncrasies make the Moon a one-of-a-kind feature of your sky.

Combined, its unique characteristics give the Moon a somewhat mysterious quality. The Moon, often, was considered a female god by those who deified the luminaries. (Whereas, the Sun was frequently male and paramount.) This is likely because the period of the lunar[1] phases is coincidentally close to that of the human menstruation cycle. The Moon also was associated with moisture, since at least its connection with Earth tides was recognized.

Notes from cultures around the world

There is a Chinese tradition that the Full Moon's light is beneficial to fertility. Why the Full Moon?

The phases of the Moon before Full represent a young girl; the Full Moon, a woman's maturity; and the phases after Full Moon, her post-child-bearing years. Of course, inasmuch as the Moon simply reflects a portion of the light from the Sun that strikes it, moonlight (Full or otherwise) has no property that is not possessed by sunlight.

China also is home to the famous mooncake served at a Mid-autumn Festival.

---

1 from Luna, Roman goddess of the Moon

https://doi.org/10.1515/9783111441245-012

12.3 Does the Moon affect the Earth?

The Moon normally is the closest astronomical body to you. And its size and mass are not small. Does it physically affect you? Perhaps by way of its gravity?

No. The gravitational attraction between a massive object, and one that in comparison has little mass at all (you), is extremely minor. Moreover, gravitational force falls off quickly with distance. The Moon is close by astronomical standards, but not by human standards.

Misconception alert

Does the phase of the Moon affect life on the Earth? Yes, some life forms respond to changing moonlight. However, this subject is one in which lore goes far beyond reality.

While certain *almanacs* traditionally provide folk with planting instructions based upon the appearance of the Moon, there is no theory to explain how the two are in any way connected.

Imagine that you are told to plant tomatoes on the third Full Moon after the Vernal Equinox and that the Full Moon might occur on a date between 17 May (for northern-hemisphere tomatoes) and 17 June. It may well be that planting on any of those days might result in local-fair-grade tomatoes. But it is the time of the year, and not the time of the month, which is important in determining when to sow. It is also unclear how a tradition that medicinal herbs must be picked at mid-month for full potency might work, too.

Some fishermen say that shellfish should be collected when the Moon is Full so that they will be fattest. Some ranchers swear that sheep should be sheared when the phase of the Moon is "growing" so that their livestock's wool will grow back faster. These practices are meaningless but harmless.

Less so are myths that might get in the way of proper medical treatment. These include ones that state that people are more prone to bleeding during a Full Moon and that quality of sleep is affected by the lunar phase.

12.4 Tides

There is one way in which the Moon does affect what takes place on the Earth. It is the **tides**. The global blue is very extensive. At the same time, fluid water changes configuration with ease.

The gravity of the Moon and Sun create ocean tides. The Sun and Moon are the only bodies both massive enough and close enough to the Earth to produce any tidal effect, and it is the nearby Moon that is responsible for most of it. (The ratio of gravitational tidal effect is about 2/3 Moon and 1/3 Sun.) While the Sun is much more massive than the Moon, it is very much farther away from the Earth.

The high seas all are connected. Tides happen because the gravitational force between the Moon and the ocean on the side of the Earth nearest to the Moon, the gravitation force between the Moon and the solid Earth, and the gravitational force between the Moon and the ocean on the side of the Earth farthest from the Moon all differ. The result is differential forces that produce oceanic bulges on both sides of the Earth.

Therefore, as the amount of seawater remains essentially constant day-to-day, two briny troughs form in-between. We speak of the **sub-lunar lobe** and **anti-lunar lobe**.

Such a differential force affects the hard Earth very little. It is about a thirty centimeters stretch, something you will never notice. So, as the exposed solid part of the Earth rotates once, it does so into and out of both the oceanic bulges and troughs caused by the tides.

Imagine standing on dry land, a little above mean sea level (say, on the coast or an island's edge). From your point of view, the water level rises and subsides in a period of time, a little greater than half-a-day: twice every **tidal day**.[2]

Figure 12.1: "Incoming" tide. Courtesy of Sharon Mollerus.

The anticipatable delay that results in your longer-than-usual tidal day is due to the Moon's revolution about the Earth in the same direction as the Earth rotates. Consider a hypothetical Moon directly overhead. It will take one tidal day for it to once again be directly overhead.

The sub-lunar and anti-lunar lobes are not expected to sit upon a line connecting the center of the Earth and the center of the Moon. The rotation period of the Earth is much faster than the revolution period of the Moon. The Earth pulls the lunar lobes with it. They should arrive about fifty minutes ahead of the actual Earth-Moon line.

---

2 or lunar day

Actual tide times and amplitudes are affected by the shape of your coastline and the depth of offshore water. For this reason, no two ports have exactly the same tidal pattern.

The **luni-tidal interval** is the time from the Moon's culmination on your celestial meridian and the next high tide you experience. It can vary from location to location by up to an hour. In some places, it even can be negative! (High tide occurs just *before* celestial meridian culmination.)

Figure 12.2: The changing tide at Portishead Dock, Somerset, England. This site, on the Bristol Channel, has one of the highest (and lowest) tides in the world. Courtesy of F. M. Soer.

When the Earth, Moon, and Sun are more-or-less lined up, it is called a **syzygy**.[3] At one of the syzygies, the differential gravitational force of the Moon and the differential gravitational force of the Sun combine to produce a particularly strong tide: Low tide is especially low, and high tide is especially high. Together, tides under these circumstances are called **spring tides**.

When the Moon and Sun are at right angles to each other on the Celestial Sphere, their differential gravitational forces are completely out of synch – the Moon produces an oceanic trough while the Sun produces an oceanic bulge – and the two effects partially cancel out. The result is an especially weak tide: High tide is lower than average tide, and low tide is higher than an average tide. Together, tides under these circumstances are called **neap tides**.

Not all spring or neap tides are equal. Their amplitude is affected by the declination of the Moon and the small variation in its distance about a mean. (See below.)

Notice that tides are a very large-scale phenomenon. For tidal effects to apply to (say) the fluid within such a tiny length as a plant stem, the Moon's gravity would have to be immense. It would certainly kill you! After all, human beings have bodies made up of a great deal of liquid.

---

3 Greek for "yoked together"

Figure 12.3: When the Moon, Earth, and Sun are aligned, in either possible order, the gravity of the Sun and of the Moon work together to produce higher-than-normal tidal extremes (spring tide). When the Sun-Earth and Moon-Earth lines are at right angles to each other, the tides produced by the Moon and Sun are out of synch; the lunar tide produces a tidal bulge at the same time as the Sun produces a tidal trough and *vice versa*. The two effects partially cancel, leading to lower-than-normal tidal extremes (neap tide). Not to scale. Artwork by Christine Tarte.

Notes from long ago

In the 1960s, 10,000-year-old (or more) worked bird-bone fragments were found in France. This age places them in Paleolithic times. American journalist Alexander Marshak saw the bones in a museum and thought that the markings on them might be purposeful (not just, say, scratches made while sharpening of a knife). He was the first to interpret them as monthly day counts, instead of something arbitrary like a hunting score, akin to the modern notches on a gun belt.

If this interpretation is correct, it makes Marshak's find the oldest record of sky watching over time, known to date. And the phenomenon being watched is the changing Moon.

There is a pattern to the incisions. Indeed, they were not made all at once. A microscope reveals that sets were cut by different tools at different times.

Why, though, jump to the conclusion that they are the work of a person(s) keeping track of the Moon day-by-day? Why assume that the marks refer to the number of days in a month (reckoned any way)? Why assume that the notches have anything to do with timekeeping at all?

Suppose that you count the month as the length of time from the appearance of one phase to its repeat appearance. Say that you decide to use Full Moon to Full Moon. Sometimes, you get 29 nights and other times, 30. It depends upon the exact moment you decide that the Moon is Full (or some other phase). (See Chapter 13.) This is precisely what is seen in the bone tallies: groups of 29 and 30.

Unfortunately, Marshak did not share a similar analysis of any bones in which the scratches did *not* fit his calendrical theory. So, our definition of a theory is not fully met.

12.5 Albedo

The Moon's brightness can dominate the night sky. The Earth's only **satellite** in your sky may limit you to seeing stars no dimmer than fourth magnitude. The Moon produces no light, though. Its brilliance is borrowed from sunlight shining upon it.

Astronomers call the ratio of light reflected by a surface to the amount of light that the surface receives, its **albedo**. Albedo is a unitless quantity between zero and one. A hypothetical, perfect looking glass would have an albedo of exactly 1. Regardless of how much light shines upon it, all of that light is reflected. When thinking about a hypothetical, perfect black body, picture black velvet or coal – only much darker. It has an albedo of exactly 0. No matter how much light shines upon it, none is reflected back.

Figure 12.4: The low albedo of the footway[4] causes it to absorb more solar energy during the day. After a nighttime snowfall, the still-warm pavement melts away the snow. Photograph by the author.

---

4 sidewalk

It may surprise you to learn that the Moon's albedo is closer to the second example than the first. Our satellite is a bad mirror. The Full Moon's average albedo is only 0.07. It looks so bright at night because {1} it has little competition and {2} the Sun is so bright that even a low-albedo surface will reflect a great deal of sunlight.

Notes from yesterday and today

Shiny silver is associated with the Moon. A custom is to turn a silver coin in your pocket when you first see a sliver of Moon some night. The idea is that as the silver Moon appears to increase in size (begin its cycle of phases), the amount silver in your pocket will increase too! Your author attests that this does not work.

12.5.1 Lunar features

While it is convenient to do so, what is usually called the disk of the Moon belies the fact that this body is actually a sphere. The proper term for the portion facing you is the lunar **nearside**. This is the hemisphere of the Moon that is closest to the Earth. Its opposite is the lunar **farside**. Using "nearside" acknowledges the sphericity of the Moon. Similarly, the term limb is preferable to "edge."

The Moon's albedo varies with location. Higher albedo regions seen on the nearside of the Moon are called the lunar **highlands**. (It turns out that their mean elevation is a little greater than that of other terrain.) The lower albedo regions are called the lunar **maria**[5]. Maria are smoother, lower, and more vaguely round than the highlands.

In solar-system history, maria were formed more recently than the highlands by the remelting of rock: Dark lava welled up from inside the young, still internally hot Moon, after a mare basin was excavated by the impact of some other solar-system body. Such objects, among many similar ones that wandered through the early Solar System, are small compared to the Moon but large enough to permanently effect the lunar surface in a collision. (See Chapter 19.) About a third of the Moon that you see from the Earth is mare.

Interestingly, both hemispheres of the Moon were not created equally. If you could see the lunar farside from the Earth, you would notice that, even to the naked eye, it is covered in fewer maria.

Similar to maria are **impact craters**[6] on the Moon. Craters are round lunar features, formed by smaller-scale impacts. (Again, see Chapter 19.) They are ubiquitous on the Moon, but ordinarily unresolvable. There are exceptions to this, though.

---

5 Mare is singular, maria is plural

6 The adjective distinguishes them from volcanic craters, though those planetary features are not visible beyond the Earth to the unaided eye; in context, simply "craters" is sufficient.

Figure 12.5: Lunar maria and highlands. Courtesy of USGS.

Naked-eye-visible lunar craters are not necessarily so because of their size. They can be resolved because of two other characteristics.

One of these is a dark crater floor. Contrast aids resolution.

Another is the presence of **rays**. Rays are subsurface material excavated by the impact and scattered radially outward. The pattern is much like that seen when a lump of flour falls on your kitchen countertop. The low lunar gravity allows rays to extend far from their source.

The Moon has no atmosphere; that is why its limb is not at all fuzzy. Any sort of solar radiation strikes the lunar surface unimpeded. The result is that, over long periods, the lunar surface darkens.

Fresh material (of a ray), newly exposed to the Sun, has a higher albedo. This still-fresh albedo pattern of rays surrounding a comparatively new crater makes the crater easier to see.

Dark-bottomed craters that still have their rays are good candidates for spotting with the naked eye. There are half-a-dozen or so that meet this criterion.

Figure 12.6: Selected lunar craters. Courtesy of Bob King.

Look for lunar craters within a few days of Full Moon. Full Moon is equivalent to "high noon" on the Moon. Shadows, which might confuse the view of features identifiable by albedo, are minimized.

At other phases, shadows may give away a feature that has relief (unlike crater rays). This is especially true when the shadow of a lunar feature is longer than the feature is tall.

A good example of a naked-eye crater is Grimaldi, seen within the lunar highlands close to the middle of the Moon's eastern limb. It is very dark, its rim is high, and it is big as craters go. (Perhaps, it should be called a small mare.)

Mountain ranges on the Moon may be made out if the terminator is nearly upon them as you observe. Our brain "likes" straight lines. If the lunar terminator is not "straight," we notice it. If it is ragged, this is because of the shadows cast by lunar mountains.

Sometimes, an elevated lunar feature remains in sunlight even though it is technically on the unilluminated side of the terminator. The result is a light-and-shadow show, or **clair obscure**.

Figure 12.7: Golden Handle. It is just visible, but only if caught under exactly the right angle of solar illumination. Courtesy of Frank Bernard.

An example is the Golden Handle. It is a sunlit arc on the Moon where there should be only darkness. The peaks of the lunar Jura Mountains catch the Sun's light several hours before the Sinus Iridum (a mare) below, which remains enshrouded by night.

12.5.2 What do *you* see on the Moon?

It is the maria that create the silhouette figures we imagine on the lunar disk with the naked eye. For instance, Mare Serenitatis is the Man in the Moon's left eye, and Mare Imbrium is his right eye.

These imaginary scenes are harder to see during the glaring contrast caused by a Full Moon against a totally black sky. They are better picked out in a twilight sky.

Such imagined shapes are something more than a game but far from reality. They are an example of **pareidolia**, your brain's desire to find meaning, even if there is not any.

The Man in the Moon is a product of the Christian era. The Moon Man shows up in one of the most popular – and reprinted – astronomy textbooks of all time, *De Sphaera Mundi* (*circa* 1200), by John of Hollywood (also known as Sacrobosco). The Man likely is still older than that. John's contemporary, German schoolman Albertus Magnus (Albert the Great), saw a complicated scene: a dragon under a tree, with a man leaning against the trunk!

Selenography[7] is the mapping of the Moon. Selenographer Ewen Whitaker catalogued scenes that people have seen in the Moon over the ages. These include a rabbit, an elderly man carrying a bundle of sticks, and a lady at a spinning wheel. Elsewhere across the world, we find a buffalo, a cook working over a stone stove, a crow, a frog, and even a moose. Some Moslems read on the Moon the *inscribed name* of Ali Ibne-Abi Taib (the Prophet's son-in-law and deputy).

12.5.3 Moon as mirror

For over two millennia, one supposedly more naturalistic hypothesis concerning the features seen on the Moon was that they were reflections of Earthly features. In other words, the Moon was a big speculum in the sky. Canadian Geographer Philip Stooke traces the notion as one held by some for more than two thousand years – right up to the twentieth century.

---

7 "Selene" is the Greek word for the Moon

Figure 12.8: Lunar pareidolia. Courtesy of Magnus Manske.

There even exists a sixteenth-century European map of the world that shows parts of the Earth, then yet unexplored by westerners. These portions look a lot like a mirror-reversed image of the Moon!

The argument against the reflection hypothesis was the same in the past as it is now. If it were correct, the Moon rising in the East should reflect different terrestrial land masses than the Moon setting in the West. Yet, no such change in the appearance of the lunar nearside is ever seen.

12.5.4 Color of the Moon?

Subtracting out atmospheric effects, the naked-eye Moon is colorless. As it reflects whitish sunlight, it looks white to our eye most of the time. While images of a Sun god often were adorned with gold, silver was reserved for that of a Moon god. Remember the silver coins? (See the Box above.)

12.6 How much of the Moon's sphere you can see?

Tableaus on the Moon are made much easier to recognize by the fact that when you look at the Moon, you always see the same hemisphere. Your view is forever of the lunar nearside.

The farside remained unknown until the Space Age, when robot probes flew past the Moon and photographed it from the other side. Finally, at least images of the farside were available.

Yet, for the earthbound naked-eye observer, a large portion of the Moon must remain *terra incognita*. From this, many people conclude that the Moon does not rotate. This is false. So, why are there definable lunar near- and farsides?

The Moon rotates on its axis a single time in the same amount of time that it takes to revolve about the Earth once. This is called **synchronous rotation**. It is the

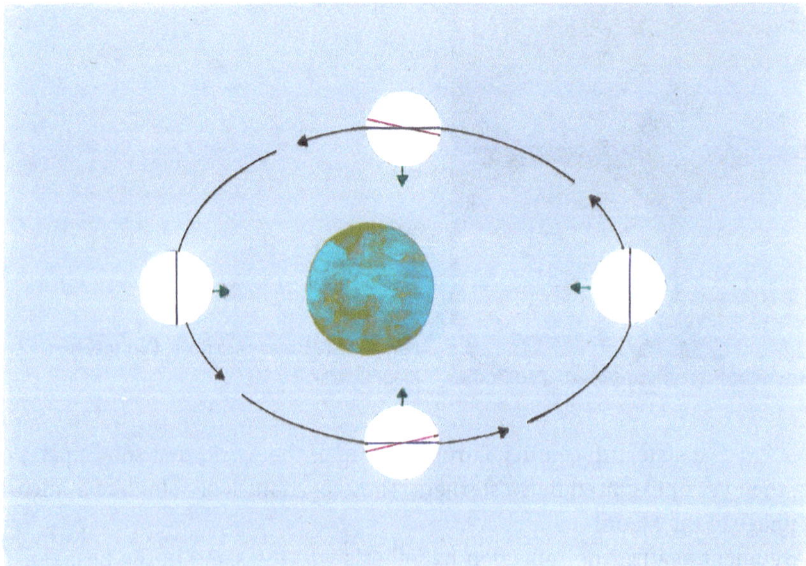

Figure 12.9: As the Moon revolves around the Earth, the direction in which the arrow *on* the Moon changes the direction in which it points, like the hands of a clock. If the Moon did not rotate, the arrow would always point in the same direction. Not to scale. Artwork by Christine Tarte.

consequence of the center of mass for both the Earth and Moon not being exactly at the geometric center of each of these bodies.

This results in both having a lever arm from geometric center to center of mass. The gravity of the more-massive Earth affects the Moon the most. As the Moon rotates, the Earth pulls the Moon's lever arm backward or forward until it is pointing directly at the Earth all the time, and lunar synchronous rotation is achieved.

12.6.1 Geometric (optical) librations

The observational effect of synchronous rotation should be that you see only fifty percent of the Moon from the Earth, the familiar nearside. The entire other hemisphere should be unavailable to you. In reality, you can see slightly more than half of the Moon.

Let us use the Full Moon as an example. If you observe it just after it rises and again just before it sets, you are doing so from two different places in space. During the time between observations, you travel a distance up to as great as the diameter of the Earth, due to diurnal rotation. The Moon is close enough that this affects what you can see. Compared to the fraction of the Moon facing you in the middle of the night, in the early morning, you see a small segment further to the West. Similarly, in the evening, you see a small segment further to the East.

This effect is called **diurnal libration**.[8] It is a libration in longitude. However, its magnitude is one degree, at maximum, and it is questionable whether it matters in naked-eye observations of the Moon.

The discovery of diurnal libration may belong to the famous seventeenth-century Italian physicist Galileo Galilei, the first observer to ever seriously study the Moon using a telescope. Of course, this instrument gave him an advantage in making the discovery, over naked-eye observers.

You can imagine observing the Moon from the North Pole; jumping on a high-speed aircraft, which flies you to the South Pole; and observing the Moon again. You might barely detect a minor north-south difference in the view of the Moon by a similar mechanism to that which produces the minor east-west difference in the case of diurnal libration. There is no record of anybody ever doing this – for obvious practical reasons!

More interesting is another kind of libration. The Moon's rotational axis is not parallel to the Earth's rotational axis. Therefore, over one months' time, you can peak over the lunar pole by a small angle, both north and south.

---

8 or "parallactic libration"

What is more, the Moon's orbit takes it a little bit above the ecliptic and a little bit below the ecliptic each month. (See Chapter 16.) The resulting alternate views over the lunar poles are even better.

Together, these particular point-of-view effects are called **libration in latitude**. Libration in latitude allows you to look an extra 6.7°, past the lunar poles, into the farside.

12.6.2 Physical libration

Both diurnal and latitudinal libration are the result of point of view. There is a **physical libration** too.

The Moon's *average* speed of revolution around the Earth is always the same. However, like the Earth about the Sun, it travels more slowly at the point in its elliptical orbit, furthest from the body that it revolves about, that is, the Earth. This point is called **apogee**.

The Moon travels fastest at the point in its orbit closest to the Earth. This point is called **perigee**.

On the other hand, the Moon rotates at a *constant* rate. Thus, at apogee, the Moon's-rotation puts a given lunar longitude a little ahead of where it would be if the Moon's-rotation and revolution were totally in sync with one another. At perigee, the Moon's rotation puts a given lunar longitude a little behind where it would be if the Moon's rotation and revolution were totally in sync.

This physics gives you a small view around the Moon's average limb (west) and then a small view around its average limb (east). Combined, these equal an extra 7.9°-in-longitude glimpse at the lunar farside.

The red limb in Figure 12.9 demonstrates the effect of physical libration. Physical libration was discovered by Galileo's younger contemporary, the Polish amateur selenographer Johan Hevel (or Johannes Hevelius). He did so based on naked-eye observations.

There are additional **true librations**[9] that are not an optical perspective effect. These are real wobbles of the Moon, either back and forth or in a circle. However, their timescale and angular amplitude make them irrelevant in this book.

Not withstanding, the three librations described make it possible for you to see 59% of the Moon, without leaving the Earth.

---

9 These often are called physical librations instead; context usually prevents confusion between the two different definitions of physical libration.

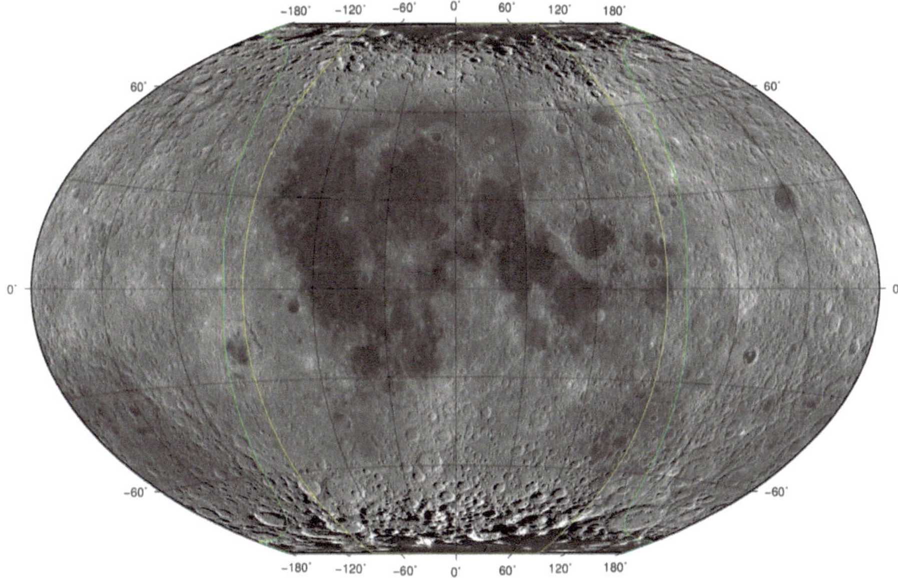

Figure 12.10: This projection of the Moon shows the extent of lunar longitude visible due to physical libration. Courtesy of NASA.

12.7 Observed change on the Moon?

There are no relatively fast **endogenic processes** currently at work on the Moon. These include volcanoes, tectonics, and water or wind erosion.

The face of the Moon is essentially changeless to us. However, **exogenic processes**, those changes caused from beyond the Moon (*i.e.*, mare and crater formation), must continue today – just not at the same frequency as in the geologic past. They *might* still cause change visible to the naked eye. But would this happen often enough so as to occur during human history?

18 June 1178. Five monks in Canterbury, England, witnessed something extraordinary. Here are the words of Gervase, their chronicler: "[The] upper horn [of the Moon] split in two [and from the division point] a flaming torch sprang up, spewing out, over a considerable distance, fire, hot coals, and sparks."

Planetary geologist Jack Hartung interprets this story as evidence for a huge, explosive impact on the Moon. Others have pointed out that, statistically, this is improbable. The formation of a large impact crater, caught in the act anywhere on the Moon within the last millennium or so would be a fantastic coincidence. What may be common over spans of geologic time can be extremely rare over shorter intervals.

It is troublesome that, as far as is known, no one else on the entire Earth saw this display. What if it were a local sight? A brilliant meteor that just happened to occur between the monks and their view of the Moon also requires serendipity, but such an event is more likely. (See Chapter 19.)

On the other hand, more-modest impacts on the Moon have been observed using a telescope. So, the thought of a naked-eye impact – if not in the past, in the future – cannot be disposed of into the realm of fantasy.

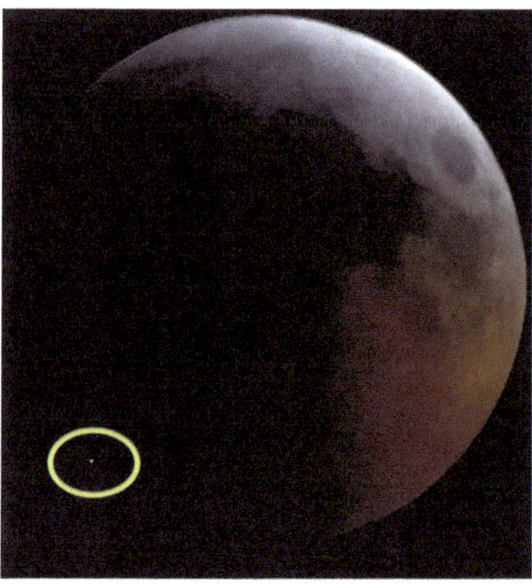

Figure 12.11: An impact on the Moon (lower left, near the limb) taking place during a lunar eclipse. (See Chapter 16.) This particular event was not visible to the naked eye. Courtesy of Ed Krupp/ Griffith Observatory.

12.8 Representation of the naked-eye Moon

Artistic representations of the Moon date back to Paleolithic rock art. However, with one speculative exception, these figures show the phase of the Moon only.

Amazingly, there exists no evidence that a western artist painted or drew the Moon as it authentically looks – -"blotches" and all – until the Renaissance. (Some pre-Columbian New World art arguably shows the lunar rabbit.) If the drawing supposedly made by the Greek/Roman author Plutarch in the first century actually existed, it is long lost.

For some time, it was thought that the celebrated Renaissance artist Leonardo da Vinci was the first to render an authentic Moon. His famous notebooks depict the Moon as the master saw it between 1505 and 1508.

Art historian Scott Montgomery points out, though, that the Flemish painter Jan van Eyck portrayed a realistic Moon, more than a generation before Leonardo. The oil painting is *The Crucifixion* from circa 1420. It includes a Moon, some days past Full. It is a small vignette, but with lunar maria of correct shape and position still possible to make out. Realistic Moons can be seen in other van Eyck works as well. Artistically van Eycke went beyond Leonardo's rough sketches.

Figure 12.12: The Moon appears to the right of the third cross in this diptych's left panel. The original van Eycke artwork resides in the Metropolitan Museum of art.

Tools of the naked-eye observer

It was not until around 1600 that English physician William Gilbert created a map of the Moon. That is, Gilbert drew a stylized Moon but also added a cartographical grid and feature names. More detailed selenography would have to await the resolution provided by the telescope.

Today, maps of the Moon are commonplace. You should be able to use one in order to match up some of the features potentially visible to the naked eye with their fanciful or arbitrary proper names attached to them, beginning in the seventeenth century.

Lunar craters are named for people. Most are famous scientists, but there are plenty of exceptions. Lunar mountain ranges borrow from those on the Earth.

Mare names represent weather or states of mind. Mare Tranquillitatis, where the first Apollo astronauts landed, is the Sea of Tranquility. The large Oceanus Procellarum is the Ocean of Storms.

Four hundred years ago, it was thought that seas seen from above might have low albedos, hence the use of "mare," Latin for "sea." In reality, there is no significant water on the Moon at all.

12.9 Chapter summary

Our planet's satellite, the Moon, is the closest astronomical body to the Earth. However, except for the tides, it has no significant gravitational effect on life on Earth. Tides are the product of the difference in the Moon's gravitational pull upon opposite sides of the whole planet.

The impact-related features on the lunar disk that are visible to the naked eye include the maria and highlands (because of differences in albedo) and the occasional crater or mountain. We tend to connect these features into pictures, due to pareidolia.

Synchronous rotation is responsible for the fact that the nearside of the Moon always faces us, and the farside always faces away. Yet, you can see more than half of the Moon on account of librations. The most significant of these is due to the Moon traveling faster at perigee and slower at apogee, while its speed of rotation remains constant.

12.10 Chapter review

1. What causes the tides? Draw a picture and explain why two high tides happen per day.
2. When the Moon was formed, it was much closer to Earth. How might tides (on land and in the ocean) have been different?
3. What phases of the Moon correspond to the highest tides? Why?
4. Given that the Moon's albedo is quite low, why does the Moon appear so bright in your night sky?
5. How do maria, craters, and rays affect how we see the Moon?
6. What accounts for why people often see shapes and images (*e.g.* the "Man in the Moon") on the lunar disk?
7. Why can you see in your sky very little "real estate" belonging to an entire hemisphere of the Moon?
8. Explain how we can see nearly 60% of the Moon rather than just 50%.
9. Why do you think western artists of the Middle Ages were at first reluctant to represent a realistic-looking Moon?

13 From New Moon to Full Moon

The month is the orbital period of the Moon. (The word "month" comes from "Moon.") The month is a handy unit of time, much longer than that other celestially defined unit – the day – but shorter than the similarly defined unit of the year. Without the Moon, your 12-month calendar, the 12 signs of the zodiac, and so forth probably would not exist.

A lunar[1] calendar (as opposed to a solar calendar) is useful to, for instance, hunting communities who rely on the Moon's light to illuminate their prey. Agricultural and herding communities might find more utility in a solar calendar, because of the importance of seasons to their food supply. Both are reasonable choices.

13.1 Chapter learning outcomes

After reading this chapter, you will be able to:
- distinguish between the sidereal and synodic month.
- recreate the phases of the Moon using models.
- test an explanation for the Moon Illusion.

13.2 Two kinds of month

There are at least five different definitions of the month! Two are the most common.

First, a bit of quantitative business: The mathematical term "commensurate" means "divides evenly into." The quantity 4 is commensurate with 12 (=3), but not 13 (=0.30769230769 . . .).

Part of the reason for the variety of months is the continual desire to make the number of months in the solar year commensurate with the number of days in the solar year. Ultimately, it is hopeless to fulfill this desire inasmuch as the Moon's revolution period and the Earth's revolution period about the Sun are independent of one another.

Rotation and revolution are independent of each other. Such a successful numbers game as commensurability would be a tremendous coincidence. Nonetheless, we map humankind's attempts.

---

1 adjectival form for the Moon

https://doi.org/10.1515/9783111441245-013

13.2.1 Sidereal month

Perhaps the most obvious thing to do, in order to measure the length of the month, is to proceed similarly to the way you measured the sidereal year: using as a beginning/ending reference point (*i.e.*, the stars) on the Celestial Sphere.

The Moon moves on the Celestial Sphere about 13° per day. Imagine that you start your stopwatch as the Moon passes over a certain star. Or, more generally, it is at a certain angular position on the Celestial Sphere. You wait until the Moon once again occupies that same celestial location, then stop the watch. The length of time you measure is the sidereal month. It is equal to 27.322 days.

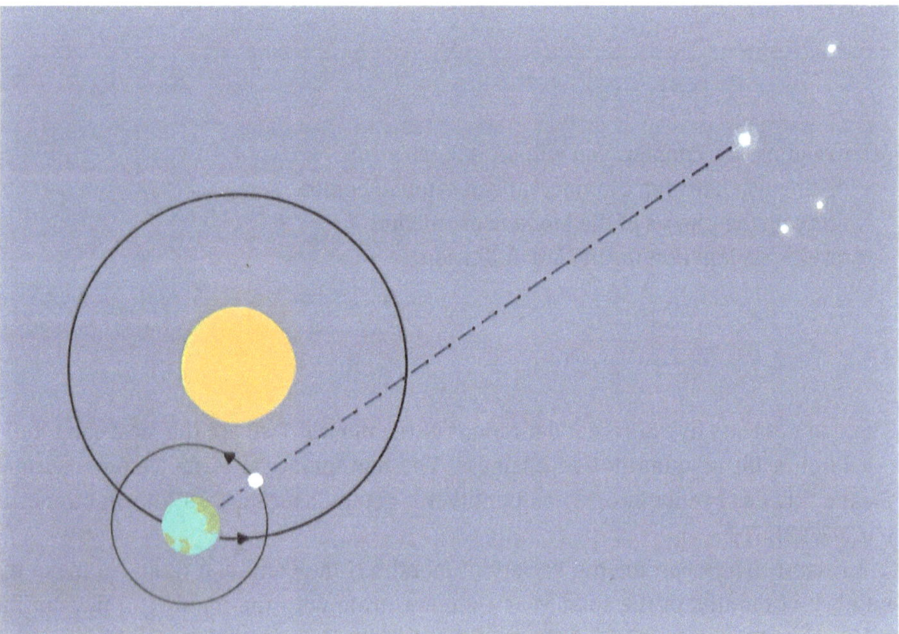

Figure 13.1: The Moon revolves through one sidereal month. Not to scale. Artwork by Christine Tarte.

Notes from long ago

In Mayan texts, we read that Moon Woman sometimes is preceded by a herald. Barbara and Dennis Tedlock (an American anthropologist and archaeologist, respectively) equate this leading character with a constellation of stars.

Such a constellation would rise just before the Moon. At other times, Moon Woman is accompanied by a spouse; they walk side-by-side. If the "spouse" is a constellation, the Moon and it rise together. Finally, Moon Woman may carry a "burden," beings who ride on her back. If we again interpret such a character or characters as a constellation(s), it/they rise(s) just after the Moon.

The only slight complication is that precisely when the Moon encounters a particular star or stars depends upon your location on the Earth. The geometrical effect of **parallax** causes a comparatively near celestial body like the Moon to appear at a different place against a distant background (*i.e.*, the Celestial Sphere) depending upon where you observe it.

Parallax is familiar to you if you have had a tall person sit in front of you in a theater. Moving slightly one way or the other will cause the view-obstructing individual to appear to move to the left or right. You now see the stage, even though he or she did nothing.

In the case of the much farther away Moon, parallax amounts to only a degree or so. If attempting to time the sidereal month exactly, simply do not move thousands of kilometers while in the act.

Despite the minor nuisance of parallax, the sidereal month is a straightforward way to measure the revolution period of our satellite. Nevertheless, 27⅓ days does not sound like the month you are used to. Most people expect 31 or 30 – or at least 28 or 29, in February – days in their month.

For 5,000 years East Asians have divided the Celestial Sphere into 28 zoomorphic asterisms[2]; the Moon occupies, on average, one of these lunar **mansions**[3] per day. However, this is not quite the same thing as a sidereal-month calendar.

The lunar mansions are arranged around the Celestial Equator, not the ecliptic, which the path of the Moon better approximates. As far as "monthlong" intervals are concerned, most cultures do not keep track of the sidereal month.

Notes from cultures around the world
The Borana, nomads who inhabit Ethiopia and Kenya, do use a sidereal-month calendar, rounded to 27 days in length, in which each day corresponds with the Moon's relationship to a particular constellation. Such a calendar is well-suited to those who live near the equator insofar as celestial objects rise and set nearly perpendicular to the horizon at low latitudes. When the Moon rises or sets alongside one of the Borana calendar constellations, the luminary and stars are north and south of each other and form a line parallel to the horizon.

---

2 Hsu, in Chinese
3 or lodges

13.2.2 Synodic month

Beside the sidereal month, another way to measure the Moon's revolution period is to use as a reference point the other luminary, the Sun. Suppose that the Moon is a certain angular distance away from the Sun. The time between this occasion and the next when the Moon has the same relationship with the Sun is called a **synodic month**.

The difference between the sidereal and synodic month lies in the fact that the Sun appears to move on the Celestial Sphere. Measuring the month in this manner is like timing a footrace that is run on a circular track. The starting line also is the finish line. During the race, though, somebody picks up the start/finish line while the racers run and moves it farther around the track. When the athletes reach the place from which they began, they must continue further to break the tape. The duration of the race is necessarily longer.

The Sun and Moon appear to move in the same direction on the Celestial Sphere. During the course of the synodic month, the Sun shifts approximately one-twelfth of the way around the ecliptic. It is a moving target. The Moon must catch up with the Sun.

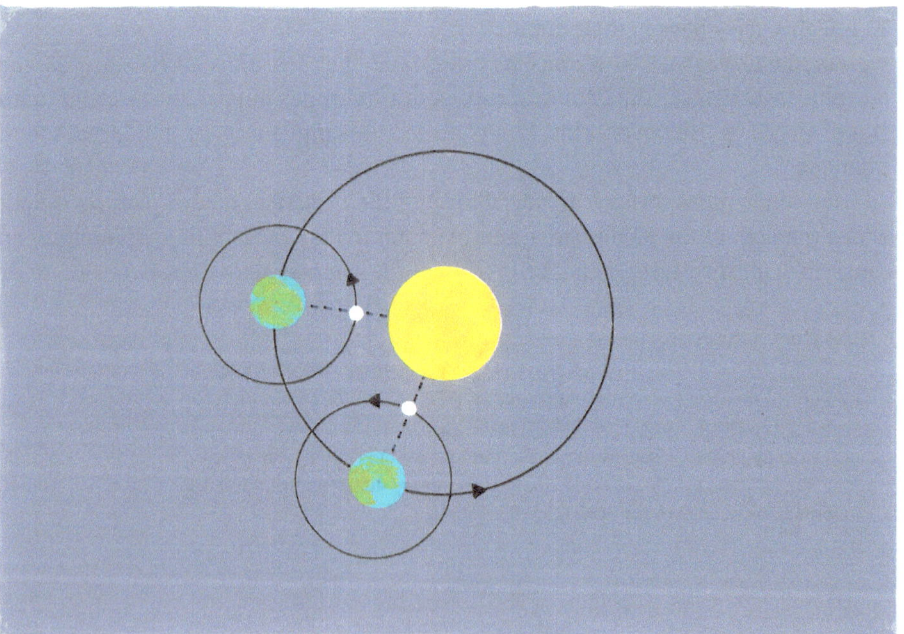

Figure 13.2: The Moon revolves through one synodic month. Notice that it has traveled farther and longer than it does in one sidereal month (as illustrated in Figure 13.1). Not to scale. Artwork by Christine Tarte.

The synodic month is 29.531 days long. In the earliest written calendar, the Mesopotamians toyed with this number and constructed a civil calendar of 29.5 days each month over an interval of three solar years. They then had to insert into the last month of the cycle an extra half day so as to make it 30 days long. The idea again was to construct a civil year of equal length, whether counted in a number of months or days.

The Mesopotamian scheme resulted in a year of 12 months or 364 days – not too bad. They eventually finagled a calendar made up of an integer number of months that ended up as 364.5 days; this is an even more realistic match.

However, months that include half days are inconvenient. It works better to round 29½ up to 30 – or perhaps 31 – days. You end up making one month an exception (e.g., February) but still keep it longer than 27⅓ days. The result converges (historically, after a lot of false starts) to 365 days and the civil year of months familiar to most of us.

Starting with a month length made up of a number of days with the fraction one-half in it is a bit easier than trying to deal with one that has months ending in one-third. Yet, ultimately, that is not why most people prefer the synodic month.

The synodic month is preferred over the sidereal month because that most unusual behavior of the Moon (lunar phases) repeats over an interval of one synodic month. It is not coincidental that a Greek Moon goddess further represented the personification of change.

13.3 What causes the phases of the Moon?

It is likely that most people pay attention to its phases more so than the Moon's location on the Celestial Sphere. Moreover, there is every reason to believe that this has always been true. Lunar phases may be the subject of an ornamented disk from Germany that appears to have been made 3,600 years ago.

It is easy to distinguish the appearance of the Moon at different phases. It is harder to keep track of the constellation on which the Moon is superimposed. There is no universal agreement on what the constellations are, at least in the same way that there is on the shapes of the lunar phases. If you know their order, lunar phases make timekeeping simpler on the order of a month.

13.3.1 You see the part of the Moon's illuminated hemisphere that is facing you

Until one reaches the distance of the stars, no other celestial object produces light of its own except for the Sun. All the other objects in our Solar System shine by reflected sunlight only. It is the same reason why any of us "shines" (outdoors).

Figure 13.3: Nebra Disk. Located in the State Museum for Prehistory in Halle, Germany. The inlays are of gold. Courtesy of Roger Jones.

This is no less true than for the Moon. As is the case for all spheres placed reasonably far away from a light source, half of the Moon's spherical surface is illuminated at all times. The other half is self-shadowed.

So, just like the Earth, one hemisphere of the Moon always is in daylight while the other half is in self-shadowed night. Locations on the Moon experience an average of about two weeks of daytime and two weeks of nighttime. (For now, we ignore the minor effect of eclipses while discussing diurnal consequences on both the Earth and Moon; see Chapter 16.)

How much of the daylit Moon you can see from our point of view on the Earth determines the phase of the Moon.

Misconception alert 
The Moon's phases do not require that a shadow be cast on the Moon by any other object. That would be a special situation known as an eclipse.

The only other object in the vicinity of the Moon is the Earth. However, the Earth can only cast a shadow upon the Moon when it is between the Sun and Moon. As we will see, this is when the Moon, from your point of view, is at its most illuminated.

Another lunar misconception: The New Moon is not the time of the synodic month at which you can first see the Moon. It is when you cannot see the Moon at all! (See below.)

13.3.2 Phases of your own

Take any sphere, say an orange. Hold it in the light of a desk lamp or other light source. In this model, you play the role of the Earth. Move the orange/Moon around your head; just do not let it pass into your head's shadow. You can recreate every phase of the Moon that you have ever seen in this simple manner. (It is OK to twist your head in order to see the phases.) Moreover, they will occur in exactly the same order as they do in the real sky.

Tools of the naked-eye observer
A physical model of the Solar System is called a *planetarium*. If the model represents just the motions of the Sun, Moon, and Earth, it is called a *tellurium*. (We are used to the Greek prefix "geo-" referring to the Earth; "tellur-" is an alternate such prefix.) The principal use of a tellurium is to demonstrate lunar phases.

13.4 New Moon

Traditionally, the synodic month begins with **New Moon**. When, exactly, does New Moon occur?

This is tricky to determine observationally. While the Moon is nearest the Sun in our sky at New Moon, it is lost to the glare of the much brighter luminary. You cannot see it but know that it must be there!

New Moon is officially the instant when the three bodies – the Sun, Moon, and Earth, in that order – are closest to syzygy during a particular synodic Month. Practically speaking, New Moon is the two or three days when the Moon appears absent from your sky.

A synodic month that starts at New Moon is commonly called a lunation. Each lunation, since a particular one starting on 17 January 1923, has been assigned a consecutive number by astronomers. Those that occurred before the arbitrary reference date are given negative numbers.

Figure 13.4: Relationship of the Earth and Moon to the Sun at New Moon. Not to scale. Artwork by Christine Tarte.

Here we go: Imagine an angle measured on the Celestial Sphere, around a great circle marking the Moon's orbital path. Zero degree (and 360°) are equivalent to the Moon's closest apparent approach to the Sun. The angle is called the Moon's **elongation**.

The lunation begins, then, when elongation equals 0°. Consequently, the **age of the Moon** is the elapsed time of the lunation since the geometrically defined moment of New Moon.

13.5 Crescent Moon

A couple of days into the lunation, the Moon is far enough from the Sun in your sky that most people will be able to distinguish it and agree that the next lunar phase is now visible. This is admittedly vague. The next phase usually occurs after the Sun has set and the sky has darkened somewhat, but while the Moon yet remains slightly above the western horizon. This lunar phase appears as a thin **Crescent Moon** because, from your point of view, the lunar orb is still mostly backlit by the more-distant Sun.

Figure 13.5: Relationship of the Earth and Moon to the Sun at Crescent Moon. Not to scale. Artwork by Christine Tarte.

13.5.1 Sighting the Crescent Moon

Especially because of their significance in the calendar used by the world's Moslems, it is useful to be able to predict on which evening/morning a given lunar phase will occur. For instance, take the first Islamic month, al-Muharram[4]. This religious month starts with the first observation after New Moon of the Crescent Moon. Without a means of predicting this appearance ahead of time, the faithful do not know when to commence the New Year.

The traditional rule is that this Crescent Moon must be spotted by two "trustworthy" individuals, in the evening, with the celebration starting the next morning. Thus, different people at different places easily can end up observing the holiday that is the first date in this sacred synodic month on a different day.

13.5.2 The Moon grows "older"

There is not much Moon visible when the lunar crescent first becomes visible. Seeing it is a challenge and predicting its first appearance is harder than the task sounds.

---

4 from the Arabic

Lunar visibility is a function of how much light the Moon reflects our way. Whether we can make out that light also is governed by things that control contrast. The latter includes twilight and extinction by air mass.

Simply waiting a number of hours past New Moon – 24 is a typical value – does not work. A ten-hour-old Moon on the ecliptic has almost the same brightness as a (nearly) zero-hour-old Moon that is 5° away from the ecliptic. The angle between the Sun and Moon is the important variable; but even at a fixed Sun-Moon angle, the width of the lunar crescent appears to vary, depending upon how near or far the Moon is to the Earth at that time. (See Chapter 16.)

There also are latitudinal and seasonal effects. In fact, there does not appear to be any easily parameterizable theory that allows prediction of crescent moon visibility. Notwithstanding, the current record for naked-eye sighting of the Crescent Moon is at age 15 hours, 32 minutes, and was set by American astronomy-journalist Steven O'Meara.

The young Crescent Moon sets quickly, but on successive nights it is higher in altitude (and more southerly in azimuth) at sunset. Moreover, the crescent widens and becomes brighter, night after night, as the angle of illumination changes. Still, less than 50% of the Earth-facing lunar nearside is illuminated.

Figure 13.6: Crescent Moon and western horizon.

The Crescent Moon, unlike the New Moon, is not technically a single instant of time. Nor does it represent a single, particular elongation. The Moon is a crescent for an interval of time and over a range of elongations that is a subset of 0–90°. Its appearance changes night to night as it works its way eastward in your sky, but the lunar terminator remains curved in the same direction as the illuminated limb.

13.6 First Quarter Moon

The Moon moves further eastward in your evening sky. Eventually, at an age about one week after New Moon, half of the lunar nearside is sunlit and half is not. The lunar terminator is a straight north-south diameter bisecting the Moon. We call this phase **First Quarter**.

Figure 13.7: First Quarter Moon. Courtesy of Soly Moses.

Figure 13.8: Relationship of the Earth and Moon to the Sun at First Quarter Moon. Not to scale. Artwork by Christine Tarte.

13.6.1 Why the phrase "Half Moon" is incorrect

Clearly, half of the Moon's disk is visible to you! Yet that is not where the name First Quarter Moon comes from. At First Quarter, the Moon's elongation is ninety-degrees east, a geometry called **dichotomy**.[5] This happens when the Moon is one quarter of the way around its orbit.[6]

Having risen at noon, at sunset, the First Quarter Moon is near your celestial meridian in the early evening. It will set at midnight.

13.6.2 Aristarchus's attempt to measure the lunar distance

Sometime in the third century BC/BCE, a Greek geometer named Aristarchus of Samos had a clever idea: The Earth, Moon, and Sun form a great triangle in space. When the Moon is a Quarter, this triangle is a right triangle.

This means that the side joining the Earth and Moon, and the side joining the Moon and Sun, are at ninety degrees with respect to each other. Aristarchus under-

---

5 from the Greek, dicha, meaning "apart" and tomos, meaning "cutting"

6 Such vocabulary leads to confusion, with people using the term "Half moon" when they mean Quarter Moon.

stood that all right triangles behave in the same way. Their sides are proportional in a ratio fixed by one of the acute angles in the triangle.

If he knew one of those angles, Aristarchus would know the ratio, too. He attempted to measure the angle between the Earth-Moon side and the Earth-Sun side of the triangle (elongation). Doing so, he was poised to determine for the first time how much farther away from the Earth the Sun is than the Moon (the length of the Earth-Sun side of the triangle).

The problem was that the required angle was very difficult to measure accurately with the naked-eye instruments at Aristarchus's disposal. His angular measurement was 17 times too large. At the same time, he probably underestimated how difficult it is to decide, using only the naked eye, when the lunar terminator is completely straight. And the moment of dichotomy is at hand.

Aristarchus's result was wrong. He concluded that the Sun is only 18–20 times farther from the Earth than from the Moon. That is much too small a number. However, his ingenious method for making this determination was, abstractly at least, completely sound.

13.7 Gibbous Moon

Our lunation continues past First Quarter. Now, more than 50% of the nearside is illuminated by the Sun. The appearance is a cookie-cutter reversal of the Crescent Moon – the lunar disk looks as if a crescent-shaped piece has been removed.

Figure 13.9: Gibbous Moon and western horizon.

The name of this phase is known to fewer people than the number who are familiar with the name of the Crescent Moon. This is somewhat surprising because this newly introduced phase can appear high in the night sky. This is unlike the Crescent Moon, which, if seen outside of astronomical twilight at all, appears to be hugging the horizon. When the Moon's elongation is more than 90° (but not 180°), it is called the **Gibbous Moon**.

The Gibbous Moon appears in your western sky during early evening. On successive nights, its shape becomes wider and its brightness increases, just as was the case for the Crescent Moon. In this phase, though, the lunar terminator always is curved in the opposite direction from that of the illuminated limb. The Gibbous Moon is at a higher altitude and more southernly azimuth night to night.

Figure 13.10: Relationship of the Earth and Moon to the Sun at first Gibbous Moon of the month. Not to scale. Artwork by Christine Tarte.

This width and brightness become greater and greater, the same as those properties of the Crescent Moon, as more and more of the sunlit Moon becomes visible to us Earth dwellers. Still, less than 100% of the nearside is illuminated.

The Gibbous Moon, unlike the First Quarter Moon and like the Crescent Moon, is not technically a single instant of time. Nor does it represent a particular elongation; it has a range of elongations. The Moon is gibbous for an interval of time and over a range of ages, as it makes its way further eastward in your sky.

13.8 Full Moon

The lunar age is about two weeks since our lunation began. Next comes your second syzygy of the synodic month. The Moon is as angular distant from the Sun as it can get. The Moon is halfway through its path around the Celestial Sphere, but the term "Second Quarter" is never used.

Figure 13.11: Full Moon. Courtesy of Nui Malama.

Finally, the lunar terminator corresponds to the lunar limb. The entire nearside is illuminated; a completely sunlit disk is visible from the Earth (though see below).

The geometry of a 180° elongation technically lasts but a moment in time. However, the same situation applies again as it did, for instance, when attempting to decide that the lunar terminator was exactly straight (the observational definition of First Quarter). Practically speaking, there are typically a couple of days during the lunation, over which the casual viewer will no longer discern anything out-of-round about the Moon and will not call it gibbous. It is the **Full Moon**.

Figure 13.12: Relationship of the Earth and Moon to the Sun at Full Moon. Not to scale. Artwork by Christine Tarte.

13.8.1 You are most likely to see a Full Moon

Did you notice that the Crescent Moon and Gibbous Moon just described were above the horizon at sunset, so they set before sunrise? The Full Moon rises as the Sun sets and sets as the Sun rises.

The Full Moon is on or near your celestial meridian at midnight. With the Moon "up" all night, your chances of randomly catching it in the sky are at a maximum compared to other lunar phases. This, and the fact that the Moon is at its brightest in the sky when Full, make it more likely that the casual observer notices this phase more often than any other lunar phase.

Notes from yesterday & today

If the intent in media is to imply something menacing or frightening, the trope is a Full Moon. But in the modern civil calendar, divorced as it is from the synodic month, the Full Moon can appear on any date during the modern month of October.

There need not be a Full Moon at Halloween, contrary to many television and movie scripts. It takes 372 solar years for any given phase of the Moon to occur on the same date as well as the same day of the week.

On the other hand, the Full Moon can be comforting – as a great nightlight in your sky. In still another connotation, the Moon often symbolizes sadness in song. An example is the frequently used lyric "blue moon."

13.8.2 How full is full?

The Moon orbits the Earth in a plane different from the Sun-Earth plane (the ecliptic). So even when it is geometrically 180° away from the Sun in the Moon's orbit plane, the Moon likely may be a little above or a little below the Sun-Earth line, which exists in the ecliptic plane.

This causes a **defect of illumination**, to varying degrees, to the Full Moon. Such a deviation from true "full" is virtually impossible to detect with the naked eye.

Occasionally at Full Moon, you are treated to the sight of both circular luminaries at the same time. Because of atmospheric refraction, it is just possible to see the Full Moon appear to rise before the Sun has appeared to set totally and vice versa. Atmospheric refraction raises the image of both the Sun and Moon in altitude. Each is near an opposite horizon, but the two luminaries look a little bit angularly closer than 180° apart.

People do not, in fact, become moonstruck at Full Moon. Objective statistical studies of records such as hospital emergency-department admissions and police blotters show no correlation with Moon phase. Sadly, people do "crazy" and do criminal things to themselves and others every night of any kind of month. (That said, burglary under the light of a Full Moon seems especially fool-hardy.)

Where did that come from?

Luna was the Roman Moon goddess. For better or worse, "lunacy" and "loonie" also are derived from this word.

13.9 Moon Illusion

Particularly at Full Moon, you are confronted with a phenomenon that rightly causes you to question what you see with your own eyes. Starting with Chapter 1, you have been urged to trust that which you observe yourself. However, there are exceptions. One is the imagined physical size of stars. Another is the **Moon Illusion**.

13.9.1 Distance from the Moon anywhere on the Earth does not affect lunar apparent size

The rising Full Moon is beautiful. Because of the convenient times of day at which it occurs, you frequently see Full Moon risings and (perhaps somewhat less often) settings. You comment on how big the Moon looks. It gives the impression of being, perhaps, as much as twice the diameter it has when it is far from the horizon.

Actually, measuring the apparent size of the Moon, though, proves that your impression is incorrect. In media, the Full Moon is routinely portrayed as overly large, compared to other astronomical objects. This might be done for truly artistic reasons. Here, though, we discuss the naked-eye Moon Illusion. It is a nearly universal trick of nonoptically-aided perception produced in our minds. Indeed, it is absent in photographs.

It may be possible to judge the change in the Moon's apparent size due to the eccentricity of its orbit. The most extreme case would be comparing the appearance of the Moon at apogee to that at perigee. However, the eccentricity of the Moon's orbit is slight, and you can never subject views of the Moon at each of these points to a side-by-side comparison. You must compare one to the memory of the other. Observation of the Moon's varying apparent size (from 29.4–33.5 minutes of arc) escapes all but the most dedicated Moon watcher.

Figure 13.13: Half the apparent lunar disk at apogee compared to half the apparent lunar disk at perigee. Courtesy of NASA.

Besides, the Moon's apogee and perigee would not explain why, in the Moon Illusion, the Moon always looks bigger near the horizon. In truth, theoretically it should look smaller. The Moon is slightly farther from us then – by one Earth-radius – compared to when the Moon is at the zenith. In reality, this one-earth-radius difference in distance (2%) has a generally imperceptible effect on the Moon's apparent size.

What is going on? The atmosphere does not magnify the Moon.

13.9.2 Potential causes of the Moon Illusion

The overly large appearance of the Full Moon near the horizon has no explanation in physical reality. It is a mistake made inside our brain. Very few are immune to it.

The same thing that is taking place in the Moon Illusion may affect your perception of the Sun's apparent size, too. It even can influence that of compact star groups, such as the Pleiades. Once upon a time it was called more generally the Celestial Illusion.

What do these objects have in common? They are far enough away that your binocular vision, by which you normally judge distance – on the Earth, at any rate – fails.

Does the Moon Illusion have to do with color? Objects close to the horizon look redder due to the effect of the thick atmosphere through which you see them.

Do you see redder objects as bigger? No. High in the sky, it is true that you do not see this color in the Full Moon. However, if you look at a high-altitude Full Moon through red-tinted glasses, it does not seem to suddenly grow in size.

Is the Moon Illusion due to atmospheric refraction? This effect may make the Moon look out-of-round (oval), but it does not magnify it.

The truth of the matter is that the Moon Illusion is not fully understood. A key to it seems to be the fact that it is most commonly observed, not just when the Moon is on the horizon, but when the horizon rim is not quite dark (dusk or dawn). That is, when nearer, terrestrial objects also can be seen on it.

This is a reason why the illusion is strongest with the Moon. Faint stars are difficult to see under such conditions; the Sun's glare makes it hard to observe other objects in the line of sight.

Apparently, we think that the Moon is bigger when there are other familiar horizon irregularities in the foreground: houses, trees, or wind turbines, perhaps even silhouetted on the horizon. High in the sky, there are no such familiar objects nearby to mislead the eye. We see the Full Moon, as the apparent size it truly is, at midnight – not on the horizon.

This does not totally explain the Moon Illusion. Sailors looking at the Moon ever a featureless sea have experienced it. So have astronauts orbiting the Earth. But it is the best theory we have.

An exercise: Try looking at the rising or setting Full Moon upside down. This can be accomplished by turning your back on the Moon, bending down, and peering between your legs. A "huge" Moon magically shrinks. You have disrupted the view your brain expects to see: a horizon and horizon features below the Moon, not "above" the Moon. So, it fails to produce the illusion.

There will be more to say about the Full Moon in the next chapter.

Figure 13.14: Thinking about the Moon Illusion the other way around: Which Full Moon looks *smaller*? Most people will choose the one that has an unexpected background, even though both are the same size. Artwork by Christine Tarte.

13.10 Chapter summary

Most people are more familiar with the synodic month than the sidereal month because it is the former that governs the change of lunar phases. The faction of the illuminated side of the Moon that you see from the Earth causes the phase, not a shadow cast upon the Moon. As the Moon "ages," and elongation increases, its phase progresses from New to crescent to First Quarter (occurring at dichotomy) to gibbous to Full. This sequence represents the first half of the synodic month. Most of the time, the Moon is not New, at a quarter phase, or Full; it is crescent or gibbous.

Aristarchus made the first serious attempt to measure the relative distances of the Moon and Sun.

The idea of unusual human behavior associated with the Full Moon is a myth. The defect of illumination in the Full Moon is real. The overly large appearance of the

Full Moon when near the horizon is a misinterpretation made by our brain. For this reason, it is called the Moon Illusion.

13.11 Chapter review

1. List a pair of numbers that are commensurate. Why would people who make calendars value that their calendars are commensurate?
2. What is the difference between the sidereal and synodic months? What is an example of a culture that used the sidereal month? What is an example of a culture that used the synodic month?
3. What causes Moon phases? Draw a picture of the Earth, Moon, and Sun to help you explain.
4. Many people wrongly believe that the Earth's shadow is what causes the darkening of the Moon that creates lunar phases. What actually is causing the shadow?
5. "Young" and "old" are used to describe the Moon. What do these terms mean in regard to the Moon?
6. Why is the phrase "Half Moon" incorrect? Why do we use First/Third Quarter Moon instead?
7. What is the cause of the Moon Illusion?
8. Some Moon phases technically occur at a specific time. Others occur over a longer period of time. Which phases belong to which category?
9. Despite the fact that a crescent Moon occurs for more days of the month than does the Full Moon, why might the casual observer, glancing up toward the sky at random times, be more likely to see a Full Moon?
10. Match the lunar phase on the left to a good time and direction at which it to see it on the right.

waxing Gibbous Moon	early evening in your western sky
First Quarter Moon	on your celestial meridian at sunset
waning Gibbous Moon	early morning in your eastern sky
Full Moon	on your Celestial Meridian at midnight

14 Full Moon to New Moon

Our lunation is only half-way complete. The phases of the Moon now repeat, but in reverse order.

On the Moon itself, the terminator relentlessly proceeds at an equatorial speed of fifteen kilometers per hour. This is fast enough that Apollo-program astronauts on the Moon, even driving their Lunar Rover Vehicles, could not outrun it.

> **Notes from yesterday and today**
> Studies show that motorcycle accidents are more frequent on nights of the Full Moon! Apparently, riders are distracted by it. This is one of the few statistically verified cases where the Moon affects human behavior in any less-than-obvious manner.

14.1 Chapter learning outcomes

After reading this chapter, you will be able to:
- define the term "Blue Moon" in two different ways.
- draw a diagram in order to explain earthshine.
- provide examples of different types of luni-solar calendars.

14.2 Only Full Moons each have their own names

The more well-known of Full Moon names come from native-American traditions of northeastern North America. Of course, these vary. However, here are some frequently repeated examples. Each now is used to refer to the only (or first) Full Moon in our **civil month**.

January = Wolf Full Moon. Presumably, this is because these predators howl in complaint as they find less prey during the middle of winter.

February = Snow Full Moon. Climatologically speaking, large swaths of North America experience their greatest snowfalls during this civil month. Obviously, there may not be any Full Moon in the civil month of February.

March = Worm Full Moon. As soon as the ground begins to thaw (from top down), worms often tunnel onto the surface.

April = Pink Full Moon. A preponderance of early blooming, North American wildflowers are pink.

May = Flower Full Moon.

June = Strawberry Full Moon. The harvest of this berry begins in June for North America.

https://doi.org/10.1515/9783111441245-014

July = Buck Full Moon. This is a roughly the civil month in which male deer typically shed their antlers.

August = Sturgeon Full Moon. This is a good civil month for fresh-water fishing.

September = Harvest Full Moon. Crops are brought in. (Technically, the Harvest Moon is the Full Moon associated with the Autumnal Equinox.)

October = Hunter's Full Moon. Hunting is easier now that the harvest has provided game animals, with less opportunity to hide.

November = Beaver Full Moon. There are conflicting explanations.

December = Cold Full Moon.

If the fish do not appear, the next Full Moon can be named the Sturgeon Moon too. This time, it is an embolismic month. (See below.)

No other monthly set of phases (*e.g.*, Quarter Moons) has its own set of names that have become ensconced in popular culture. However, the second New Moon after the Winter Solstice *is* well-known as the traditional Chinese New Year[1] (notwithstanding that it is celebrated throughout Asia). This date often is the closest New Moon to the Vernal Equinox. It falls between 21 January and 20 February on the modern civil calendar.

> **Notes from long ago**
>
> The original Roman calendar did not number days of their civil month. Instead, it referenced certain significant dates and counted *backward* from them. For instance, 1 April (hypothetically corresponding to the date of the New Moon) was *Kalends*[2]. So, the last day of March would be called "one day before Kalends" and so on.
>
> The seventh or the fifth day of the early Roman month was *Nones*, depending upon the length of the civil month. It corresponded, at least hypothetically, to the First Quarter Moon.
>
> The most famous of these reference dates on Rome's calendar was the *Ides*, mid-month (or, in some cases, the fifteenth day of the civil month), which corresponded, at least hypothetically, to Full Moon. The Ides of March is immortalized as the date upon which Julius Caesar was assassinated. This 44 BC/BCE date was one that changed the direction of Roman history and, therefore, that of the western world.

14.3 Angle of illumination

At Full Moon, the brightness of the Earth's satellite is at its greatest by far. This is not just because its maximum area is reflecting sunlight back toward you. The Moon has different average albedos at different phases.

At Full Moon, it is midday in the middle of the Earth-facing lunar nearside. The physical surface of the Moon is rough though. Once the angle of illumination is no

---

1 or Lunar New Year, even though it is a luni-solar calendar holiday; see below

2 the word from which "calendar" comes

longer vertical, a myriad of little shadows is produced by lunar relief features, large and small.

These shadows add up and in total, must be subtracted from the amount of surface available to reflect light toward the Earth. Shadowing increases rapidly, as a trigonometric function, with increased angle of illumination. The overall brightness of the Moon decreases markedly with phase, as one "travels" either forward or backward in time, during the synodic month, from Full Moon.

As daylit lunar surface decreases, shadowing increases. The average brightness of a Quarter Moon is only about 8% that of the Full Moon.

14.4 What is a Blue Moon?

In the sixteenth century, when it first appeared in written form, a **Blue Moon** was synonymous with an absurdity. "Absurd" evolved into "never." In the nineteenth century, the definition changed again to "a rare, and not entirely predictable event."

A real connection between the phrase and the Moon (or, at least, the calendar) has a slippery etymology. Canadian folklorist Philip Hiscock has tracked it down.

Blue Moon was used in the context of the calendar on the Public Radio program *StarDate* in 1980. By 1985, it appeared as a calendrical expression, having to do with the calendar in a popular children's book. With its calendrical definition, the Blue Moon showed up in the popular Trivial Pursuit game produced in 1986. By about 1988, it was firmly associated with the calendar.

To trace the connection earlier than this, it is necessary to go back to what is today a somewhat obscure publication called the *Maine Farmers' Almanac*, which used "Blue Moon" in the 1930s. This would seem to be the earliest published instance of a Blue Moon, with anything like its modern definition.

Though not quite: According to the *Almanac*, the third Full Moon in a season with four Full Moons is the Blue Moon. Today's usage is slightly different. It refers to the second Full Moon that occurs in a modern civil month.

Apparently, the author of an article in a popular astronomy magazine, describing the Blue Moon concept, misunderstood the rule that the almanac authors used. Yet, it is this second-Full-Moon-in-a-calendar-month meaning that has become accepted.

Why do such second Full Moons happen at all? It is because we use a civil month of 30 or 31 days. The Full Moon, occurring once every synodic month, will become placed earlier and earlier in each successive civil month.

Pretend that a Full Moon happens on, say, 1 November. The next Full Moon will occur 29.5 days later, on 30 November. It is still November. We say that the Full Moon of 30 November is a Blue Moon.

Blue Moons defined this way are not all *that* rare. In a 31-day month, it is even easier to have a Blue Moon. The first Full Moon could occur on the first or even the second day of the calendar month. Alternatively, the Blue Moon could take place on the last or even second-to-last day of the calendar month. The only month in which a Blue Moon is impossible is February.

Figure 14.1: August 2023 featured a calendrical Blue Moon. Courtesy of Ed Krupp (Griffith Observatory).

There is a Blue Moon every two-and-a-half civil years or so.

Nobody seems to take an interest in the equally likely occurrence of two *New* Moons in a single civil month.

Totally unrelated to the calendar is the question: Can the Moon literally turn blue?

Yes! This sounds bizarre because you are most familiar with a Moon – if it appears to have any color at all – being yellow, orange, or red due to atmospheric effect. These colors are on the opposite end of the spectrum from blue.

Normally, the particles in the Earth's atmosphere scatter blue light. If the Moon is anything other than white, then, it is because blue light reflecting from the Moon is scattered out of your line of sight. This leaves only the redder light to pass straight from the Moon to you.

Nevertheless, occasionally, particles in the air are just the right size to cause things to happen in reverse: Red light is scattered, and blue light passes directly through from the Moon to you.

An example is when there are micron-sized particles injected into the air by volcanoes. The phenomenon appears to have been first noted when the infamous volcano Krakatoa (Indonesia) violently erupted and polluted the stratosphere globally in 1883. The Moon turned blue.

Incidentally, it can happen to sunlight, too. The Sun can be turned blue under even rarer, extreme conditions. It has been reported to have occurred during a sandstorm.

Figure 14.2: Gibbous Moon. Courtesy of Ed Uthman.

14.5 Second Gibbous Moon of the month

You continue through the rest of the lunation. The lunar elongation is between 180° and 270°. Now at dusk, the Moon has not yet risen. For a time, there is no Moon in the sky.

When the Moon does rise, it is in the gibbous phase again. The geometries that created the phases during the first half of the month now repeat, in reverse order, during the second half of the month.

> **! Misconception alert**
> Artists frequently depict the Moon as a crescent. It has a traditional aesthetic appeal.
> Notice, though, that the Moon phase is seen as gibbous slightly *more* than it is crescent (because of the difficulty in seeing the crescent Moon near New Moon). Nevertheless, this closer-to-Full phase is almost absent in the visual arts.

On successive nights, the gibbous Moon appears later and more southerly. It grows thinner, though more than 50% of the lunar nearside still remains sunlit. The Sun's illumination recedes on the opposite side of the Moon from that upon which it encroached during the previous gibbous Moon.

Either gibbous Moon is a good time to note what is largely an effect of contrast. However, it also has to do with the angle of illumination on the rough lunar topography.

Figure 14.3: Relationship of the Earth and Moon to the Sun at the second Gibbous Moon of the month. Not to scale. Artwork by Christine Tarte.

Figure 14.4: Relationship of the Earth and Moon to the Sun at Third Quarter Moon. Not to scale. Artwork by Christine Tarte.

There is nothing as dark in the sky as the darkness of empty space. Compare the brightness of the illuminated limb to the illuminated portion of the Moon, just past the terminator. Often, the limb looks oddly brighter.

14.6 Third Quarter Moon

The Moon eventually is 270° from its starting point at the beginning of the lunation; it is a second dichotomy. At first glance, the third-week **Third Quarter Moon**[3] looks like the First Quarter Moon: the lunar terminator appears straight, and half of the lunar nearside shines in sunlight. The difference lies in the fact that it is the opposite semicircle of the nearside that is illuminated (compared to the First Quarter). After rising at noon and before setting at midnight, the Moon sits on the celestial meridian at sunrise, not sunset.

Figure 14.5: Third Quarter Moon. Courtesy of Michael Seeley.

The Third Quarter Moon is dimmer than the First Quarter Moon, even though we see equal areas on the Moon at both times. The western half of the Moon's nearside happens to be intrinsically darker than the eastern. This lower albedo means that the Third Quarter Moon does a poorer job of reflecting sunlight than its doppelganger earlier in the month.

By now you will notice that, during the second half of the month, where the "action" involving the Moon takes place has changed, it is east of the celestial meridian,

---

3 or Last Quarter Moon

Figure 14.6: Relationship of the Earth and Moon to the Sun at the second Crescent Moon of the month. Not to scale. Artwork by Christine Tarte.

and not west of it, anymore. Language may become more useful that describes what the Moon and its phases are doing in the early morning, at sunrise, at dawn, *etc.*

Your lifestyle is more likely to be on a schedule such that you are out and about in the early evening than in the early morning. For this reason, you tend to be more familiar with the phases during the first half of the month than those of the second half of the month.

14.7 Second Crescent Moon of the month

It is time in the lunation for the Crescent Moon again. It rises still later than any of the phases before it. You see it hanging low over the (south) eastern horizon before sunrise brightens the sky and causes the Moon to quickly disappear. If you do see the Moon in the daytime, it is likely to be when it is at an elongation, not too far from 180 degrees, somewhere between First and Third Quarter.

During the several nights belonging to the second appearance of the Crescent Moon, elongation between 270° and 360°, a thinner shape (and, therefore, dimmer Moon) is seen on each successive night. (However, the illuminated lunar nearside technically does not reach 0%.) You say that the Moon is becoming old.

Figure 14.7: Crescent Moon. Courtesy of Ed Uthman.

Figure 14.8: Difficult-to-make-out, twenty-four-hour-old Moon. It was photographed from Minnesota, USA, in May 2010. Courtesy of Bob King.

14.7.1 Cusps of the Moon

The "horns" that the visible shape of the Moon exhibits at other than New and Full Moon phases are called its **cusps**. Geometrically, the outer curve of the crescent Moon always should extend 180 degrees around the circumference of the lunar disk. However, in practice, the naked eye cannot resolve it at its thinnest.

In the last century, French astronomer André-Louis Danjon investigated a phenomenon that likely had been observed casually long before: thin crescent Moons that appear to extend, cusp-to-cusp, less than 180°. In other words, a line connecting the tips of the two lunar cusps does not always pass through the center of the lunar disk.

The effect manifests itself in very young and very old Moons. Experts debate the cause of this phenomenon. But they agree that, because of it, there is a theoretical minimum elongation at which (or maximum elongation after which) no crescent can be observed. This angular distance is called the **Danjon Limit** and is about 7°.

An exceptionally thin crescent Moon might exhibit breaks before the Danjon limit is reached at the cusps. These are due to the topography of the Moon. However, such a sight requires a rare confluence of circumstances in order to manifest itself to the naked eye.

Figure 14.9: These are sixteen consecutive Full Moons. However, the images were made at different times of night. Note the isolated, elliptical-appearing mare at the top. Think of it as the hand of a clock. Notice that it changes "time." This illustrates different position angles of the Moon. Courtesy of Theo Wellington.

Lost at night? If the Moon is crescent, draw a line in your mind through the cusps of the Moon. Where that line intersects the horizon is south. That goes for the northern hemisphere; in the southern, the line intersects the north horizon.

Insofar as the visible part of each lunation begins and ends with a crescent, it is not so odd that two-horned beasts are historically and culturally associated with the Moon. In Paleolithic rock art, the horns on animals often are painted to look like crescent Moons. The Roman god Luna frequently was portrayed riding a chariot pulled by bulls; and as American planetarium-director/astronomy-educator Ed Krupp reminds us: In the old nursery rhyme, "The Cow Jumped Over the Moon."

14.7.2 (Lunar) Terminator illusion

An adage about the Moon says that you can forecast the weather by whether the cusps of the Moon appear to be pointed more upward with respect to the horizon or more perpendicular to the horizon. "Rainy down, fair skies up." The upturned cusps metaphorically hold the water back.

The **position angle** of the Moon refers to its disk orientation in your sky. It is the angle that true north *on the Moon* makes with respect to the celestial north. The Moon's position changes as it moves daily across your sky. The degree to which it does so is a function of latitude.

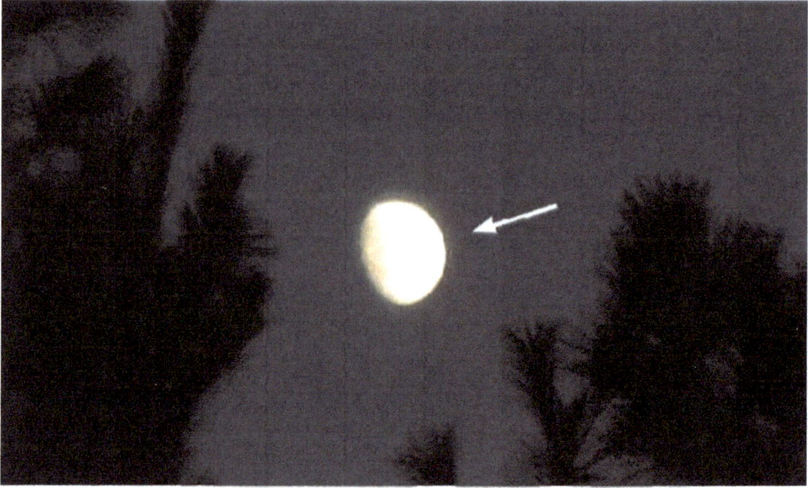

Figure 14.10: While not of a crescent Moon, this image illustrates the perceived direction of the Sun (white ray) just after sunset. Courtesy of Michael Berry.

Position angle should correlate with how the Moon's cusps appear to be pointed. Depending upon the way you choose to define "more" in this case, the cusps might be expected to appear pointing more toward the horizon or more parallel to the horizon, equally often.

There is of course the obvious truism that the Moon is best seen at night. Therefore, the crescent Moon is more often noticed as a "smile" in your sky. The appearance of a "frowning" Moon, with respect to the horizon, is a daytime event.

Nonetheless, in reality, the quoted aphorism quoted earlier would seem to tell you nothing more than that there are equal chances of rain and sunshine. In some climates, often, this is true. However, there is no causal correlation between the directions in which the Moon's cusps appear to point and the weather.

Notwithstanding the above, note the use of the word "should." Something odd can happen as you study the orientation of the Moon in the sky. It is most striking if the Sun is still in the sky, or a little below the horizon, so that you can infer where it set (alternately, just in the sky or a little below the horizon, so that you can infer where it will rise).

Your perception of how the Moon's cusps point with respect to the imaginary Moon-Sun line, or – more generally – how the lunar terminator is oriented with respect to what you perceive as this line, may differ from what it really is. This is a very common visual misperception.

You know that the imaginary line connecting the cusps, or the similar line tangential to the terminator, ought to be perpendicular to the Moon-Sun line in your sky. However, your brain may not correctly take into account the spherical projection of the scene. The orientation of the imaginary cusp or terminator line may appear differently; the Moon's position angle, suggested by the cusps or terminator, may look as if it is less than what it actually is. The discrepancy can be more than 10°. It is a lunar tilt illusion.

Figure 14.11: Earthshine, as seen in Cambridge, England. Courtesy of Brian Josephson.

14.8 Earthshine

It is sometimes possible to see more of the lunar nearside illuminated than its phase
would dictate. This illumination is dim but is sufficient to provide a complete view of the
Moon's Earth-facing hemisphere, even if the Moon is a crescent. This means that, in addition to seeing part of the lunar dayside – the phase, horn to horn – you also see part of
the lunar nightside. This usually happens during a crescent phase, and when it happens,
the phenomenon is called poetically "the New Moon in the Old Moon's Arms." It was
first documented by the Mesopotamians. Today, we call it **earthshine**.[4]

Figure 14.12: Path of the sunlight that produces earthshine. Not to scale. Artwork by Christine Tarte.

14.8.1 What causes earthshine?

If the Moon produces no light of its own, where does the supplemental irradiance
originate, if not from the Sun? Galileo thought that the Moon might be translucent!

No, earthshine comes from the only other object in the vicinity: the Earth. The
Earth mirrors (unfocused) sunlight just as the Moon does. The Earth has a higher albedo than the Moon and provides a comparatively large reflecting area. It can be
bright.

---

4 or ashen light, Da Vinci glow

Some of the sunlight striking the Earth shines off of it, and a portion of this redirected light reaches the Moon (the whole nearside, not just the sunlit part producing the phase). This Earth light is reflected *again*, by the Moon, toward you.

By now, the strength of the original sunlight is well attenuated by a double reflection; much of it has been absorbed either by the Earth or by the Moon. Still, it is just brilliant enough to produce noticeable contrast with the pitch black of the night sky.

Earthshine was first correctly explained by the Italian polymath Leonardo da Vinci. It is about a thousand times fainter than normal "moonshine."

Earthshine comes from the dayside of the Earth, obviously, though it is too faint to see from the Earth during the day. However, we have seen how at certain geometries, part of the lunar nearside is lit by the Sun and part is not. (The New Moon and Full Moon are excluded.) These are the lunar phases.

Let us turn things around. It is not too difficult to envision that, at these geometries, the disk of the Earth that is facing the Moon is partly illuminated by the Sun (in daytime) and, by the separation of a terminator, partly not. It is therefore possible to sit on the nightside of the Earth watching the Moon while it is illuminated by the dayside-produced earthshine. This is light reaching and then reflecting off of the surface of the Earth, perhaps some thousands of kilometers away from you, measured eastward or westward along its circumference.

14.8.2 How & when to see earthshine

To see earthshine, all you have to do is be on the side of the Earth facing the very young or very old Moon, at night (maybe, outside of twilight).

At this point it may be useful to think of the phase of the Earth as seen from the Moon. The situation is completely reciprocal – ask an astronaut who has stood on the Moon looking up – except for the fact that terrestrial phases (as viewed from the Moon) are one half-month out of synch with lunar phases (as viewed from the Earth). For instance, a crescent Earth takes place during a gibbous Moon and vice versa. So, there is more Earth light available during a crescent Moon.

Starting not-too-many days after New Moon, not only is the glare from the brightening Moon becoming significant enough to make seeing earthshine difficult, but less and less of the gibbous Earth is sunlit and available to produce the redirected light that will become earthshine. It quickly becomes impossible to discern earthshine. For completeness, note also that, until not-too-many days before New Moon, the glare from the crescent Moon remains significant enough to make seeing earthshine difficult.

Earthshine is a phenomenon of the thin crescent Moon. How long before or after New Moon you can see it, or whether you can see it at all, is a function of local factors but also varies due to what is taking place on the Earth's daylit hemisphere. A greater percentage of ocean (as opposed to land mass) means greater-than average reflectivity; less means lower-than-average reflectivity. More clouds mean a higher-than-average albedo; fewer means less. As a rule, the Moon should be less than four days old, or more than 25 days old.

In the northern spring, the North Celestial Pole is pointed closer to the Sun than away from it. Yet, there likely still is snow and ice on the high-latitude ground and sea. (Think Siberia.) This high-albedo surface makes the Earth brighter than average. Thus, earthshine is brighter at this time of year too.

There is a smaller albedo peak in the southern spring. This mainly is due to the presence of Antarctica. Either way, Spring is the season for earthshine.

Long-term, the Earth's albedo, and subsequently the brightness of earthshine, *may* be affected by the sunspot cycle. (See Chapter 19.) It even is possible that climate change affects cloud cover and, thus, terrestrial albedo.

14.9 Return of the New Moon

Eventually, the Moon is New once more. Our lunation has ended.

There may be as many as four nights (centered on New Moon) between the last visibility of the Crescent Moon in one lunation and the first visibility of the Crescent Moon in the next lunation. However, if you have a good eye, you may have to wait little more than a day-and-a-half between the appearances of the crescents. If the instant of New Moon happens to occur at midday, it is sometimes just possible to catch sight of the Crescent Moon on the morning of the day before, and then again in the evening of the day after, with the New Moon occurring in-between.

The opportunities for such back-to-back Crescent Moons are rare. Sky-observing conditions must be ideal.

Figure 14.13: Twenty-eight phases of the Moon (2021). Pexels.com.

Incidentally, some scholars think that old-time cross-quarter-day celebrations in Europe did not take place exactly on the cross-quarter day itself. Instead, they occurred at the New Moon (or, perhaps, Full Moon) closest to these days.

14.9.1 Waxing & waning

How do you tell one lunar phase from another? Nobody will confuse a New Moon with a Full Moon. However, what about the First and Third Quarter Moons? What about the pairs of Crescent and Gibbous Moons?

If you keep track of the synodic Month, of course, there is no difficulty. It is even possible to purchase a clock-like timepiece that does this for you. However, these suggestions apply to few of us.

Where and when the Moon is in the sky, and which part of the lunar disk (east-facing or west-facing) is illuminated, help to sort it out. The most straightforward way, though, is this: Are you seeing less and less of the sunlit lunar nearside night after night? If so, it is the latter half of the synodic month, and the older Moon is said to be **waning**. Are you seeing more and more? If so, it is the beginning half of the synodic Month, and the younger Moon is said to be **waxing**.

In the northern hemisphere, the Moon always waxes and wanes from west to east. In the southern hemisphere, the Moon always waxes and wanes from east to west.

These adjectives, "waxing" and" waning," make each phase distinct, when placed in front of the word "Crescent" or "Gibbous." First Quarter Moon is a waxing phase too. Third Quarter is a waning phase. Since lifestyle causes more people to look at the evening sky than at the morning sky, you are more likely to see the waxing phases of the Moon than the waning ones.

Among those who practice such things, waxing Moon rituals tend to thematically involve beginning, creating, or acquiring. Waning Moon rituals tend to thematically involve ending, destroying, or disposing.

14.9.2 Practice!

You can now recite the eight phases of the Moon, in order, this way: New Moon, Waxing Crescent, First Quarter, Waxing Gibbous, Full Moon, Waning Gibbous, Third Quarter, and Waning Crescent.

1582	OUTUBRO				1582	
Dom	Seg	Têr	Qua	Qui	Sex	Sáb
	1	2	3	4	15	16
17	18	19	20	21	22	23
24	25	26	27	28	29	30
31						

Ao ser o calendário Juliano corrigido pelo Papa Gregório, em 1582, foram dele eliminados dez dias, como aí se vê. O dia que teria sido sexta-feira, 5, ficou sendo sexta-feira, 15. A continuidade dos dias da semana não foi alterada nem interrompida, nem o foi em alteração nenhuma do calendário.

Figure 14.14: *Il Calendario Gregoriano Perpetuo* by Bartolomeo Dionigi (1582).

14.10 Fixing the calendar

How important is a lunar (synodic month) *calendar*? So far in this book, we have dwelt on a Sun-based calendar. However, for hundreds of millions of people, a lunar calendar is very important. Consider the ninth Islamic month of Ramadan. It represents the holiest synodic month (out of twelve) in a strictly Moon-based religious calendar.

The observance of Ramadan occurs completely independently of the time of solar year and shifts backward through the seasons. Muslims celebrate Ramadan twice in a single, modern, civil year, every 33 years; it will likely begin on 6 January 2030 and *again* on 26 December 2030.

You can produce a calendar based on either the Sun/year or the Moon/month. The trick comes when a civilization wishes to fit both calendars into one **luni-solar calendar.** Remember that the early Mesopotamians and Egyptians did not worry too much about doing this, but it has been a quest of civilizations ever since their time.

14.10.1 The luni-solar calendar

Twelve synodic months, equaling 354 days, are close to, but short of, the (approximately) 365 days in the solar year. The age of the Moon on the first day of the solar year is called the **epact**. Even if you managed to start counting your solar year at New Moon (epact equals zero), the Moon would be older on the next New Year's Day (would have a positive epact) and would be still older on the next (greater positive epact).

Each year, it will be necessary to start counting the twelve-synodic-month cycle over again, earlier and earlier. Eventually, you will be at the end of month twelve, and there will be more than 29 ½ days left in the year! In other words, the epact for the next solar year is greater than the length of the synodic month itself.

One can imagine things getting so far out of hand that the *first* month of your attempt at a twelve-month lunar-solar calendar occurs in the *middle* of the solar year. Things have become completely out of synch. Traditional seasons (associated with a given month) no longer match up with climate.

So as to keep this from happening, it becomes necessary to occasionally **intercalate** – insert an extra thirteenth month into the calendar to reset the epact. If this is done wisely, the first month in your twelve-month cycle always remains early in the civil year, and the last month stays late in the civil year. The epact keeps reasonably small.

A civil month that exists because of intercalation is called an **embolismic month**. For example, in the Jewish calendar, the twelfth month of Adar sometimes is not by the first month of Nisan; it is followed by a thirteenth embolismic month called Adar (again). A year that contains an embolismic month is an **embolismic year**.

Eskimos do it the other way: Their thirteen-month calendar *drops* a month during the arctic winter, every few solar years.

Intercalation is based upon a human decision, and it can be misused (*e.g.*, for political purposes). The *Koran* expressly forbids intercalation. This is why the Islamic calendar bears no relation to the Sun-based calendar, and lunar-based special days (such as Eid) can occur at any time of solar year.

Attempts were made to codify intercalation in order to avoid intercalation by fiat. The Classical Greeks decided to automatically intercalate a synodic month once every eight solar years, as early at the sixth century BC/BCE. It worked, but only for 149 solar years.

Meton of Athens established *circa* 432 BC/BCE a more nuanced pattern that functioned better. It is the nineteen-year-long **Metonic Cycle**. Metonic intercalation occurs every third, then third, then second, then third, then third, then third, then second, solar year – a pattern which is then repeated. Such a system results in an average year 365.26 days-long. This is very near the desired goal, which we know to be 365.24 days, the true solar year.

Callippus of Cyzikus went so far as to suggest, in 330 BE/BCE, the following: If four Metonic cycles are followed, but then *one day* of the last resulting solar year is dropped, the result is even closer to reality: 365.25 days per solar year. Admittedly, the 76-year-long **Callippic Cycle** is awkward.

14.10.2 The Julian calendar

A major calendar innovation was made in the days of the Roman Republic. Imagine again a graph of the rate of the Sun's change of position on the Celestial Sphere versus time. Recall that the two inflection points that are the Solstices, occur when the graph is flattened. The Sun does not change its location on the Celestial Sphere, north to south, very fast. It is not easy to observe a change in the Sun's altitude at this time of a solar year over the span of a few days.

It is easier to establish the exact time of solar year at an Equinox, when the Sun is at an inflection point on a graph such that the Sun curve is steepest. The change in solar altitude day-to-day is much more obvious than at a Solstice.

Thus, the new year traditionally began in March, "home" of the Vernal Equinox. This notwithstanding, the Romans *changed* the beginning of their civil year! The result was the now familiar January (a time of year when there is less going on).

Much later, it was Julius Caesar who discovered that within what was now his empire, there still was a variety of competing and conflicting calendars in use. He asked his astronomers to standardize the calendar and to one based upon the fact that the solar year is closer to 365.25 days long, not 365.

They did so using integer-long months of different lengths, with little attempt to match the length of a synodic month. A civil month with, say, 30 days (too long) was a **full month**. One with 30 − 1 = 29 days (too short) was a **hollow month**.

The result was the **Julian calendar** of alternating full and hollow month. But the key to it was a single month at the end. This civil month, but only this one, was full or hollow (now meaning 29 or 28 days), depending upon a leap-year rule.

The Julian calendar was a major change of strategy. Beforehand, intercalation was a means of correcting an error. Leap years are a means of prohibiting an error from occurring in the *future*.

The Julian calendar is anticipatory. We call it a **proleptic** calendar. The calendar makers fatalistically accept that their calendar always will be "wrong" and take steps to minimize its divergence from reality. In doing so, the calendar is applied to the past as well as the future.

This was the western calendar throughout the Middle Ages. However, nearly right away, it was decided that Julius should be honored with his own civil month, that of his birth. It was impolitic not to give his powerful heir Augustus, the next civil month. But could an emperor have anything other than a 31-day month? This is how a regular pattern of full and hollow months was disrupted. In other words, it is why you likely are compelled to memorize some tool of memory, for instance, the familiar English rhyme that begins, "Thirty days hath September . . ."

14.10.3 The Gregorian calendar

The modern Julian calendar received a final tweak during its separation from the real Moon. To five places, the solar year is 365.24 days long. The difference of one-hundredth of a day, between this value and 365 ¼ days, added up to enough time by the Renaissance so that the seasons were no longer matching their proper dates.

Of great importance to many in the West, the paramount Christian religious holiday Easter was threatened with no longer being an early spring holiday. Pope Gregory decided to solve the problem once and for all.

Dropping a day from the civil year every 125 years was considered. Ultimately, Gregory instead decided to finagle the existing leap-year rule slightly so that a troublesome slippage would not occur again until the far, far future. At the same time, he reset the count so that the Spring Equinox occurs on 20 March or 21 March.

For example, the year 2000 was noteworthy not only for being the end of a millennium, it also was a leap year! That is not, by itself, surprising. You learned in elementary school that an extra day 29 February is to be inserted into every civil year that is divisible by four.

However, according to the proleptic **Gregorian calendar** of 1582, based on the correct 365.24-day solar year, you must skip a leap year every hundred civil years *except* in years divisible by 400. The year 2000 was a leap year, but it also was divisible by 400, which made it special. So was the year 1600.

The Gregorian calendar is the one most used today. Even Pope Gregory's calendar eventually will go wrong, though We likely will have to drop a leap year in 4000, nearly two thousand years from now.

The change in the leap-year rule was subtle, but the more obvious damage due to the Julian calendar already was done. In order to reset to the Equinox, Gregory decreed that Thursday 4 October 1582 was to be followed by Friday 15 October 1582. People were not happy. It was as if time had been stolen from their lives. Would they be paid for a whole month of wages? Would they die ten days sooner? In that age, only the Pope could have pulled off what needed to be done.

Despite the long-term advantage of the Gregorian calendar, Great Britain did not adopt the Gregorian calendar until 1752. In Greece, it was 1923. In Ethiopia . . . they remain Julian to this day.

14.10.4 Other calendars of Interest

Omar Khayám, better known today as a poet, was foremost a mathematician. In the eleventh century, he proposed a calendar with eight leap years every 33 civil years. While complicated, it is *more* accurate than the Gregorian calendar. Nobody took him up on it though.

> **Notes from cultures around the world**
>
> The nomenclature associated with the western civil month divides it into approximate quarters (weeks). Yet, quarters are not the only way in which to apportion a civil month.
>
> In Hindu society, the half-month (after New Moon) of waxing phases (up to Full Moon) is given its own name, the Suklapakṣa. The half-month (after Full) of waning phases (up to New) is the Kṛṣṇapakṣa. The Batak of Sumatra split their civil month into ten-day thirds!

You might be tempted to think that the modern civil calendar, which no longer pays attention to the phase of the Moon, is a new invention. Yet, such a calendar was first adopted by the Egyptians more than six-thousand years ago.

Their "Sothic Year" was even simpler than ours: twelve thirty-day civil months, plus five extra days of festival at the end of the year. A 360-day calendar is called a **perfect calendar**, even though most of us would think of it as far from "perfect." On the other hand, ending the civil year with a long party seems remarkably modern.

14.11 Chapter summary

Full Moons are unique in that people have given them names. During the second half of the synodic month, the lunar phase continues to progress past Full to Gibbous to Third Quarter to Crescent to New Moon again. A waxing Moon is becoming closer to Full, a waning Moon is becoming closer to New. The naked-eye visibility of the crescent Moon is influenced by the Danjon Limit.

Earthshine is real and occurs when sunlight is doubly reflected, once by the Earth and once by the Moon. The perceived orientation of the lunar cusps and terminator may not be real.

A Blue Moon seldom happens. More often, the term is used to refer to a quirk in the civil calendar: a second Full Moon in a civil month.

The civil calendar familiar to most people today is proleptic and applies leap years, assigned by an algorithm, so as to remain close in length to the solar year. This is the Gregorian calendar.

14.12 Chapter review

1. Even though the First and Third Quarter Moons are roughly "half-full," why are they only 8% as bright as a Full Moon?
2. What does the phrase "Blue Moon" mean today? How does this differ from what "Blue Moon" meant in the past?
3. What are some of the things that could happen on the Earth that would make the Moon literally appear blue?
4. Why does the Moon often appear more red/orange/yellow near the horizon than when it is higher in your sky?
5. Why does the Third Quarter Moon appear dimmer than the First Quarter Moon, despite the fact that in those two phases, equal fractions of the lunar disk are visible?
6. Why is it that sometimes you can see the part of the lunar nearside that is not experiencing daytime?
7. Why might earthshine be more dramatic during a Crescent Moon phase?
8. "Earthrise" is a famous photograph taken in 1968 by William Anders, an Apollo 8 astronaut. (See below.) If you were standing on the Earth looking at the Moon at the time this photo was taken, what Moon phase would you have seen?
9. How would you go about deciding if a lunar phase is waxing or waning?
10. Describe, using a bulleted list, how we arrived at the civil calendar nearly all of us use today.
11. Why do you suppose there was once a desire to start the civil year in March?

Figure 14.15: The iconic "Earthrise" (for use in answering Question 8). Courtesy of NASA.

15 Further lunar cycles

Like the Sun, the Moon is a wanderer. It is one of the seven planets of the ancients. Today, though, we label the naked-eye planets as a group separate from the luminaries.

The physical Moon is a satellite of the Earth. It can be seen to move in its orbit, eastward against the background of the Celestial Sphere. It does so at a rate equal to its own diameter in an hour – more quickly than any other periodic object.

15.1 Chapter learning outcomes

After reading this chapter, you will be able to:
- recognize different times of the month, year, and the 18.6-year lunar cycle based upon the Moon's rising, culmination, and setting.
- place lunar standstills, and lunar maximum and minimum standstills, hierarchically according to frequency of occurrence.

15.2 Lunar retardation

Moonrise and moonset have a lot in common with sunrise and sunset. In the case of the Sun, you know that the diurnal times of these events correlated with the Sun's declination through the solar year. So, it is with the Moon. However, because the Moon's apparent angular velocity on the Celestial Sphere (13° per day on average) is much greater than that of the Sun's, its declination changes more rapidly.

The time from sunrise/-set to the next sunrise/-set differs with regard to the Celestial Sphere's apparent rotation period by mere minutes. This daily delay in the Moon's debut above the horizon is called lunar **retardation**.

The average length of the Moon's retardation is 50 min. The same is true, of course, for its setting. Clearly, retardation varies much more than that of the more minor change in the times at which you witness the corresponding behavior of the Sun.

15.3 Lunar altitude

You have learned that the ecliptic can be nearly perpendicular to the horizon or quite oblique to it. It depends upon your latitude and the time of solar year. The Moon's

https://doi.org/10.1515/9783111441245-015

orbit is tilted 5° 9' to the orbital plane of the Earth about the Sun. Thus, its projection onto the Celestial Sphere may be more than 5° away from the ecliptic.

15.3.1 Moonrise/moonset

The direction of the Moon's diurnal motion in the sky changes accordingly. The component of the Moon's daily motion perpendicular to the horizon is what determines how much later moonrise and moonset will occur each day.

Consider a moonrise, one which occurs near the onset of nighttime, as an archetype. The declination of the Moon is about 5° less in northern summer and fall than is the declination of the ecliptic. When the Moon is near the horizon, the relationship between the ecliptic and the lunar orbital plane is such that little of the Moon's daily orbital motion is perpendicular to the horizon. Most of it is parallel. So, in each successive 24-hour interval, the Moon does not change its altitude much.

The change is mainly in azimuth. Moonrise during one daytime or nighttime is only slightly later than it was during the daytime or nighttime before. The same is true for moonset.

We tend to notice this phenomenon more, as well as nearly everything else that has to do with the Moon, at Full Moon. Near the Autumnal Equinox, lunar retardation may be as little as 23 min. There will be two, or even three, nights when a nearly Full Moon rises approximately at sunset. This is the famous Harvest Moon: the Full Moon closest to the second annual Equinox.

There is some ambiguity regarding the timing of *the* Harvest Moon: Is it the first Full Moon after the Equinox? The last before? More commonly, and by the definition used here, it is the one nearest in time to the Equinox (before *or* after). So technically, the Harvest Moon can occur in October, not September.

Notwithstanding, this Full Moon provides several nights of bright moonlight after sunset. This phenomenon traditionally gives farmers more time of visibility to harvest their crops, if the farm is at latitude such that harvest occurs during this time of year.

Still consider our archetype Full Moon. Six months later, the Moon is in a very different place with respect to the horizon near nightfall and daybreak.

The declination of the Moon is about 5° greater in northern winter and spring than is the declination of the ecliptic. The relationship between the ecliptic and the lunar orbital plane is such that less of the Moon's daily orbital motion is parallel to the horizon. Most of it is perpendicular. So, in each successive 24-hour interval, the Moon does not change its azimuth.

The change is mainly in altitude. Moonrise on one daytime or nighttime is much later than it was on the daytime or nighttime before. The same is true for moonset.

You may well see a rising March Full Moon. However, if you look for it at approximately the same hour the next night, it will not yet have risen. Retardation at the

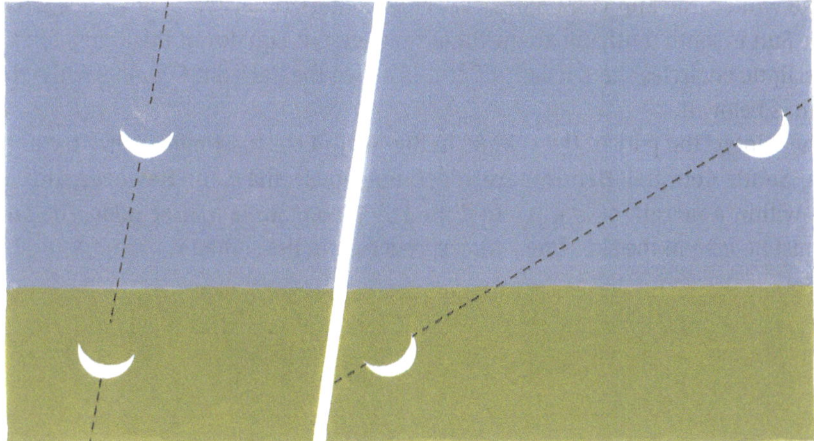

Figure 15.1: In these two panels, the Moon is shown starting at the same negative altitude in each. Both panels then show the rising Moon in the sky after the same amount of time has passed. Nevertheless, the altitude that the Moon achieves is different on the left and right. The panel on the right represents a winter moonrise, while the panel on the right represents a summer moonrise. Artwork by Christine Tarte.

Equinox may be as much as 75 min. The times between successive moonrises can be longer near the Vernal Equinox than near the Autumnal Equinox by about an hour.

This difference is less at more equatorward latitudes but greater at more poleward latitudes. Keep in mind the reversal of the pattern in the southern hemisphere.

> **Notes from yesterday & today**
> People tend to notice the Moon nearer the horizon than when it is high in altitude, enough so that it is associated with views on the skyline. This is part of the reason why certain northern people associate the Full Moon more with Hallowe'en Jack O' Lanterns than with, for instance, Easter Eggs. They tend to see the Full Moon more often on the fall horizon than on the spring horizon.

15.3.2 Lunar culmination

Consider two variables of the ecliptic's orientation with respect to the horizon, and the orientation of the Moon's orbital path with respect to the ecliptic. These also determine at what altitude the Moon will culminate at, on a given evening (morning). Why do these season-dependent behaviors of the Moon take place? Here is a way to think about it.

Again, let us dwell on a Full Moon; it is the easiest example. Recall that at Full Moon, the Moon is approximately opposite the Sun with respect to the Earth on the Celestial Sphere. It is on the other side of your sky from the direction of the Sun.

In the winter, the Sun culminates closer to the horizon. As you know, this is because the Sun is south (antipodean north) of the Celestial Equator at this time of year. Yet the ecliptic encircles the Earth. Half of it is above the Celestial Equator, while the other half is below it.

In the winter, the part of the ecliptic in the North Celestial Hemisphere (for antipodeans, South Celestial Hemisphere) does not house the Sun. However, this is where – within a margin of 5° – we find the Full Moon. So, 12 h after noon, the Full Moon, must be *high* in the sky. Everything is reversed in the summer.

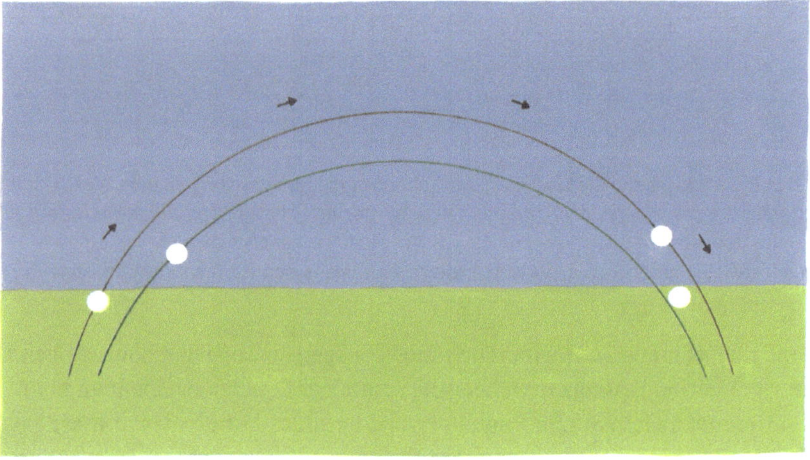

Figure 15.2: Diurnal path of the Moon in wintertime (top) and summertime (bottom). Artwork by Christine Tarte.

If you are a northerner from a climate that includes snow in the winter, you might conjure the memory of a January evening: It is likely clear and cold – the kind of night when you can hear your footsteps crunch upon fresh fallen snowflakes. All the snow crystals glisten. Your eyes may be drawn upward to the cause: a high, bright Full Moon somewhere overhead.

Parts of Australia, New Zealand, Argentina, and Chile also experience snow. There, this is a story about July.

You also may remember a summer's night, one when the Full Moon spends its time seemingly slinking close to the horizon. Yes, you normally may pay more attention to the Full Moon when it is near the horizon. However, what if the sky is hazy, not unusual in summer? The Full Moon is dimmed. Close to the horizon, the odds of its being obscured by clouds are increased. There are more or less the same numbers of Full Moons in the summer as there are in the winter. Practically speaking, though, you may be *less* likely to notice them.

To some tropical-dwelling peoples, the dates on which the Sun is at the nadir are important. You cannot see the Sun at the nadir. However, the opposite relation between the Sun and the Full Moon may be exploited. The Full Moon may act as a "stand-in" for the Sun. If the Full Moon is seen at the zenith, the Sun must be near or at the nadir.

15.4 Lunar azimuth

So much for the Moon's altitude. What about the Moon's azimuth throughout the sidereal month? Just as with the Sun, you are most likely to notice changes in the Moon's azimuth as it rises or sets.

The Moon's place on the horizon varies over an even greater range of azimuth than does the Sun's. This is again because the Moon's orbital plane is inclined to the ecliptic plane.

For half of the sidereal month, when the Moon's path on the Celestial Sphere takes it into the North Celestial Hemisphere, it will rise in the northeast and set in the northwest. For the other half of the month, the Moon's path on the Celestial Sphere takes it into the South Celestial Hemisphere. It will rise in the southeast and set in the southwest. The range in azimuth may be as great as 57°.

15.4.1 Standstills

When the Moon is as far north as it will travel during a given sidereal month, it rises at its northern standstill. When the Moon is as far south as it will travel during a given sidereal month, it is at its southern **standstill.** Of course, comparable standstills occur at moonset, too.

> **Where did that come from?**
> To parallel the term "solstice," a lunar standstill really should be called a "lunastice." Blame the 1971 book Megalithic Lunar Observatories. Alexander Thom, its author, used the English-derived term "standstill." "Standstill" caught on, "lunastice" did not.

At a lunar standstill, the Moon appears to change directions. Every day it rises and sets more and more southernly. Then it reaches its standstill. It will thereafter rise and set more and more northerly.

The Moon does this until it reaches a second standstill. At that time, it will begin to rise and set more and more southernly again. Notice that the Moon rises due east or sets due west only twice during the month.

Think of the azimuth of moonrise or moonset as another swinging pendulum. Each half of the symmetric cycle is about two weeks apart. The Sun's azimuthal variation between the Solstices, over the course of the solar year, is mimicked by the Moon over the course of each sidereal month. And like the Sun, the Moon changes azimuth faster in the middle of its pendulum-like cycle than it does at the extremities.

15.4.2 Maximum & minimum lunar standstill

So far, the Moon's behavior on the eastern and western horizons seems analogous to the Sun's. Now, though, we arrive at where it becomes more complicated.

The orbital plane of the Moon twists in a circle, counterclockwise as viewed from north, around our planet. It does so every 18.613 solar years. This is a multi-body gravitational effect. The process is somewhat like the Earth's axial precession. It is called the lunar **nodical cycle**[1].

As the Moon's orbital plane projected onto the Celestial Sphere is tipped 5° to the ecliptic, these two effects conspire to make the apparent path of the Moon in your sky periodically closer to or farther from the Celestial Equator. It ranges from declination +18.134° to +28.725° and –18.134° to –28.725°.

When the Moon's apparent path in your sky is closest to the Celestial Equator at all times during the sidereal month – the two opposite, geometrically necessary intersections of these two apparent circles are the exception – the southernmost and northernmost standstill azimuths of the rising/setting Moon do not differ much. They remain close to that of the Celestial Equator (due east and due west). These two standstills, stray the least from the Celestial Equator compared to any other time during the nodical cycle. They are named the minimum southern standstill and the minimum northern standstill.

When the Moon's apparent path in your sky is farthest from the Celestial Equator at all times during the sidereal month – the two opposite, geometrically necessary intersections of these two apparent circles are the exception – the southernmost and northernmost standstill azimuths of the rising/setting Moon differ greatly from one another. The Moon rises and sets far to the southeast and southwest. Half a month later, it rises and sets far to the northeast and northwest. These two standstills now stay farther from the Celestial Equator than they will at any other time during the nodical cycle. They are named the maximum southern standstill and the maximum northern standstill.

Every 9.30 (18.613/2) solar years, there is a synodic month in which the Moon arrives at either a minimum or maximum lunar standstill. Unlike the sunrise and sunset, though, both moonrise and moonset hover in the vicinity of one of their maxi-

---

1 or Regression of the Nodes

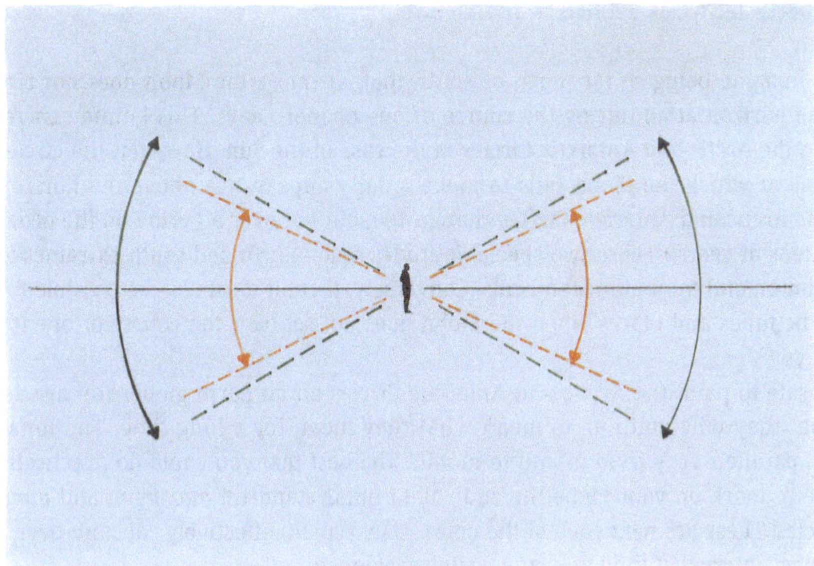

Figure 15.3: In this illustration, the horizon surrounds the reclined observer. The zenith is perpendicular, out of the page. At given latitude, and during a random month, the Moon rises and sets at azimuths symmetric with the continuous east-west line. It does so somewhere between the dashed lines.

The black dashed lines represent the maximum possible angular distance between moonrise and moonset during *any* month (standstill maximum). The red dashed lines represent the minimum possible angular distance between moonrise and moonset during *any* month (standstill minimum). Not to scale. Artwork by Christine Tarte.

mum/minimum inflection points on the horizon for about three years. The swinging behavior of the "pendulum" is different.

The last maximum lunar standstill (prior to this writing) was in 2025.

Of course (deviating from our example), roughly half of those moonrises or moonsets will occur in the daytime. The Moon may be washed out. It might be near New Moon. The odds are that some standstills will happen during cloudy weather. Regardless, the scene will not repeat until the same window of opportunity opens one lunar nodical cycle later.

Thus, in reality, catching the Moon at a standstill maximum or minimum is tricky. Like a Solstice, it is easier to determine that it has passed than it is to decide that it is happening. The only thing making the required observations easier is the slowing of the lunar pendulum near a standstill. Then there are, in practice, several nights in a row during which the Moon is near its standstill rising or setting azimuth.

We saw that the **inclination** of the Moon's orbital path in the sky to the Celestial Equator also determines at what altitude the Moon culminates on the celestial meridian. The lunar nodical cycle affects the maximum and minimum altitude of the Moon, too.

15.4.3 Special latitudes defined by the Moon

You can imagine being so far north or south that, at times, the Moon does not rise above the horizon at all during the course of one or more days. This latitude corresponds to the Arctic and Antarctic Circles in the case of the Sun. However, the circles of latitude at which the Moon fails to make a daily appearance above the horizon vary. (The Arctic and Antarctic Circles change, too, but not over a period on the order of mere tens of years.) These two special latitudes reach north and south extremes at the maximum and minimum standstills. Conversely, then, it must also be possible for there to be times and places when the Moon does not set over the course of one (or more) days.

It is safe to paint the Arctic and Antarctic Circles on an Earth globe. You are assured that they will continue to mean what they mean for a long time. The lunar standstill parallels vary from month to month. The best that you could do practically would be to mark on your globe the equivalent lunar standstill *maximum* and *minimum* circles. These are near each of the poles. They remain effectively constant over a much longer interval of time than any ordinary standstill.

You also can imagine indicating special parallels of latitude defined by the Moon that correspond to the roles of the Tropic of Cancer and Tropic of Cancer. These would apply to the tropical behavior of the Moon in the sky. The Moon may appear at the zenith near the equator, something never seen at mid-latitudes. If such parallels were made part of our imaginary globe, again, it would make sense only to highlight them at their maximum and minimum latitude values. There would be twice as many of these newly delineated circles of latitude as there are Tropics.

However, your globe is crowded enough as it is! The utility of noting all these lunar-based circles of latitude is much less than that of the Arctic and Antarctic Circles, as well as the Tropics. Encoding them on a globe is never really done.

15.5 Who pays attention to the changing path of the Moon?

Your ancient ancestors clearly studied the motion of the Sun in the sky. Did they do so as diligently for the Moon? There is little in history to help you answer this question. Looking for extant, prehistoric, architectural alignments that definitely refer to the Moon is difficult.

15.5.1 Difficulties encountered while attempting to observe lunar alignments

Whatever the range of azimuths at which the Sun may rise or set, as we observe from a given location, the range of azimuths at which the Moon rises or sets will be greater.

Thus, the potential for coincidental orientations with horizon features is greater in the case of the Moon.

There are practical impediments to observing moonrises as well. Unlike sunrise, where dawn provides some warning, moonrise takes place suddenly at the horizon. One must use other signals provided by sky phenomena in order to prepare.

Even so, what do you measure when surveying a moonrise or -set? Do you assign azimuth based upon the lunar mid-disk? The middle may be difficult to define unless the Moon were to be always Full!

Is it the limb you pay attention to? Which one?

The lunar terminator is constantly changing its location with respect to the geometrical center of the Moon's disk. The angular difference between the apparent lunar disk's center and this edge is small. Yet it is one more factor that makes any precise determination of standstills harder.

Discovering the interval itself between lunar standstill maximums and minimums is challenging. Standstill records would have to be kept for many solar years. At the same time, even more easily than those for the bright Sun, moonrise and moonset records could be interrupted by poor weather. Such recordkeeping is hard in a preliterate society.

At best, timing would take 19 solar years. This was a sizable interval of an average human lifetime until only recently in the history of our species.

The identification of lunar standstill alignments *has* been claimed. Of course, if both lunar and solar suspected alignments exist, you might *expect* to find more of those that have to do with the Moon than those that have to do with the Sun.

There are only two Solstices. There are four interesting lunar standstill azimuths. These correspond to standstill-maximum moonrise, standstill-maximum moonset, standstill-minimum moonrise, and standstill-minimum moonset. Coincidence becomes a greater distraction.

There is an additional element that confuses the verisimilitude of lunar alignments. It may or may not have been constructed by our ancient forebears. Because of a geographic serendipity, they may be the same as certain solar alignments. Today's community of anthropologists is much more convinced of alignments in stone having to do with the Sun than they are of such architectural features having to do with the Moon.

15.5.2 Example: Colorado, United States

Remains of an Ancestral Pueblo community's Great House in Colorado (*circa* 1000) are situated within sight of a two-spired butte. It is named Chimney Rock and Companion Rock. Visiting it, American solar physicist J. McKim Malville noticed something extraordinary. There, at the northern lunar standstill maximum, the Moon rises spectacularly between the two tower-like formations as viewed from the ruins of what was a major pueblo. It takes between five minutes and fifteen minutes to do so. One date that archeologists consider to be a possible building date for the pueblo is the early 1090s. At that time, the inter-butte moonrise took place.

Figure 15.4: In 2012, US President Barack Obama established Chimney Rock as a National Monument. Courtesy of USDA.

The apparent gap between the "chimneys" is wider than the apparent size of the Moon. Those lunar standstills, centered in time upon maximum azimuth, are far enough northward for about three solar years such that the Moon will fit within the chimneys upon rising for some portion of the month.

The prehistoric inhabitants near Chimney Rock may have recognized the phenomenon. They even may have celebrated it. Nonetheless, it is another thing entirely to suggest that they *purposely* situated their settlement where they did. Did they really do so, so as to align with Chimney Rock and the lunar maximum standstill?

Whatever our predecessors thought about it, today we are fascinated by what happens at Chimney Rock. "Pilgrims" filled available space to see maximum standstill moonrise in 2024. Furthermore, it was livestreamed.

Notes from long ago

Another potential lunar alignment may exploit an oddity only possible far to the North. It is situated in the Scottish Hebrides Islands.

The site named Callanish consists of a Stonehenge-like circle of *mehirs* (orthostats). These "standing stones" were placed on the ground, longest direction up, *circa* 2900 BC/BCE. There are others radiating out from the center of the monument. Most conspicuous is a parallel set of radial uprights called the Avenue.

The southern maximum standstill is significant at Callanish – more so than some other proposed lunar-aligned sites, including Stonehenge itself. This is because Callanish has high latitude.

The connection was described by British archeoastronomer Clive Ruggles: As viewed from the end of the Avenue, a (dramatic) Full Moon, near both the Summer Solstice and its maximum standstill, just barely appears. It does so briefly, skimming the horizon. And it does so behind the stone circle.

At Callanish, latitude and the local horizon conspire to produce a one-of-a-kind effect. Such a thing is more difficult to dismiss as accidental, than just an alignment of two stones to a standstill.

Why, exactly, should your predecessors have cared about lunar phenomena, such as standstills, at all? The practical answer to the parallel question regarding Solstices and the Sun is obvious. It is less so for the Moon.

There is one possible explanation for a hypothesized, early human interest in standstills. That would be for the purpose of attempting to predict eclipses. (See Chapter 17.) However, there is no evidence whatsoever that alignments were used for this sophisticated purpose.

Misconception alert

Trying to decide whether people with whom you cannot communicate, due to the barrier of time passed, recognized and marked this or that celestial phenomenon (or did not) is hard. Consider this: The potential for utility is only one factor to keep in mind. It is difficult to place your way of thinking into someone else's head. Perhaps your ancestors spent some time and effort – as we all do – on matters that have no *practical* application at all.

15.6 Changing brightness of the Full Moon

As always, the Full Moon has a special attraction for us. Full is our satellite's brightest phase. However, there is bright and then there is bright.

Periodically, a Full Moon occurs when our satellite is at or near perigee. (See Chapter 16.) This happens in cyclical clusters and did so half-a-dozen times in the years 2023–2025.

These events often grab the attention of news media. They call them "supermoons." A 30% increase in brightness is advertised over that of a Full Moon at or near apogee.

It is true that anything closer will be brighter, everything else being equal. The perigean Full Moon is indeed 29% brighter than the apogean Full Moon. However, are you likely to make that comparison in reality? It is more probable that you compare the brightness of a supermoon to the memory of a closer-to-mean-distance Full Moon. This is the Moon you presumably have seen more often.

An apogean Full Moon occurs as often as a perigean Full Moon. Yet have you ever heard someone look up and exclaim, "See how dim the Moon is!" (This would be a "micromoon.") No. It is simply not our way.

The difference in brightness between the perigean Full Moon and the average Full Moon is only 16%. While a 29% difference should be recognizable, it is unclear than the human eye/brain will notice a 16% difference. This is especially true inasmuch as the comparison is not side-by-side. It is rather from a memory of at least a synodic month before. Air mass and transparency should be taken into account, too.

15.7 *Abandon* the calendar?

A last word on calendars, lunisolar or otherwise: As the calendar has evolved and differed from place to place, it can become difficult to unambiguously cite the date of a particular celestial event. It can become more difficult to cite the duration of, or time lapsed since, such an event. Calendrical conversions become unavoidable.

One tactic is to throw up your metaphorical hands. Forget about traditional calendars. Such a move never will be societally popular, but the much smaller cadre of astronomers sometimes does just that.

The alternative, **Julian date** system [JD], could not be simpler. A date in the far past is chosen. Then, each successive day since that date is counted. Julian date 0 is 1 January 4713 BC/BCE. The next Julian date, 1, is 2 January 4713 BC/BCE. And so on until the present. Julian dates technically begin at noon (following the long-time convention of astronomers) at the longitude of the Greenwich Observatory. (It still is the custom of astronomers to use Greenwich as a fundamental fiducial mark for terrestrial references.)

Differences, sums, products, and ratios of time turn into simple arithmetic on a number line. You even can convert other diurnal units (*i.e.*, hours and seconds) into a decimal following the JD integer.

The system was invented by Italian physician Julius Scaliger in the 1580s. Hence, the "Julian" does not refer to Julius Caesar, though Scaliger still used the Julian calendar.

Today, we call the Julian start date 24 November 4714 BC/BCE. This is the same date as Scaliger's, only using the Gregorian calendar.

How did Scaliger come up with day 0? It is a date on which several time cycles, including ones involving the Moon and Sun, coincided. But they also include the Roman taxation interval! The meta-cycle lasts 7,980 years.

By now, Julian dates reach into the millions. A quicker-to-write notation subtracts 2,400,000.5 days from each more recent Julian date. So, JD=24062674.5 is the same as the now modified JD=62674 (21 June 2030).

15.8 Chapter summary

Lunar retardation and altitude vary season to season. The Moon's rising and setting azimuth fluctuates monthly but also over the lunar nodical cycle of 18.6 years. This latter periodicity is due to the inclination of the Moon's plane of orbit about the Earth to the ecliptic.

At a standstill, the Moon changes direction from northward to southward or southward to northward. Standstills result in days upon which the Moon never rises, or never sets, at certain latitudes. These latitudes change with the lunar nodical cycle.

When people began to pay attention to lunar maximum and minimum standstills (and to what degree) is controversial. There appear to be some alignments that put the date in prehistory. Yet opportunities for "false positives" exist. The difficulty in making the required measurements – particularly of moonrise – and the ambiguity of motive, further muddy the waters.

The perigean and apogean Full Moon vary in brightness. It is unclear how often this will be recognizable by the naked eye.

15.9 Chapter review

1. The Earth's rotation makes the Moon appear to travel from east to west. What direction, from your perspective, does the Moon revolve about the Earth? How does this explain lunar retardation?
2. Why might measuring the azimuth and time of moonrise be more difficult than moonset?
3. Why does the Moon not always rise at the same azimuth?
4. Other than at moonrise and moonset, why is the Moon at a higher altitude in the winter than in the summer?
5. What is a lunar standstill? What causes it to happen?
6. If you want to observe the Moon continuously for a 24 hour period, where do you go on the Earth?
7. What is the significance of Chimney Rock in regard to the Moon?
8. Draw two pictures that illustrate the cause of a brighter-than-average Full Moon and a dimmer-than-average Full Moon.
9. What is the value of a Julian Date? Explain, in simple terms, how it works.

16 Eclipses

Eclipses are likely the most spectacular occurrence in the naked-eye sky that you may see. Few people have witnessed a total eclipse of the Sun and been unmoved.

Some say that humans first recorded a total solar eclipse in enigmatic Irish rock art, dated with enough precision such that it could relate to the eclipse of 3340 BC/BCE. More trustworthy are Chinese registries of eclipses inscribed into (for instance) tortoise shells, starting at least with an eclipse in 1226 BC/BCE. A Mesopotamian-documented eclipse appears to be specifically that of 5 March 1223 BC/BCE. Written descriptions of total lunar eclipses exist from not long after. Of course, it is absolutely unbelievable that eclipses went *unnoticed* before these times.

16.1 Chapter learning outcomes

After reading this chapter, you will be able to:
- describe events during the progress of a total eclipse of the Sun using the nomenclature of eclipse contacts.
- illustrate the geometry that causes a total solar eclipse, partial solar eclipse, annular eclipse, hybrid eclipse, total lunar eclipse, partial lunar eclipse, and penumbral eclipse.
- determine the type of eclipse based on eclipse magnitude.

16.2 A total eclipse of the Sun begins

First, the dark silhouette of the Moon encroaches upon the brilliant Sun. When more than half of the Sun has been extinguished, you might begin to notice a dullness to the blue sky. Shortly thereafter, the quality of light gives everything below a grey, metallic cast.

The western sky is clearly darker now, regardless of where the Sun is on your sky dome. This darkness, the shadow of the Moon, expands and rushes toward you. It always comes from this direction. As the remaining Sun is reduced to a point source of light, shadow edges below become sharp, leaving only the abrupt transition between adumbration and no shadow at all.

The entire sky darkens. A tour d'horizon reveals 360° of sunset/sunrise colors.

You may be lucky enough to see the atmospheric cells of air above you produce **shadow bands** racing across the ground. It is as if you stood on the bottom of a fish tank looking up, and a giant stirred the waters.

https://doi.org/10.1515/9783111441245-016

Figure 16.1: Notice its sharpness as the photographer takes a picture of his own shadow, moments before a total solar eclipse. Courtesy of Catalin Baldea.

Figure 16.2: You are familiar with a colorful late afternoon sky at low altitudes in the West. Or a colorful early morning sky at low altitudes in the East. However, this same chromatic variegation occurs on *every* horizon at the same time during a total eclipse of the Sun. Courtesy of Alexandru Barbovschi.

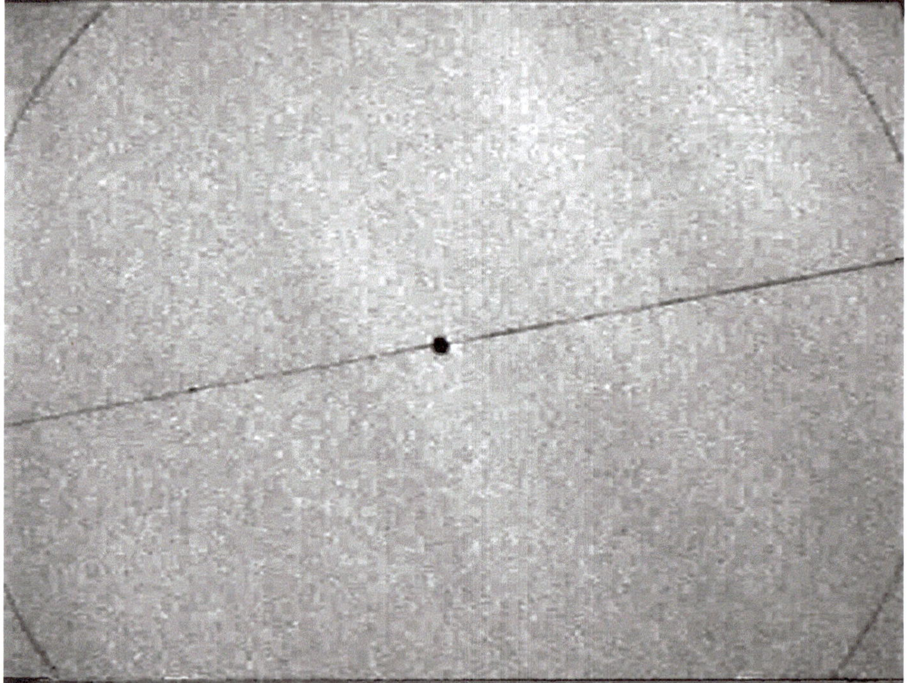

Figure 16.3: Shadow bands. 10 July 1972, Baker Lake, Northwest Territories, Canada. Courtesy of Jack Newton and Phil Harrington.

16.3 Totality

The bright disk of the Sun, called the **photosphere**, disappears in a flurry of one-or-more sparkling **Baily's Beads**. These glints of sunlight are seen through lunar valleys and between lunar mountains on the rough limb profile of the Moon. They are named for English stockbroker Francis Baily, who observed and explained them in 1836.[1] If the last bead is spectacular enough, it may produce a **Diamond Ring Effect** out of the otherwise eclipsed Sun. It only lasts a few seconds. (Coincidence occasionally yields a "Double" Diamond Ring: two last beads of similar brightness.)

Once the photosphere is completely blocked from view, what you see is **totality**. During the few minutes of totality, all is dun, with light from the sky reduced to the level of deep twilight. Planets and bright stars (down to magnitude 3 or 4) are visible, even though it may be the middle of the day.

---

1 or Halley's Beads (Edmund Halley described the phenomenon in 1715)

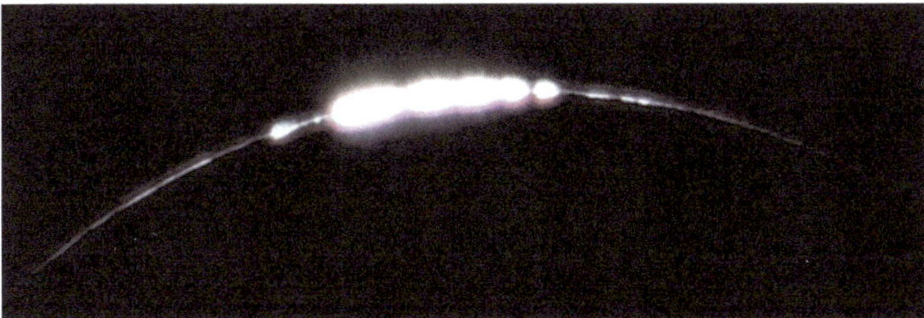

Figure 16.4: Baily's Beads serve as proof that the Moon is not smooth, like a mottled glass paperweight, but is instead a rough body. Courtesy of Alan Sliski.

Figure 16.5: Computer-aided impression of the 2019 Total Solar Eclipse, as Seen from Chile. Courtesy of M. Druckmüller, P. Aniol, K. Delcourte, P. Horálek, and L. Calçada/ESO.

The temperature might go down. Breezes may commence, change direction, or cease. Insects swarm. Birds fly home to roost. Farm animals return to their stables. Dogs bark or are uncharacteristically silent. Cats slink away.

The only solar illumination may be from the thin, characteristically red-glowing **chromosphere** above the photosphere, and quite obviously, from the amorphous **corona**, the faint outer layer of the Sun. The chromosphere is hot, thin, glowing hydrogen. The broader, but still more muted and delicate-appearing corona includes even hotter and rarer gas. Normally, both are visually lost in the glare and the blue sky produced by the now-blocked photosphere.

Sometimes, you see with your naked eye red tongues of what look like flame erupting from the chromosphere into the corona. They are **prominences**, which are really huge loops of hot gas following the lines of force produced by the Sun's magnetic field.

During certain total eclipses of the Sun, long creamy streamers radiate from the corona. They may be several Sun-diameters long. The Sun is no longer circular.

> **Misconception alert**
>
> It may be counterintuitive to think that the middle of a total solar eclipse may be observed *without* a filter or some sort of indirect viewing apparatus. Nonetheless, it can. Indeed, you must remove anything designed to protect your vision in order to see the phenomena of totality. For example, the solar corona is merely the approximate brightness of the Full Moon.
>
> Solar eclipse totality is the ONLY time it is safe to look in the direction of an otherwise dangerously bright Sun without eye protection.
>
> There is nothing *different* about the quality of sunlight associated with a solar eclipse. Contrary to myth, there are no unusual "eclipse rays."
>
> There are sad stories from the 2017 total eclipse of the Sun, seen in North America, of school children kept indoors by their principals during totality. They were forced to miss what for most of them would be a rare, perhaps unique, opportunity to see what a total eclipse really is about.

Figure 16.6: Most photographers point their camera upward during a total eclipse of the Sun. This rarer view of the landscape shows the darkening caused by a midday solar eclipse. Courtesy of Yuliana Ivakh.

The entirely shadowed Moon sits where you normally expect to see the solar photosphere, looking like a misplaced black hole in the sky. Fellow witnesses instinctively hush. Or cheer.

Notes from long ago
The corona has been documented in writing since at least the tenth century, but what exactly was it? (The word itself was not coined until 1803.) In the seventeenth century, Imperial astronomer Johannes Kepler was arguably the first modern astronomer (despite his title). It was he who correctly hypothesized that the corona is actually part of the Sun. It is not, say, part of the Moon or an optical illusion produced *by* an eclipse.

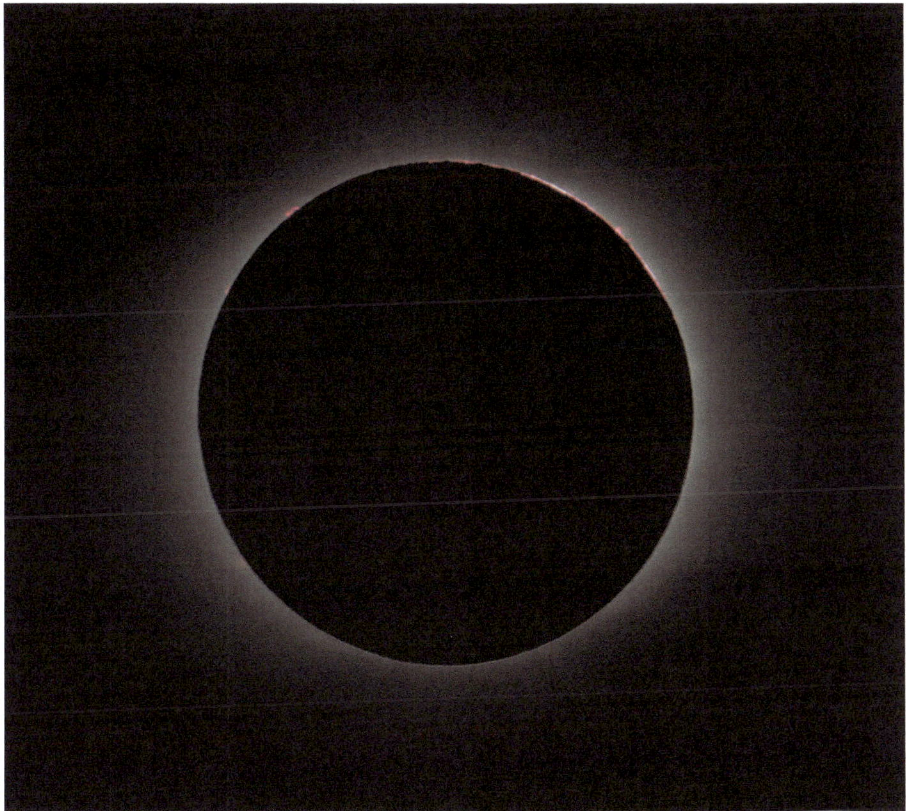

Figure 16.7: The chromosphere is at most only 1% of the solar photosphere's radius. Thus, its visibility at the very beginning and end of the total eclipse may be extremely brief. Courtesy of Catalin Baldea.

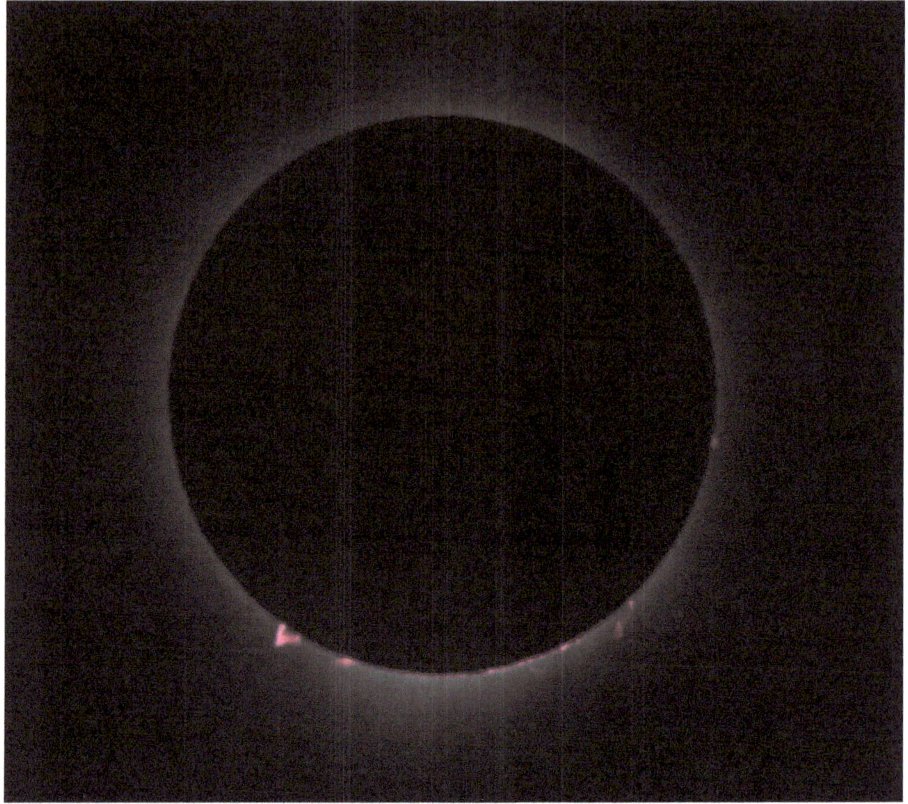

Figure 16.8: Prominences are at most some minutes of arc in apparent height. They are only visible during a total eclipse of the Sun when they happen to be at the solar limb. Courtesy of Catelin Beldea.

How dramatic are the phenomena visible to the naked eye only during a total eclipse follows the approximate 11-year cycle of sunspots. (See Chapter 19.) For instance, the corona is less circular and looks more like the wings of a butterfly near the sunspot maximum. Notwithstanding, every total eclipse of the Sun looks different.

16.4 A total eclipse of the Sun ends

The abrupt end of totality is signaled again by sparkling Baily's Beads or, perhaps, another single, bright Diamond Ring Effect. The total eclipse of the Sun has been a visceral experience.

It is often quoted (without attribution) that only one in ten thousand of the world's citizens have seen a total eclipse of the Sun. It is true that just fourteen total eclipses will be visible in North America during the 21st century, eight in Europe. Dur-

Figure 16.9: The solar corona as seen during the 2017 total eclipse of the Sun in Wyoming, USA. Courtesy of Alexandru Barbovschi.

ing the Middle Ages and early modernity, London went 837 civil years without such an eclipse.

Notes from yesterday & today

In the United States, much attention was given to the fact that the latitudes and longitudes belonging to the state of Illinois experienced two total eclipses of the Sun in only seven civil years (2017 and 2024). However, southern Papua New Guinea witnessed a total eclipse on 11 June 1983 and also on 22 November 1984 – an incredibly brief wait of less than one-and-a-half civil years.

16.5 What causes an eclipse of the Sun?

A complete cycle of lunar phases happens every synodic month. Remember that New and Full Moon occur when the three bodies, Earth- Moon-Sun, form a syzygy.

It does not have to be a perfect line-up. In fact, most of the time, the Moon is a little above or below a line running from the center of the Sun through the center of the Earth.

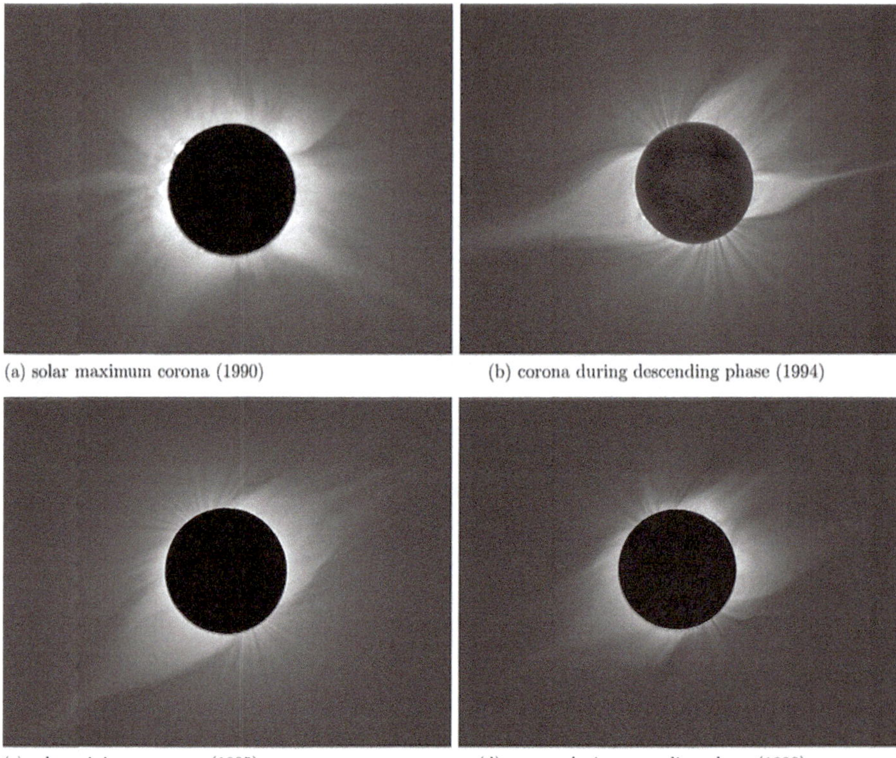

(a) solar maximum corona (1990)

(b) corona during descending phase (1994)

(c) solar minimum corona (1995)

(d) corona during ascending phase (1998)

Figure 16.10: Solar corona at the sunspot maximum (left; top and bottom) and sunspot minimum (right; top and bottom). (See Chapter 20.) Courtesy of NOAA.

16.5.1 Cosmic coincidence

What happens when these bodies *are* perfectly lined up? For instance, if they do so in the order, Sun-Moon-Earth?

Imagine shooting an arrow and striking all three spheres through the middle. Then, and only then, does the Moon physically prevent light emitted by the Sun from reaching the Earth. It is a total eclipse of the Sun.

By now, you have seen that the celestial realm is a busy venue. Objects pass in front of other things on your sky dome all the time. The Moon covers up a star, but nobody notices one star's worth of light missing. A planet, such as Venus, crosses the solar disk; at best, it is a barely detectable dark spot. (See Chapter 19.) These are less-than-awesome events because the bodies involved are of such disparate apparent sizes.

By serendipity, the Sun and the Moon are the same apparent size in the sky. Such a thing happens nowhere else in the Solar System. Other planets have natural satellites. However, the distances of these planets from the Sun, and the distance of any

Figure 16.11: The Diamond Ring Effect has been called that since at least the 1912 total eclipse of the Sun visible in France. It was commented upon using that name by a woman in Paris. Courtesy of Dimitry Rotstein.

satellite from its planet, result in the satellite having an apparent size too small to cover the Sun, or so big that it covers the Sun too well. The strange outer layers of the Sun, one of the things that draws you to a total solar eclipse, are blocked from view as well as the photosphere. Additionally, the physical size of the satellite itself may be too big or small, given a fixed distance, to produce a satisfying eclipse.

The Moon is about the angular extent of a pea held at arm's length, half-a-degree. We tend to think that the Sun is bigger. Yet, in reality, you blot it out exactly with that same pea.

Here is the coincidence: The Moon is much smaller than the Sun, of course. The ratio of the two radii is about 400 times. More specifically, the diameters of the two bodies are 3,475 kilometers versus 1.393 million kilometers. If the diameter of the

Figure 16.12: Only two syzygies are possible. Not to scale. Artwork by Christine Tarte.

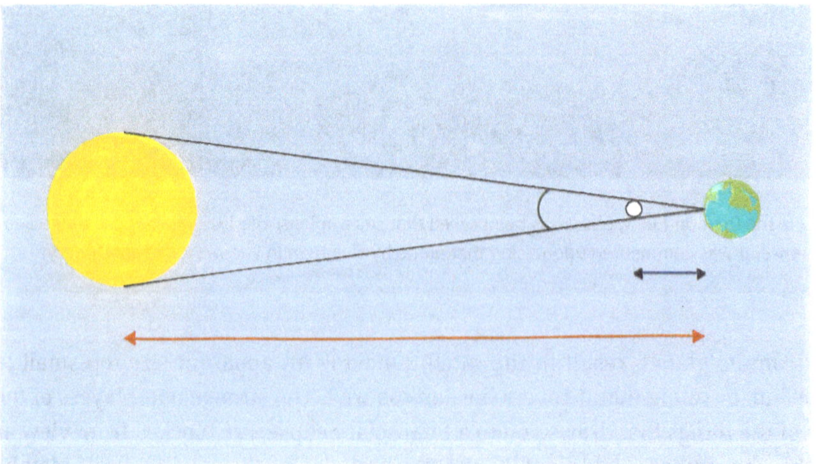

Figure 16.13: The apparent sizes, in your sky, of the Sun and Moon subtend the same angle. Not to scale. Artwork by Christine Tarte.

Moon was much smaller (or the diameter of the Sun much greater), you would never see the total solar eclipse that you see in reality.

Moreover, the ratio of the Moon's average orbital radius (as it revolves about the Earth) and the Earth's average orbital radius (as it revolves about the Sun) is also about 400 times. Notice that it is the same number. More specifically, these average radii are 384,400 kilometers versus 152,100,000 kilometers. If one or other of the Moon or Sun was much closer to you (or much farther from you), you would not see a total eclipse of the Sun either.

Your opportunity will not last forever. The Moon is working its way away from the Earth: Its orbital radius increases by 3.8 centimeters every sidereal year. Eventually, the Moon never again will appear large enough to cover the Sun. The final total solar eclipse takes place 1.2 billion years from now!

16.5.2 Lunar nodes

Returning to the present, why does a solar eclipse not occur every synodic month? Recall that the Moon's orbit is inclined to the plane of the Earth's orbit around the Sun, the ecliptic. (The similarity of the words "ecliptic" and "eclipse" is not coincidental.)

There are only two places in the Moon's orbit where it is in the plane of the ecliptic, that is, two places where it *crosses* the ecliptic. Each of these points is called a **node**.

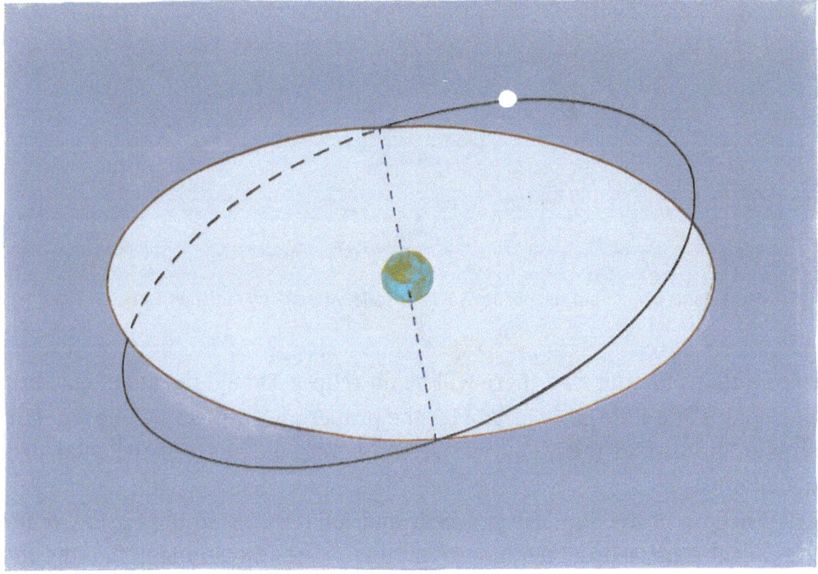

Figure 16.14: Because the Moon's orbit is inclined to the ecliptic, the two planes can intersect at two and only two points. These nodes, opposite each other in the lunar orbit, are the only places at which an eclipse can take place. Not to scale. Artwork by Christine Tarte.

For a total eclipse of the Sun to occur, there must be a New Moon. At the *same time*, the Moon must happen to occupy one of the nodes of its orbit. As the interval it takes the Moon to go from node to node is different and unrelated to the time it takes to go from any phase to the same phase again (the synodic month), for these two events to occur simultaneously, it requires yet another coincidence.

Figure 16.15: Circumstances of a total solar eclipse. Not to scale. Artwork by Christine Tarte.

If the two conditions are not met, there will be no eclipse. Should the Moon be a little off of a node point or a little early or late for the proper phase, there will be a **partial solar eclipse**: The disk of the Moon will seem to "bite out" part of the solar disk, though not all of it.

A partial eclipse of the Sun also proceeds and follows a total. During any occurrence of a partial solar eclipse, the Moon's outline makes the Sun look like the title character in one of the original video games, Pacman.

Having said that, if you are unaware that a solar eclipse is taking place, ordinarily nothing out of the ordinary will draw your attention to it until 90% of the solar disk is occluded – unless you happen to glance upward at the Sun for unrelated reasons.

Here is an exception: You may have noticed small circles appear underneath a leafy tree when the Sun is shining. The overlapping and interwoven leaves act as small, natural apertures.

Figure 16.16: 2012 Partial Eclipse of the Sun, as seen in New Mexico, USA. Courtesy of Kosmas Gazeas.

What you see actually are pinhole-projected images of the Sun. This optical effect is common enough that you may not always pay attention to it. However, during a partial eclipse of the Sun, crescents appear at your feet instead – a most uncommon sight.

16.5.3 Solar eclipse timing

It is useful to divide the "schedule of events" for an eclipse of the Sun by way of contact times; a total eclipse has four, a partial eclipse only has two.

When the limb of the Moon first begins to cross over the limb of the Sun, this moment is the first **contact**. First contact is the beginning of the solar eclipse; it is the start of the initial partial phase.[2]

Second contact is the moment at which the same limb of the Moon reaches the far limb of the Sun. The total phase begins. The time between the first contact and the second contact may be over an hour.

At the third contact, totality ends. The trailing limb of the Moon crosses the limb of the Sun. The second partial phase of the eclipse begins.

---

2 Note that the word "phase" used in reference to an eclipse is different from its definition used in regard to the synodic month.

Figure 16.17: Pinhole projections under a tree in California, where the 2017 solar eclipse was only partial. Courtesy of K. U. Chang.

Fourth contact. The trailing limb of the Moon leaves the disk of the Sun. The eclipse is over. The time between the third contact and the fourth contact may last over an hour too. In contrast, the interval between the second and the third contact is measured in minutes.

In a proverbial nutshell, a total eclipse of the Sun is simply what you see when the Moon's shadow passes over you. Shadows seem to have two parts: a darker, inner portion – the **umbra** – and a lighter, outer portion – the **penumbra**.

After the first contact, you are in the Moon's penumbra. At the second contact, you are in the Moon's umbra. The third contact returns you to the penumbra. At the fourth contact, you emerge from the Moon's shadow completely. It is easy to see how a partial eclipse of the Sun has only a first and last contact.

Figure 16.18: Umbra and Penumbra. Photograph by the author.

Think of eclipse contacts as a linear variable across the disk of the Sun. Solar illumination, and the associated eclipse phenomena, changes as the *square* of this variable. This is all to say that changes in the appearance of the eclipse accelerate up to mid-eclipse and decelerate after mid-eclipse.

16.5.4 Solar eclipse centerline

Solar eclipses may seem even rarer than they actually are because they are visible from only beneath the Moon's shadow cone, that is, within the intersection of that cone with the surface of the Earth. The Moon moves in its orbit about the Earth during the eclipse. The Earth rotates during the eclipse. These factors change the eclipse geometry during an eclipse.

Most noticeable is that the oval-like region within total eclipse gets stretched into an **eclipse path.** This path can be quite narrow.

People within the path of totality, the width of the Moon's narrow umbra, experience the second and third contact. Adjacent to the path, observers remain in the penumbra and only experience a partial eclipse throughout.

Maximum totality duration at a given longitude is to be found in the middle of the eclipse path, on the centerline. It dwindles to zero as one moves orthogonally away from the **centerline**. The greatest length of totality anywhere (during a particular total solar eclipse) occurs on a centerline, halfway along the eclipse path.

A total eclipse of the Sun is a "digital" event. Simply put, if you are in the path, even just barely, you see it. Just outside of the path, no matter how little, you do not.

This distinction was quite graphic during the 1925 total eclipse of the Sun, viewable from New York City: Witnesses say that the "edge" between the Moon's umbra and penumbra fell somewhere between 95th and 97th Street in Manhattan.

Figure 16.19: Even in 1925, eclipse prediction was not in the state of precision that it is today. The onset of the New York total solar eclipse was noticeably late, provoking anxiety among those poised to observe it.

Some scientists *intentionally* station themselves – not on the centerline of the eclipse path – but on the parallel predicted border between totality and partial eclipse. They do this to lengthen their time during which the visible eclipse phenomena, specifically associated with that brief period around the instant of the second-to-third contact, are best seen. Examples are the chromosphere and prominences.

16.6 Annular eclipses

The Moon is not *always* just the right apparent size to cover the solar photosphere and only the solar photosphere. Remember that at apogee, the Moon looks smaller than it does at perigee. Meanwhile, the apparent size of the Sun changes slightly between perihelion and aphelion.

These two effects – with that caused by the Moon's orbit being the most significant one – conspire to, during certain eclipses, make the Sun appear a little larger than average. At the same time, the Moon appears a little smaller than average.

In such an eclipse, the Moon does not appear big enough to cover the solar disk, even if it moves directly over the middle of it. The result is a special kind of partial solar eclipse, not a total solar eclipse. This version of a partial eclipse of the Sun is an **annular eclipse**.[3]

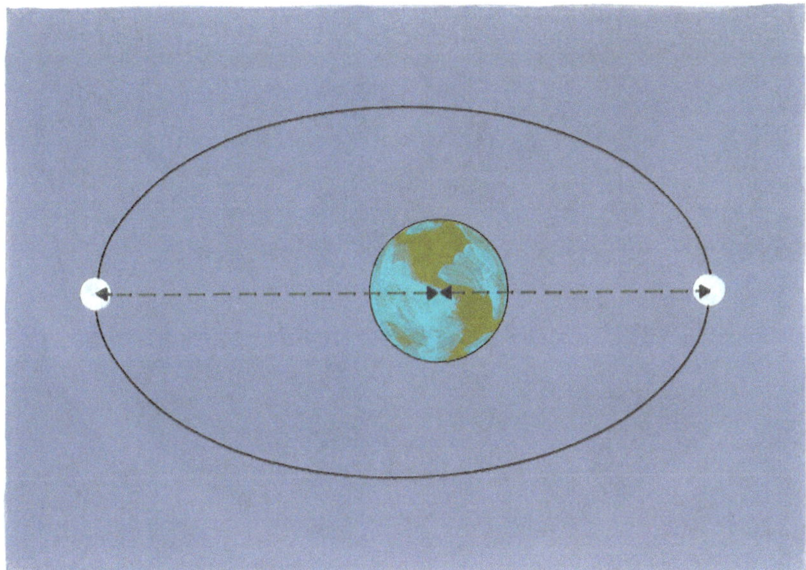

Figure 16.20: The Moon at apogee (left) and perigee (right). (The difference in distance from the Earth is exaggerated.) Not to scale. Artwork by Christine Tarte.

Think of an annular eclipse as the result of a *negative* Moon shadow, an **antumbra**, falling upon the Earth. Instead of the tip of the shadow cone (theoretically) reaching into the Earth, it ends short of the Earth's surface.

In the midst of an annular eclipse, the Moon's penumbra surrounds its umbra. The Sun looks like a round band of light in the sky.

No partial eclipse of the Sun, even an annular eclipse, is as dramatic as a total eclipse. However, the ring is not a form encountered often in nature. Therefore, its appearance in the sky is noteworthy. And, as you can see, it requires conditions that are rarer than the more common version of a partial solar eclipse. Nonetheless, two

---

3 It is not necessary to use "annular solar eclipse" inasmuch as to no other kind of eclipse does the adjective annular apply.

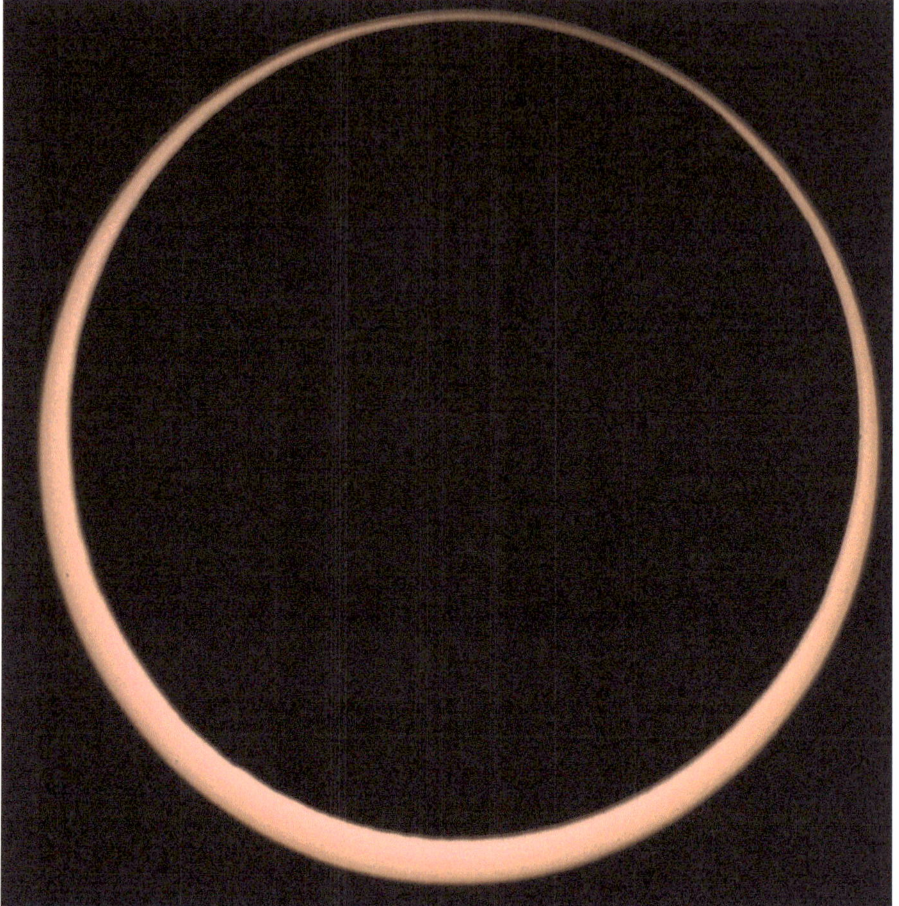

Figure 16.21: Annular eclipse of 1994. Courtesy of Rick Feinberg/AAS.

annular eclipses were visible in North America recently. These were in the civil years 2020 and 2023; they and represent an unusual frequency.

Annular eclipses have paths, centerlines, and locations of maximum duration, too. Outside the path of an annular eclipse, only a traditional-looking partial eclipse is seen. On the centerline, the solar annulus is of equal breadth all the way around.

There even are solar eclipses that are **hybrid eclipses**. They are (a very thin) annulus at the start, briefly total in the middle, and then annular again, along their paths. Such eclipses are the rarest among the different variations on the solar eclipse.

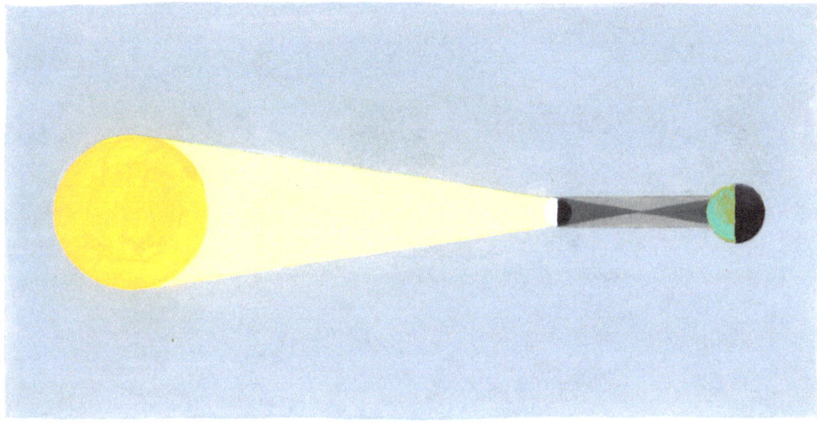

Figure 16.22: Circumstances of an annular eclipse. Notice that the Moon's umbral cone fails to reach the Earth; an antumbral cone does. Not to scale. Artwork by Christine Tarte.

16.7 Eclipse magnitude

Here is a way to keep the solar eclipse variations straight. **Eclipse magnitude** refers to the ratio of the Moon's apparent diameter to the Sun's during an eclipse. Alternately, you can choose to think of eclipse magnitude as the positive or negative percentage of the Sun's diameter that is eclipsed. Note that the variable is diameter, not area; that fraction is called **eclipse obscuration**.

An eclipse magnitude more than 1.0 indicates a total eclipse, less than 1.0 refers to a partial eclipse. Eclipse magnitude changes to a certain extent along the eclipse path; an average eclipse magnitude very near 1.0 can result in a hybrid eclipse.

Tools of the naked-eye observer

Are you not convinced that the apparent size of the Moon changes? Place a small, round hole in a rigid material. Attach the resulting aperture holder to a rod or arm (*e.g.*, a meter stick), such that it can be slid up and down. As you sight along the long piece, the smaller piece (with the hole in it) is perpendicular to your line of sight.

Aim at the Moon and adjust so that the apparent lunar disk just fits within the hole. Mark the location of the aperture holder on the rod or arm. Repeat the experiment about half-a-month later.

Are the two marks at identical locations? If not, the apparent size of the Moon was greater when the aperture was closer to you and smaller, when it was farther away.

Should you decide to perform this experiment, you are following in the footsteps of the Hellenistic Greek Archimedes of "Eureka" fame.

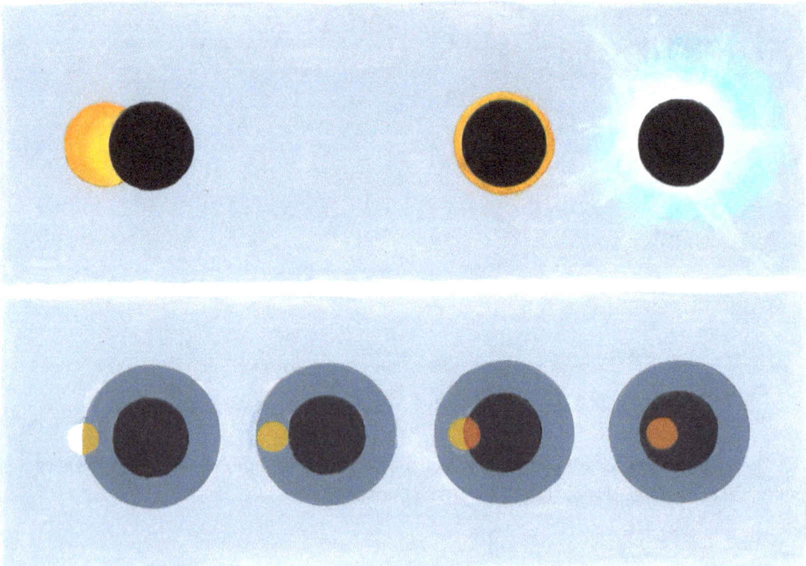

Figure 16.23: Magnitudes of eclipses. Top: solar eclipse. Left = M < 1 [partial], middle = M < 1 [partial, annular], and right = M > 1 [total]. Bottom: lunar eclipse. Notice the Earth's dark, circular umbra surrounded by its stylized penumbra (See below.). Not to scale. Artwork by Christine Tarte.

16.8 Solar eclipse eye safety

NEVER spend more than a moment watching a partial solar eclipse without taking steps to protect your eyes. If you do not heed this warning, the resulting effect on your vision can range from momentary to permanent blindness experienced in one spot within your field of view, up to your entire retina.

Your author repeats himself, but it is important: Our brains are prepossessed to avoid staring at the Sun. Indeed, there is normally no particular reason to do so. However, during any kind of partial solar eclipse, there is something particularly interesting happening to the Sun – the change in shape of its illuminating surface. You might be tempted.

Unfortunately, a partial eclipse is the same as no eclipse, as far as eye safety is concerned. Each tiny fraction of the Sun's disk is as bright as any other, and even one sliver is intense enough to cause harm.

Yet, because of the interest in the changing appearance of the Sun as the eclipse progresses, you might try to override common sense and stare at the Sun for too long. Herein lies the danger.

16.9 Eclipses of the Moon

In addition to a solar eclipse, there is another kind of eclipse. It is an eclipse of the Moon. During a total lunar eclipse, it is the Moon that becomes dark, not the covered Sun.

Figure 16.24: Total lunar eclipse of 2014. Courtesy of D. Munizaga/NSF.

As you might expect, lunar eclipses are primarily a nighttime event. And because the Moon is so much fainter than the Sun, a lunar eclipse might be missed altogether, unless anticipated.

16.9.1 What causes a lunar eclipse?

Look at what happens when the Moon, in its orbit about the Earth, is on the other side of our planet. This is the geometry opposite to that of a solar eclipse.

In a lunar eclipse, the Moon is not in the way, the Earth is. What happens when your syzygy is perfectly lined up: Sun-Earth-Moon? In other words, what if the Moon is 180 degrees from the Sun and at one of its orbital nodes (a necessary condition for all eclipses)? Then, the shadow of the Earth falls upon the Moon. Sunlight does not reach the Moon.

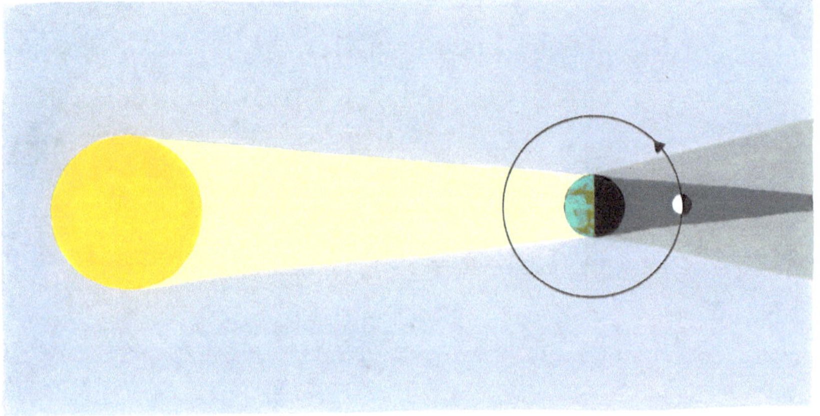

Figure 16.25: Circumstances of a total solar eclipse. Not to scale. Artwork by Christine Tarte.

Because the Moon only shines by reflected light, when sunlight is cut off, it darkens. This is all the more noticeable because the phase of the Moon at this time is Full. The event may be total or partial, depending upon whether the Moon is at any time entirely in the Earth's umbra, and upon your location.

The lengths of total lunar eclipses are determined by the time it takes the Moon to pass through the Earth's shadow. They last for up to an hour, not the minutes of totality as in a solar eclipse. The whole event, from the eastern limb of the Moon becoming shadowed to the western limb reappearing under normal sunlight – can last several hours.

You cannot see a shadow unless there is something for it to appear projected upon. In the case of the Earth's shadow, there *is* – during a total eclipse of the Moon.

Knowing the size of the Earth and assuming (correctly) that the Sun is much farther away from the Earth than is the Moon, the diameter of the shadow of the Earth at the Moon's distance must be close to the diameter of the Earth itself. Measuring how many Moon's, limb-to-limb, would span the Earth's shadow (about four) allows computation of the size of the Moon.

By the way, the Moon never *completely* disappears in a total lunar eclipse. You know that sunlight is absorbed by the Earth's atmosphere. Red light is absorbed the least. So, of the sunlight that manages to trickle around the Earth's terminator, it is this color of weakened sunlight that gives the eclipsed Moon an eerie, ruddy glow. The exact appearance depends upon the state of the atmosphere: whether there are clouds on the terminator, how much dust happens to be in the air, and that sort of thing. The darkest total eclipses of the Moon often are rare **central lunar eclipses**,[4] during which the Moon passes through the middle of the Earth's shadow cone.

An attempt at quantifying the darkness of a lunar eclipse is the **Danjon scale**.

Figure 16.26: The Danjon Scale. Courtesy of NASA. Edited by the author.

As the last bit of lunar limb enters (or leaves) the Earth's umbra, it may take on a tint of blue green. This too is an atmospheric effect. It is caused by sunlight passing through the Earth's stratospheric ozone layer.

16.9.2 Penumbral eclipses

Just as a partial eclipse precedes and follows a total eclipse of the Sun and is visible outside the solar eclipse path, so too does a partial eclipse precede and follow a total

---

4 "Central eclipse" sometimes is used more colloquially in order to refer to both a total solar eclipse and an annular eclipse inclusively.

Figure 16.27: Total lunar eclipse of 2022. Approaching totality, atmospheric ozone tints the edge of the Earth's shadow. Courtesy of Lucy Hsu.

eclipse of the Moon. A partial lunar eclipse also takes place when the entire disk of the Moon misses the Earth's umbra altogether.

There exists one more "twist" to the lunar eclipse: The Moon need not necessarily enter the umbra of the Earth at all in order for us to see a lunar eclipse. Some or all of it may pass through just the penumbra. In such a case, the Moon is still exposed to some sunlight.

The Sun is so bright that the subsequent reduction in lunar illumination is minimal. The dimming of the Moon in a penumbral eclipse is sometimes difficult to discern from no eclipse at all.

Figure 16.28: The penumbral eclipse of 2023. Courtesy of H. Raab.

A penumbral eclipse[5] is unique to the Moon. In solar eclipses, there is no such analogous situation (*i.e.,* dimming of the visible part of the Sun). The eclipse magnitude for a lunar eclipse is the fraction of the Moon's apparent diameter within the Earth's shadow; however, it is necessary to indicate whether you are referring to an umbral magnitude or a penumbral magnitude.

Inasmuch as there are three kinds of lunar eclipses, four total lunar eclipses in a row are not common. Such an occurrence is worth a special name, a **tetrad**.

16.10 Chapter summary

The average apparent size of the Moon is the same as that of the Sun.

A total solar eclipse occurs as the New Moon passes directly in front of the Sun. This happens when the Moon is at one of the two nodes in its orbit about the Earth. Totality allows you to see the outer layers of the Sun, and associated phenomena, which are normally invisible.

A partial solar eclipse precedes and follows totality. During a particular solar eclipse, in which the silhouette of the Moon is of an apparent size slightly too small to cover the photosphere, an annular eclipse is seen. Observing a partially eclipsed Sun requires *eye protection.*

A total lunar eclipse occurs when the Full Moon passes into the Earth's shadow. Light refracted around the Earth's atmosphere gives such an eclipse a faint, red hue. A partial eclipse of the Moon by the Earth's umbra is possible, as well as is a penumbral eclipse.

---

5 It is not necessary to use the adjective lunar inasmuch as penumbral eclipses are unique to lunar eclipses.

ⓘ 16.11 Chapter review

1. When you observe a total eclipse of the Sun, you have to be within a well-defined path. If you are just outside of this path, you will see a partial eclipse of the Sun. Why is this the case?
2. What phases of the Moon must it be in order for there to be a total solar eclipse? Why? What phase of the Moon must it be in order for there to be a total lunar eclipse? Why?
3. Of the three syzygies, one is impossible. Why?
4. Why are some total eclipses of the Moon darker and/or redder than others?
5. During a total eclipse of the Sun, you are in the Moon's shadow. Why, then, would your entire horizon take on colors associated with sunrise/sunset?
6. While ordinarily, you need protective eye filters to look at the Sun, why is this not the case during the totality part of a total solar eclipse?
7. Why does the Moon not completely disappear during a total lunar eclipse?
8. Match the solar eclipse contact referenced on the left with the phenomenon most likely to be seen at this time on the right.

after the first contact (but before the second contact)	Baily's Beads may turn into the Diamond Ring Effect.
at the second contact	The chromosphere is visible just before the photosphere.
at the third contact	The silhouette of the Moon appears on the Sun's east side.
before the fourth contact (but after the third contact)	A partial eclipse is seen with the silhouette of the Moon on the Sun's west side.

9. What is the difference between a total solar eclipse and an annular eclipse? What causes the difference?
10. What is the difference between a total lunar eclipse and a penumbral eclipse?
11. If you were on the Moon during a total lunar eclipse, what would you experience?

17 Predicting eclipses

Eclipses are the first subject we have undertaken in which the question of visibility merits an entire chapter. You may be inclined to skip it. However, no other subject better exemplifies what can be accomplished by naked-eye observations alone.

17.1 Chapter learning outcomes

After reading this chapter, you will be able to:
–   make certain predictions regarding solar and lunar eclipses.
–   contrast four different kinds of months.
–   list parameters that determine your likelihood of seeing an eclipse.

17.2 Factors involved in eclipse prediction

It is as if the Moon steals the Sun. In the past, a solar eclipse was startling and serious. People thought that the Sun was on fire. When eclipsed, they feared that the Sun was in danger of becoming extinguished.

A major reason for the past concern/superstition about all eclipses was their seeming unpredictability. What governs the circumstances of an eclipse was completely unknown. Legend has it that Charlemagne's son and heir Louis, emperor of the entire Holy Roman Empire, died of fright, part way through a total eclipse of the Sun (840).

Will an eclipse take place or not? Predicting an eclipse is more complicated than anticipating the phase of the Moon. It is more difficult than planning for when the Sun will reach a Solstice. Notwithstanding, eclipse prediction is possible. However, the means for doing so took a long period of eclipse record-keeping to be recognized.

There is a famous story of two Chinese court astronomers several thousand years ago. Hsi and Ho supposedly were beheaded for failing to predict an eclipse! Such a prediction was impossible at that time. It would remain so for more than a millennium. This tale is hopefully apocryphal.

17.2.1 Nodical month

Eclipse prediction would be much more straightforward if the Moon's nodes were fixed on the Celestial Sphere. Then, if a New Moon (or Full Moon) approached this spot, it would be warning that a solar eclipse (or lunar eclipse) was about to occur. However, the lunar nodes move.

https://doi.org/10.1515/9783111441245-017

Traditionally north is considered "up" on a map. When the Moon travels "up" through the ecliptic, it is at its **ascending node**. When it travels "down" through the ecliptic, it is at its **descending node**.

The line of nodes is an imaginary one passing through these two points. The line of nodes slowly shifts around the Earth. This motion takes place clockwise as viewed from north. It is due to the gravitational effect of bodies beyond the Moon.

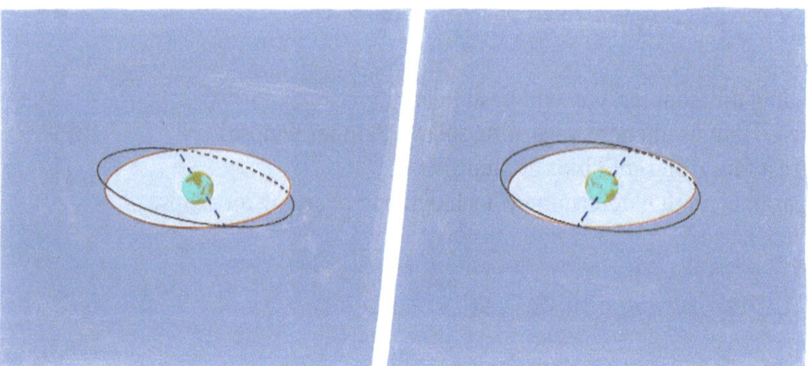

Figure 17.1: Moving nodes. Artwork by Christine Tarte.

Remember the time from the Moon's presence at its ascending node to once again at its ascending node (or descending node to once again at descending node). It is not exactly one synodic month (New Moon to New Moon or Full Moon to Full Moon). It is 27.212 days, the nodical month, the time from ascending node to ascending ode.

Because the nodes appear to drift through the ecliptic, it becomes more difficult to keep track of their locations and therefore more difficult to predict eclipses.

There is a way. First, though, let us introduce yet another month. This will complete our list of ways in which to measure this particular length of time.

17.2.2 Anomalistic month

Independently, in an eclipse-of-the-Sun scenario, it is also hard to predict the type of solar eclipse. This is because, as you already know, the apparent size of the Moon is always changing.

A line connecting the Moon's apogee point and perigee point also is called the line of apsides. This line also slowly rotates around the Earth. It does so, too, in the counterclockwise direction, as viewed from north.

Thus, the time between perigee and perigee is slightly different from either the synodic or nodical month. It is 27.555 days, the **anomalistic month**.

> **Where did that come from?**
> The nodical month is also called the "draconic month." (Or "draconitic month.") "Draconic" comes from the myth that an eclipse occurs when a celestial dragon attempts to devour the Sun. The Line of Apsides is the "Head and Tail of the Dragon" in Asia.

The time of the anomalistic month governs eclipse magnitude. It determines whether a solar eclipse is total or annular. Inasmuch as it affects the apparent size of the Moon, the time of the anomalistic month controls the width of a total solar or annular eclipse path too: If the Moon is closer to us, it casts a wider shadow onto the Earth. The time of the anomalistic month may determine whether you are on the eclipse path at all.

If the Moon were at its average distance (or farther) from the Earth all the time, there never would be a total eclipse of the Sun. From this fact, you may logically deduce that annular eclipses are a bit more frequent than are total solar eclipses. The anomalistic month also affects the duration of both solar and lunar eclipses.

Will there be an eclipse? What will that eclipse be like? Where will it appear on the Celestial Sphere? In order to predict answers to these questions, it would seem that you have to keep track of five variables. These are the times of the sidereal, synodic, nodical, and anomalistic month, as well as the time of solar year.

Moreover, knowledge of three more variables – your latitude, longitude, and local solar time – is necessary to determine whether or not *you* will see the eclipse. This is why the idea that Stonehenge, or any other prehistoric monument, was designed to predict eclipses seems farfetched.

17.2.3 Eclipse seasons

Still, there are some tricks that make eclipse prediction easier. The Moon must pass through a lunar node twice per solar year. So, there must be at least two **eclipse seasons** during which an eclipse might take place. They are 173 days apart.

Eclipse seasons are independent of the real seasons. They migrate through any solar-based calendar.

You may experience at least a partial eclipse any time when the Moon is within 12° to 9° of a node. So, you can be on the alert for an eclipse during the days between these **eclipse limits**. They are the beginning and end of an eclipse season. You need not worry about an eclipse the rest of the solar year.

You have noticed that 2 × 173 does not equal 365. The **eclipse year** is from the onset of one eclipse season to the onset of the next eclipse season to the onset of another eclipse season. It is shorter than the solar year. The eclipse year, exactly twelve nodical months, is only 346.620 days.

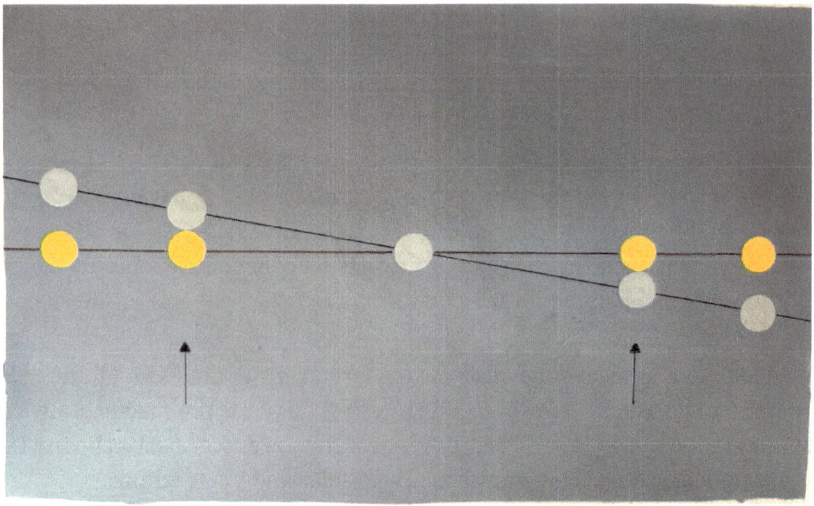

Figure 17.2: The horizontal line represents the ecliptic and the diagonal line, the path of the Moon (at an exaggerated angle). They cross at a node. Should the Moon be New between the eclipse limits, marked by arrows, a solar eclipse will occur. The arrows for a lunar eclipse (Full Moon within eclipse limits) would be spaced slightly farther apart. Artwork by Christine Tarte.

The Earth's speed varies during its elliptical orbit about the Sun. Nonetheless, the time between eclipse limits, the duration of an eclipse season, is going to be 31–37 days (depending upon the dates). There *must* be at least one solar eclipse and one lunar eclipse, each eclipse season, because the New Moon and Full Moon are only about fifteen days apart. There have to be four eclipses per solar year.

The synodic month is less than the length of an eclipse season. Therefore, it is possible to have *three* eclipses in a single such season: one or two solar, the remaining lunar. This happens if one eclipse occurs early in the eclipse season and another late in the eclipse season. The result is at least *five* eclipses in a solar year.

What is more, the first eclipse season of the solar year may take place before 18 January (or 19 January in a leap year). In this case, a third eclipse season will begin in December. At maximum, there are possibly *seven* eclipses in one solar year.

Such an "eclipse-friendly" solar year yields three/four eclipses of the Moon plus four/three of the Sun. Alternately, the pattern is two/five of the Moon plus five/two of the Sun. An average of 2.38 solar eclipses occur per solar year.

There really were seven eclipses in 1982; this kind of record-setting year will not happen again until 2038. The last time there were five solar eclipses in one solar year was 1935. The next? 2206!

17.3 Attempts at prediction

Mesopotamia was the home to the first great city builders *(e.g.,* the Babylon). It was here, fittingly, that people noticed and recorded a pattern to eclipses. What did the Mesopotamians, circa the eighth century BC/BCE, do – apparently, before anybody else? They figured out that eclipses tend to be separated by five or six Full Moons, though an eclipse was not guaranteed at that interval.

17.3.1 The saros

Much later, the Mesopotamians found that a cycle of eclipse activity repeats every (about) nineteen eclipse years. Other civilizations around the world eventually would make the same discovery.

Nineteen solar years is the amount of time after which the synodic month, nodical month, and anomalistic month are nearly commensurate with each other. In other words, 223 synodic months equal 239 anomalistic months equal 242 nodical months, almost exactly.

This period is called the **saros**[1] **cycle**. It is actually equal to eighteen solar years, 11 days. Alternately, it is an interval of eighteen eclipse years. There are 68 to 87 eclipses in each saros cycle.

You witness a total solar eclipse. The solar eclipse, a single saros cycle later, is also most likely to be total (Or maybe annular.) It is in a nearby place on the Celestial Sphere.

A similar eclipse will happen at the same points of repetition in the saros cycle. With this realization, the Mesopotamians finally succeeded in predicting a lunar eclipse that occurred in the year 612 BC/BCE.

> **Notes from long ago**
>
> An unexpected modern discovery in archaeology is a set of metal gears and dials remarkably preserved in a two-millennia-old Hellenistic Greek shipwreck. It is named the Antikythera Mechanism, after the Mediterranean island near where the ship lies on the seafloor. As far as is known, there was nothing else like it built for a thousand years!
>
> The Antikythera Mechanism seems to be a sophisticated, mechanical calendar. This one-of-a-kind artifact still is being studied. New analysis techniques reveal what looks like a saros-based eclipse prediction dial. The Antikythera Mechanism may eventually cause us to rethink ancient eclipse prediction abilities.

Though solar years apart, eclipses even can happen on the same day of the week! The longer interval required is called a **semanex**. There will be two eclipses on the same *date* after one, still longer **Grattan-Guinness cycle** (named after the Irish evangelist who discovered it as late as the latter nineteenth century).

---

1 from the Babylonian, sar means "tablet writer"

You guessed it: There is a periodicity to two eclipses occurring on the same date and on the same day of the week. It is the very long (eighteenth-century French astronomer Alexander) **Pingré cycle**.[2]

A complication inherent in a solar eclipse forecast is the next digit in the saros cycle duration quoted above. It is not eleven days *exactly*. It is eleven days and eight hours. The fraction of one-third of a day (one-third diurnal rotation) guarantees that each eclipse of the Sun, predicted by the saros cycle, will occur one-third of the way around the Earth from the previous one.

This fractional rotation may not push a long total lunar eclipse out of view. However, it does a brief-in-comparison total solar eclipse. You must wait three saros cycles to experience the same kind of eclipse at your (same) longitude.

Three saros cycles equal one **exeligmos**.[3] Even then, after an exeligmos, there is a "hitch." That "same" solar eclipse will occur a thousand or so kilometers north or south of the last, similar, eclipse. It takes 1,840 solar years for an eclipse mid-path to sit on the same latitude.

This happens because the three types of month mentioned above fail to be, not unexpectedly, *exactly* commensurate. Eventually, there will be a jump or "glitch." And the cycle will have to be reset. Perhaps, it is better to think of the saros cycle as a series of discontinuous segments of a cycle. Each one of these represents a family of eclipses and is called an individual **saros**.

A saros may begin with a partial eclipse of the Sun (We are just keeping track of solar eclipses so far.) Earlier, the Moon's umbra just missed intersecting the Earth.

Successive solar eclipse paths of the same saros work their way northward. They do so at an average displacement of about three hundred kilometers. They eventually end near the North Pole. There are one or more partial eclipses and then none at all. We say that that particular saros has "died."

The solar eclipses of the saros described have in common the fact that they are associated with the Moon's descending node. Eclipses belonging to a saros that migrate north to south are associated with the Moon's ascending node.

Regardless of its starting point, each successive eclipse over a given saros is deeper, meaning less partial and more likely total during the first portion of the cycle. The opposite happens during the second portion.

The eclipses of a particular saros can be parameterized by the somewhat ambiguous sounding parameter, **gamma** [γ]. The γ of a given solar eclipse is the distance between the center of the Solar umbra and the center of the illuminated disk of the Earth. (This is the half-of-the-Earth that is in daytime; its center is the latitude and longitude at which the eclipsing Sun is directly overhead.) The measurement is made at maximum eclipse and is expressed in units of Earth radii.

---

2 or hypersaros
3 Greek for "turning of the wheel"

A central eclipse is $\gamma = 0$. If the axis of the umbral cone passes north of the Earth's center, $\gamma =$ a positive number; and if the umbral cone passes south of the Earth's center, γ equals a negative number. When the axis of the umbral cone passes an Earth radius from the Earth's center, $\gamma = 1$.

There are about forty saros taking place at the same time. Individual saroses are numbered sequentially in time, based upon the order of mid-saros. Even numbers are reserved for northward-evolving saroses (γ increases). Odd numbers are reserved for southward-evolving saroses (γ decreases). Each saros exists for 1,226–1,550 years and represents 69 to 87 solar eclipses. The time between consecutively numbered saroses is 358 synodic months; that is, a little more than 29 solar years. This interval is called an **inex**.

There are saroses of lunar eclipses, too. However, for this kind of eclipse, those saroses that migrate southward receive even numbers and are associated with the Moon's ascending node. Those saroses that migrate northward receive odd numbers and are associated with the Moon's ascending node. Gamma may be applied to lunar eclipses as well, though its interpretation is less straightforward than for a solar eclipse.

There are other cycles that may be used for eclipse prediction. One **tritos**[4] **cycle** lasts for a period equal to the difference between that of the saros cycle and the inex cycle. There are still others. However, working with the saros cycle is easiest.

17.3.2 Success!

Predicting the details belonging to a total eclipse of the Sun, for a particular date and particular time, remained out of reach without a theory as to what actually causes an eclipse. To the Mesopotamians, it was all an abstract number game.

Nevertheless, the saros cycle is a good model of eclipse behavior. It does not yield an exact correspondence between eclipses, such that if you see one eclipse you know everything about a future eclipse. Predictions based upon it alone eventually will fail. Still, it is a tribute to ancient record keeping that the saros cycle was discovered at all, as early as it was.

What is, ultimately, the answer to the question, "Who first accurately predicted a solar eclipse?" It depends upon your criterion for success. Societies the world over have made the attempt. Most used the saros cycle. More than one succeeded in establishing eclipse probabilities: when an eclipse was more or less likely to occur.

Progress in eclipse prediction was advanced only once the Greek **geocentric** (Earth-centered) theory of the Universe was replaced. That model had envisioned celestial bodies mounted on transparent concentric spheres. The replacement was a new theory of, if not the entire Universe, the Solar System. It incorporated freely mov-

---

4 or Sarod

ing bodies, under the control of Isaac Newton's gravitation, orbiting about the Sun. This is a **heliocentric** theory.

Only since the Renaissance has there been eclipse prediction that you can count upon. And small refinements still are made.

Tools of the naked-eye observer

Since it became possible to predict eclipses, these predictions are redacted in annual *almanacs*, along with other special events of the civil year in your sky.

Eclipses were once standard observing fare for observational astronomers. They wished to keep track of the exact location of the Moon. Timing a total eclipse allows placing the Moon precisely at the exact location of the Sun.

Until accurate compasses, clocks, and electronics superseded the technique, a lunar eclipse was of direct navigational use. Such an eclipse is a simultaneous event over a relatively large area of the Earth.

It works this way: Compare the time of the eclipse's local occurrence to that printed in an almanac. The almanac is written for a known location/port. The difference in the two times is converted to a longitudinal difference, between your location and the almanac's.

17.3.3 Example: Norway

Historians find the ubiquitous records of eclipses useful in pinning down dates. An example: Haakon IV, King of the Norwegians, invaded islands claimed by Scottish King Alexander III in the thirteenth century.

Legend has it that on the date Haakon set sail, there was an eclipse of the Sun. As was typically the case, this was considered a bad omen. The King departed, anyway. Most of his ships were lost in a storm at sea. Haakon was repelled. The Orkneys remain British until this very day.

For a time, there was some debate in historiography about whether the year of this (albeit minor) battle was 1262 or 1263. In an interdisciplinary effort, astronomers computed that an annular eclipse did take place. It was visible from Norway in the year 1263. The history books have been corrected.

Notes from yesterday and today

In 1869, the Native-American Oglála and Brulé nations only had just been forced onto a reservation by the United States government. The army physician assigned to them wanted to impress upon these people the superiority of western medicine and, in general, science over the magical arts. He read of an upcoming event in his almanac and then announced to some of the principal chiefs and warriors that there would be a total solar eclipse on 7 August.

The Native Americans were impassive lookers-on until the total eclipse reached second contact. They then concluded that the exhibition had gone far enough. They must drive away the evil spirits causing it. The warriors commenced discharging their rifles in the air.

17.4 Eclipse duration

So far, our discussion has dwelt on predicting the event itself. This has meant a certain kind of eclipse occurring at a given place and on/at a given date and time. What about the amount of time that passes during the eclipse? Of greatest interest, how do you predict the length of totality (or annularity) in a solar eclipse?

Duration also affects path length; the two parameters are intertwined. For a total solar eclipse, path length is defined as the distance between the two centerline limits from where the Moon's umbra first lands upon the Earth's surface to where it leaves the Earth's surface. Path length is roughly the product of the shadow speed and eclipse duration as a function of time.

Let us concentrate on a total eclipse of the Sun. Moreover, think about an eclipse near an eclipse magnitude 1.0. Certainly, a higher eclipse magnitude will result in a longer totality. But even total eclipses of the same eclipse magnitude can have different durations for a given observer.

The shadow of the Moon travels along the eclipse path at 3,680 kilometers per hour on average. Yet at the equator, that speed is only 1,770 kilometers per hour. This is true because the surface of the Earth is traveling faster at the equator, and in the same direction as the shadow travels. This motion cancels out some of the umbral speed relative to the moving observer.

The speed of the Earth's surface, which is following along with the eclipse shadow, is simply due to the rotation of the Earth about the axis mundi. This speed is fixed. When expressed as an angular velocity, it is one rotation per day. A point on the equator must travel a 40,000-kilometer circle in the same time that a point near one of the poles travels a much, much smaller circumference.

The speed of the Moon in its orbit about the Earth, also in the same direction, is not forgotten. However, this angular velocity is one revolution per sidereal month. In other words, the length of the day is short compared to the length of the month. During a comparatively brief solar eclipse, the Moon's motion is a secondary effect. The Earth's revolution about the Sun plays a still lesser role.

Admittedly, it is true that if the Moon happens to be near perigee, its speed in its orbit is slightly greater than average. This does slow down the shadow moving across the Earth's surface. It lengthens the eclipse of the Sun just a little bit. For a Moon near apogee, it is the other way around.

All this assumes a midday total eclipse, though. If it takes place shortly after sunrise, or shortly before sunset, the direction of motion for the Earth's surface below is not entirely tangential to the Moon's shadow cone. A significant component of the Earth's rotation is at a velocity orthogonal to the Moon's shadow cone. The result is that the rotating Earth plays a lesser role in these circumstances in determining eclipse duration.

Incidentally, suppose that a total eclipse of the Sun has already begun at dawn. (It is post-first contact but pre-second contact.) You are treated to an apparent double sunrise: The sky brightens, then darkens, then brightens again. A similar, double-sunset effect occurs for an eclipse that has not ended at dusk.

Let us return to what is normally the primary effect on eclipse duration. Total eclipses of the Sun seen near the Earth's equator are longer than total eclipses in the far North or South. To make the eclipse a bit lengthier still, the Moon/Sun should be directly overhead. This is possible in the tropics.

Nonetheless, the longest total solar eclipse of all time lasts no more than seven minutes, thirty seconds. And you have to wait a while for an eclipse approaching this length. The next occurs on 16 July 2186.

Figure 17.3: Those wishing to invest heavily in an opportunity to witness a total eclipse of the Sun sometimes charter a high-speed aircraft. As the plane flies long with the umbra of the Moon, the time length of totality can be drawn out for hours. Courtesy of Catalin Baldea.

As for total eclipses of the Moon, the comparatively recent one of November 2021 lasted (first contact to fourth contact) 3 h, 28 min. This was very close to the maximum duration. It was the longest lunar eclipse since the year 1440! None will take this long again until 2669.

The duration of a total lunar eclipse theoretically can be as long as 3 h, 40 min. Totality takes place for as much as 1 h, 47 min.

Incidentally, the interval between eclipses – not having the same duration but – occurring at the same time of day is 3,310 solar years.

17.5 Eclipse paths

The path *width* of a total or annular eclipse also depends upon eclipse magnitude. The higher the eclipse magnitude, the greater the apparent size of the Moon compared to that of the Sun.

Let us call the area on the Earth that experiences either totality (within the Moon's umbra) or annularity (within the Moon's antumbra), at a given instant, the "footprint" of a total or annular eclipse. All else being equal, the greater the eclipse magnitude, the larger is the shadow "footprint" on the Earth. However, for illustrative purposes, we are comparing solar eclipses for which eclipse magnitude equals a number close to 1.0. The path width yet may still differ.

Figure 17.4: The Moon's umbra covers the Canadian provinces of Quebec and New Brunswick, plus the state of Maine in the USA, as seen in this photograph taken from the International Space Station (solar eclipse of 2024). Courtesy of NASA.

Path width also is governed by how obliquely the Moon's shadow cone strikes the Earth. As you know, the Sun and Moon always are to be found near the Celestial Equator. Perhaps, an eclipse is seen passing through your zenith if you stand near the Earth's equator. In the case of a total solar eclipse, visible at the zenith, the shadow that projects onto the

Earth's surface is a circle. (The "footprint" is, for a moment, round.) Further up and down the path, this intersection is oblique. The "footprint" becomes stretched out-of-round. The path width may be slightly wider.

But as was the case with eclipse duration, latitude is the most significant effect on path width. So, what about total or annular eclipse paths elsewhere on the Earth? What about where there is always a solar zenith angle, and the eclipse shadow must always strike the Earth obliquely?

Total and annular eclipse paths run only more-or-less west to east. The eclipse "footprint" is going to be out-of-round at some other (changing) angle with respect to direction of the eclipse path as it proceeds. Path width increases and decreases all along it.

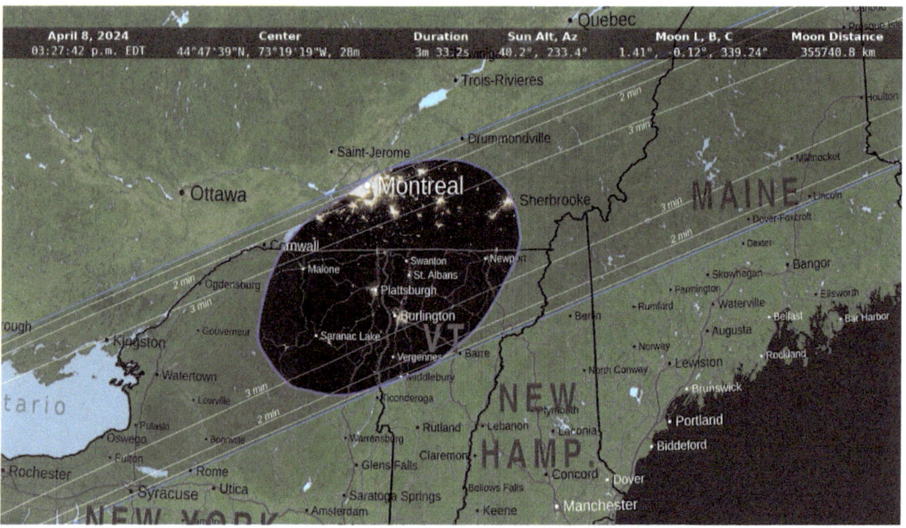

Figure 17.5: Computer simulation of a portion of the path for the 2024 total solar eclipse. Courtesy of NASA.

The "footprint" falls upon the Arctic and Antarctic very obliquely. A solar eclipse path at either of these locations may be quite wide. Indeed, there is the possibility at such high latitudes that a total eclipse centerline misses the Earth entirely. (It is above or below the planet.) Moreover, it is possible, in these circumstances, that the best eclipse that may be seen is partial. A solar eclipse path entirely on the ground is limited to about 250 kilometers in width.

Let us conclude this subject by mentioning that the Moon's shadow cone converges to a tip of zero area, just like any cone. For this reason, there is no theoretical *minimum* total solar eclipse duration or path width.

While calling the Earth a sphere is normally a very good approximation, in truth, it is a little **oblate**. Its diameter is greater at the equator than that measured through its rotation axis. It could happen that a very small lunar umbra, which would produce a total eclipse of the Sun at one latitude, misses the Earth and yields only a partial eclipse at another latitude!

Misconception alert

In the *Gospel of Luke*, what sort of reads like a solar eclipse, accompanies the crucifixion of Jesus. However, no literal eclipse circumstance matches fully with other associated events in the New Testament story.

On the other hand, *Acts of the Apostles* 2:20 reads, "The Sun shall be turned into darkness, and the Moon into blood, before that great and notable day of the Lord come." This does sound like a historical, deep partial eclipse of the Moon that took place during the period in which exegesis suggests the book was written.

17.6 Scale

Perhaps, you think that the perfect alignment required for a particular eclipse is overstated. This is understandable. Nearly every illustration you likely have seen, which involves the Sun, Moon, and Earth, is not to scale.

That is because the distances between these objects and their physical sizes exhibit such a range in arithmetical magnitude. A single solar-system scale of reasonable display length results in bodies too small to see. Alternately, it results in objects too near each other to resolve. All this happens when one tries to make a diagram to a uniform scale on a single page.

You easily can make a three-dimensional representation to a uniform scale though. Get hold of a "nine-inch" diameter Earth globe. This size is commonplace. It probably is most often manufactured after the "twelve-inch" variety. (The Earth in this model need not literally be a globe; any sphere of about the same size will do.)

Now, acquire a play ball about "two inches" in diameter. It is nearly, on the same scale, a correctly sized stand-in for the Moon. Again, this is a commonly manufactured size: A tennis ball is about right. These two spheres represent the four-to-one diameter ratio that exists between the real Earth and Moon.

It is time to place the faux Earth and Moon at their appropriate distances from one another. You do so to the *same* scale as their diameters. The distance ratio is thirty times. You need to move the Moon "globe" *seven meters* away from the Earth globe!t

17.7 Your chance of seeing an eclipse

About 32% of solar eclipses are normal partial eclipses; about 32% are annular (somewhere on the Earth); about 31% are total (somewhere on the Earth); and about 3% are hybrid. Around 35% of lunar eclipses are partial; about 36% are penumbral; and about 29% of lunar eclipses are total.

Total lunar eclipses are neither rarer nor more frequent than total solar eclipses. Notwithstanding, you are more likely to see a *lunar* eclipse. Why?

In a solar eclipse, the shadow cone is very narrow. A total eclipse is only visible within the path made by it.

These locations may be in a hard-to-get to, sparsely populated regions, like the Sahara Desert. They easily could be in the middle of the ocean.

Elsewhere, only a partial eclipse is seen. You have to be on the path, sometimes only a few kilometers across, to see a total eclipse.

You will recall that at Full Moon, when a total lunar eclipse takes place, the Moon rises at about sunset and is above the horizon until around sunrise. That involves half of the world. Indeed, because total eclipses of the Moon last so much longer than do total eclipses of the Sun, you are more likely to experience the Moon in total eclipse

Figure 17.6: Recent and future total-eclipse paths. Photograph by Aaron Spurr.

while it is rising or setting than is the case with total solar eclipses. Counting the people who do not see the total eclipse from beginning to end, even more than those who live in an entire hemisphere of the Earth have the opportunity to see the total lunar eclipse.

During a lunar eclipse, everybody for whom it is nighttime can see the eclipse. Of course, it must be nighttime, or nearly so, because that is when the Full Moon appears in your sky.

If you stay in one place, it is unlikely that it will be the right place to see any given total solar eclipse. Your odds are about[5] one in 375 years. Total solar eclipses are for travelers.

In the case of a lunar eclipse, the probability is close to "50-50": If it is night when it occurs, you see the eclipse. There is a window of opportunity of approximately twelve hours.

Total eclipses of the Sun are for pilgrims. American John Beattie is not an astronomer. He is an "umbraphile," the preferred word for a person whose hobby is observing eclipses. As applied to Beattie, it is a very serious hobby. He has regularly left his native New York and traveled the world to be at the right place at the right time to see a total solar eclipse. At last count, he has experienced totality *thirty-four* times so far.

17.8 Chapter summary

Predicting a particular type of eclipse is a challenge because the lengths of the nodical month and anomalistic month differ from each other. And both differ from that of the synodic month. Still, you know to be on the lookout for one during an eclipse season. Plus, a pattern of eclipses, known as the saros cycle, makes the prediction easier. Today, eclipses can be predicted accurately enough such that knowledge of past eclipses may be used to help better date historical events.

A total eclipse of the Moon is seen over half the Earth. It may last for hours.

A total eclipse of the Sun, though, only is seen on a narrow path over the Earth. Outside of, short of, or past this path, you see a partial solar eclipse instead. Due to the fact that the Moon is far from the Earth, compared to the size of the Earth, a solar eclipse geometry lasts mere minutes for a given observer on the ground.

---

5 Currently, someone in the northern hemisphere is favored

ⓘ 17.9 Chapter review

1. Why might eclipses have frightened people in the past?
2. List three factors involved in eclipse prediction and briefly explain them.
3. Match the time interval on the left with how it is counted on the right.

nodical month	time it takes for the Moon to go from perigee to apogee to perigee
anomalistic month	time from an early/late eclipse limit to that of the next early/late eclipse limit to that of the next early/late eclipse limit
eclipse year	time between the midpoints of two sets of similar eclipses
saros	time it takes for the Moon to go from descending node to ascending node to descending node

4. How do eclipse seasons help reduce the uncertainty of predicting eclipses?
5. What is the maximum number of eclipses that may take place per solar year?
6. Based upon what you know about solar eclipses, how many Earth diameters is the diameter of the Sun?
7. Why was not progress in accurate eclipse prediction possible until after the Renaissance?
8. What is the maximum duration of a solar eclipse? What variables impact the duration?
9. What is a major factor that affects the width of a solar eclipse path?
10. Why are you less likely to see a total eclipse of the Sun than a total eclipse of the Moon?
11. What do you think motivates an umbraphile?

18 Planets

Our eyes and brain insist on equating brightness with size. Yet the planets are no more disk-like than stars are to the naked eye. Which are the planets among so many stars?

18.1 Chapter learning outcomes

After reading this chapter, you will be able to:
- describe the motions of planets in your sky.
- identify a planet as either inferior or superior and explain how this distinction affects its potential location in your sky, direction of motion, and brightness.
- summarize unique visible characteristics of Mercury, Venus, Mars, Jupiter, Saturn, (and Uranus).

18.2 How to tell a planet from a star

Except for little Mercury, the planets are fairly bright. Mars is noticeably red. Mercury, Jupiter, and Saturn are faintly yellow. Venus is definitely white. All this helps. However, there are bright, colored stars in the sky, too.

Recall that, on average, planets twinkle less than stars. (The exception is Mercury, which is always viewed near the horizon and therefore through a large air mass.) However, on a night of poor seeing, everything twinkles. This is not a trustworthy method of discerning planets from stars, either.

Planets are to be found in the zodiac. Memorizing these twelve constellations allows you to pick out a "star" that does not belong. Confirmation must await another night, when the suspected planet is seen to have changed position with respect to the real stars.

Historically, this is the most reliable method. The fact that these five special objects, the naked-eye planets, were noticed long ago and thought to be exceptional was based upon their *motion* on the Celestial Sphere. They were not "fixed" stars, well-behaved because they stay in constellations.

Thankfully, today, there are other people to keep track of the planets for you. A published **ephemeris** provides astronomical data that can be used to yield the celestial coordinates (or the coordinates themselves) for all sorts of astronomical objects as a function of time.

https://doi.org/10.1515/9783111441245-018

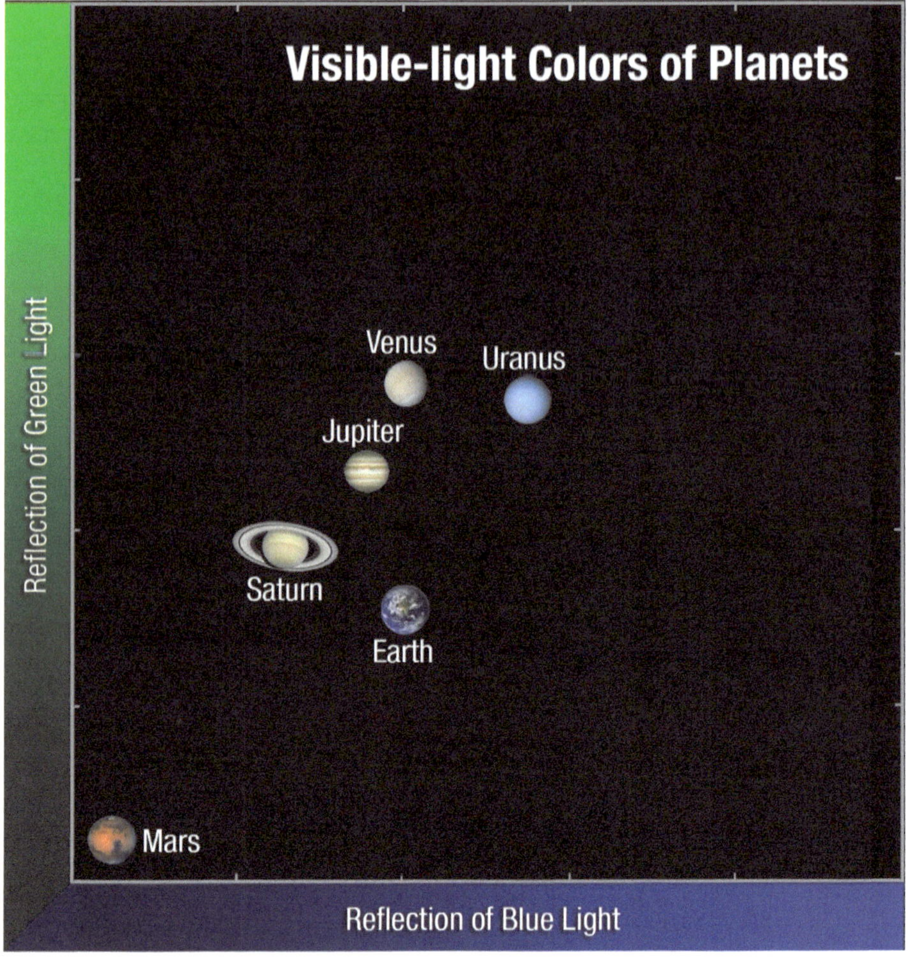

Figure 18.1: With the exception of Mars, the colors of the planets are determined by their atmospheres and surface geology (plus, in the case of the Earth, its oceans). Courtesy of NASA.

18.3 Motion of planets in your sky

The apparent motion of the planets in your sky is more complicated than that of the stars, Sun, or Moon. In the case of the stars, they are so far away that the revolution of the Earth about the Sun inconsequentially affects our point of view for them. Only the diurnal motion of the Earth causes them to appear to move on your sky dome at all.

In the case of the Sun's apparent motion, you do have to take into account both the rotation and revolution of the Earth. For the Moon, it is the Earth's rotation and the Moon's own orbital motion around the Earth. Still, apparent lunar and solar motion are (roughly) constant. These facts help to keep things simpler.

Planets, though, are at the same time close and distant enough that, in charting their apparent motions in your sky, you must take into account the Earth's rotation, the Earth's revolution, and – independently – the unique and sometimes more variable revolution of each planet itself about the Sun.

Moreover, each planet has a different orbital path. These orbits are, not surprisingly, elliptical. Each has a different size, eccentricity, inclination to the ecliptic, and orientation with respect to the Celestial Sphere.

Things get messy. For this reason, this textbook does not discuss the position of the planets (their rising and setting points on the horizon, their altitudes as they culminate, *etc.*) in as great detail as it did for the luminaries, the Sun and Moon.

18.3.1 Sidereal & synodic periods of the planets

As it orbits the Sun, each planet has two differently measured revolution periods. These are a sidereal year and a synodic year, just like the Moon. The former represents the planet's relationship with respect to the stars. It is more fairly called *the* revolution period about the Sun. The latter represents the time it takes for the planet to move from one relationship with respect to the Earth to that same relationship again.

Both the Earth and another planet orbit the Sun in the same direction. Thus, the two differently defined years (the Earth's and the planet's) can be quite different from each other.

Mercury travels fastest around the Sun among all the planets; Venus is second swiftest. Both have sidereal periods shorter than that of the Earth, while the other planets all have longer sidereal periods than the Earth.

Venus is the closest planet to the Earth, and it is traveling at a speed around the Sun not much greater than our planet's. It takes Venus a long time to lap the Earth. The synodic period of Venus, and to a lesser degree Mercury, is significantly longer than its sidereal period.

Notes from long ago
Pretty planet Venus has attracted attention for a long time. The Venus Tablets of Assyrian King Ammi-saduqa (in Mesopotamia) date from the seventeenth century BC/BCE.

In eighth-century Mexico, the Maya aligned a building at Uxmal, Yucatán, (now called the Governor's Palace) to the most southernly rising point of Venus. It is believed that this orientation with the planet is an intended one (an alignment) because the inscriptions on the structure refer to Venus.

After Venus and Mercury, the next closest planet to the Earth is Mars. It is the first one orbiting further from the Sun than our planet. Hence, it does so a little more slowly than the Earth. The situation is the same. It takes the Earth much longer than a solar year to overtake Mars. The result is a major discrepancy between Mars's sidereal and synodic period.

Beginning beyond Mars, planets move much more slowly against the Celestial Sphere. The Earth catches up with them easily. Their synodic periods are not much different from their sidereal periods.

The part of the **Planetary System**, Mercury through Mars, constitutes the inner planets. They are characterized by relative proximity and small physical size. Then come the outer planets, Jupiter through Neptune. They are characterized by great distances, from the Sun, from each other, and – significantly – the Earth. They also are of large physical size. These properties affect their appearance (or nonappearance) in your naked-eye sky.

Notes from yesterday and today

It is likely that Mercury has always been associated with speed. Statues of the Roman god often depict an athletic figure with winged feet. This is why, as recently as the year 1914, a group of florists chose this symbolism as the emblem for their new, telegraphic, rapid-delivery business model.

18.3.2 Prograde motion

The orbital planes of all the planets, while not identical, are similar. Mercury, Venus, Mars, Jupiter, and Saturn, Uranus, and Neptune (never visible to the naked eye) always can be found near the ecliptic. Another way to say this is that our Planetary System is pretty flat. Only Mercury has a significant inclination to the ecliptic.

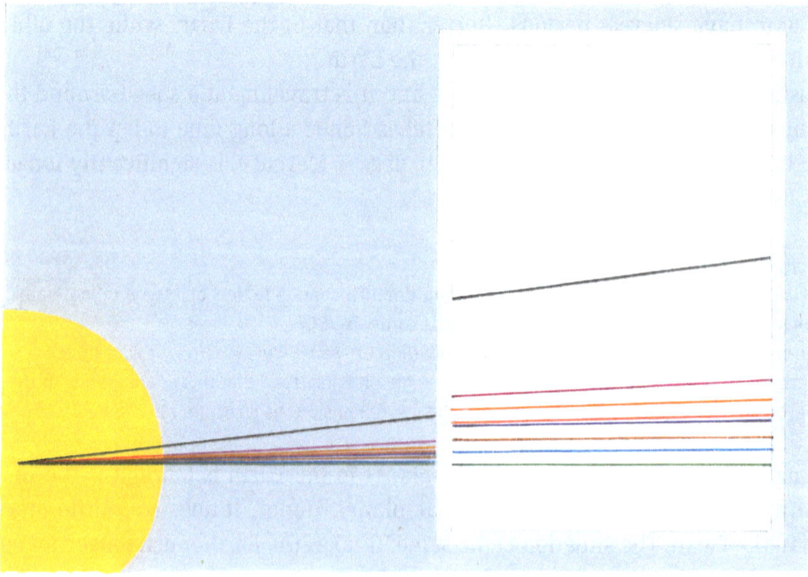

Figure 18.2: Inclination of the planets shows why those planets in our sky are to be seen near the ecliptic. Artwork by Christine Tarte.

Most planets spend the great majority of time moving predominantly west-to-east on the Celestial Sphere. (There are two exceptional instances.) This primary direction of planetary movement is called **prograde**. It is a feature of the Planetary System because all planets, including the Earth, revolve about the Sun in the same direction: counterclockwise as viewed from the direction we call north.

We all have learned that the Earth is just one more planet. It is third in order among the nested orbits of the Planetary System, counting outward from the Sun. This simple fact results in two kinds of planets in our sky.

There are planets that orbit closer to the Sun than we do here on the Earth and other planets that orbit farther from the Sun than we do. There are two planets in the first category and five in the second. This simple distinction governs how you see the planets in your sky the most.

Note that in this distinction, the Earth does not count as a planet in the same way that the other seven do. All ephemerides treat the Earth, not as a planet, but as a reference.

18.4 Inferior planets

A planet revolving about the Sun inside the Earth's orbit is called an inferior planet. Does this sound like an odd word choice? There is nothing "wrong" with these worlds. Here the term refers to smaller areas within their orbits, compared to the Earth.

Because of this positioning, these planets are never far away from the Sun in your sky. Inferior planets have orbits smaller in radius than the Earth. They are "chained" to the Sun, from your point of view. Inferior planets can never be on the opposite side of the Celestial Sphere from the Sun.

Such geometry makes it harder to catch sight of an inferior planet. If it always appears near the Sun, then most often during the time that the inferior planet is above the horizon, the Sun is also above the horizon. You cannot easily see a planet against blue sky.

If it is west of the Sun, the best opportunity to spy an inferior planet is during the short interval after it has risen, but before the Sun also rises. Alternately, the inferior planet may be east of the Sun. You might spot it just after the Sun sets, but before the planet sets as well. In both cases you are probably looking for the planet in twilight, neither in daytime nor the dark of night.

18.4.1 Planetary elongation

The two inferior planets are Mercury and Venus. Venus is the "morning" and "evening" star. These nicknames are very old. They were coined before the distinction between planets and stars was made. The terms indicate the times of day you are likely to see the planet.

Theoretically, Mercury could serve as a "morning" or "evening star," too. But consider this: Venus, which orbits farther way from the Sun than does Mercury, gets angularly farther from the Sun in your sky than does Mercury. Moreover, Venus moves more slowly on the Celestial Sphere than does Mercury. Add these facts: {1} Venus has a higher average albedo than does Mercury; {2} Venus has a greater reflecting surface than does Mercury because its physical size is larger than that of Mercury; and {3} Venus approaches closer to the Earth than does Mercury. The result is that one is more likely to see Venus than Mercury. If someone points out *the* "morning star," it is almost always Venus.

Used in this context, elongation is the angular separation between an inferior planet and the Sun, as seen in your sky. When elongation is at its greatest, astronomers say that the planet is at maximum eastern or maximum western elongation. Every synodic period, an inferior planet reaches a maximum eastern and maximum western elongation. The maximum elongation of Venus (48°) – regardless of which elongation it is, east or west – is significantly more than the maximum elongation of Mercury (28°).

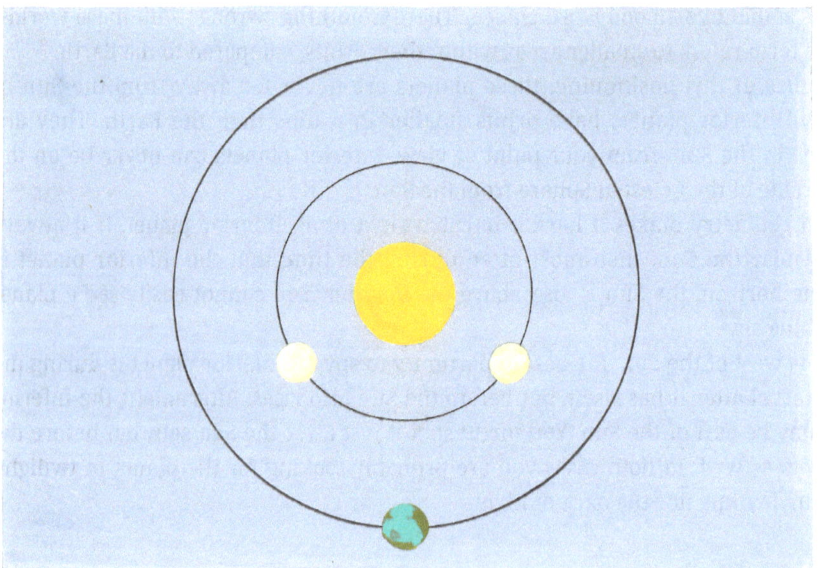

Figure 18.3: Inferior planet at maximum eastern and western elongation. Not to scale. Artwork by Christine Tarte.

18.4.2 Retrograde motion

If a planet appears to be moving east-to-west in your sky, it is called **retrograde** motion. The points of transition from prograde to retrograde motion, and from retrograde to prograde motion, are called a planet's **stationary points**.

An inferior planet swings between stationary points. Approximately half of the time it appears to move prograde, and about half of the time it appears to move retrograde.

Picture watching a wooden, carved horse while standing next to a carousel in motion. Sometimes the horse moves from your left-to-right. At other times (as you look through the carousel), it appears to move right-to-left.

When an inferior planet shifts from maximum western elongation through the far part of its orbit to maximum eastern elongation, though, it is traveling through a longer arc of its orbit than it does when it is traveling from maximum eastern elongation through the near part of its orbit, to maximum western elongation. So, its prograde motion on the Celestial Sphere does last a little longer than its retrograde motion.

Maximum elongation, as described above, is a good time to see an inferior planet in the sky. Which of the two maximum elongations you wish to look at depends on whether you want to be awake before the Sun rises or after it sets.

18.4.3 Inferior & superior conjunction

There are two times when it is all-but-impossible to observe an inferior planet at all. These occur when the Earth, Sun, and inferior planet are nearly lined up (elongation equals zero).

Think of an inferior planet and the Earth on opposite sides of the Sun. This must happen once during each of the inferior planet's synodic revolutions. To look at the planet you must also look in the direction of the Sun. You cannot make the planet out due to the Sun's overwhelming glare.

The planet and Sun are said to be in conjunction (meaning on the same "line" in your sky). Specifically, an inferior planet behind the Sun is at **superior conjunction**.

Now think of an inferior planet and the Earth on the same side of the Sun. This must happen once during each of the inferior planet's synodic revolutions. To look at the planet you must once again look in the direction of the Sun. The same problem arises. This near lining up, Earth-inferior planet-Sun, is called **inferior**[1] **conjunction**.

The distinction between whether an inferior planet is at superior conjunction or inferior conjunction is based upon whether it is farthest from the Earth at the time

---

1 Note the two different uses of the word "inferior"

(superior conjunction) or closest to the Earth at the time (inferior conjunction). Either way, inferior conjunction or superior conjunction, these are poor times at which to try to see the planet.

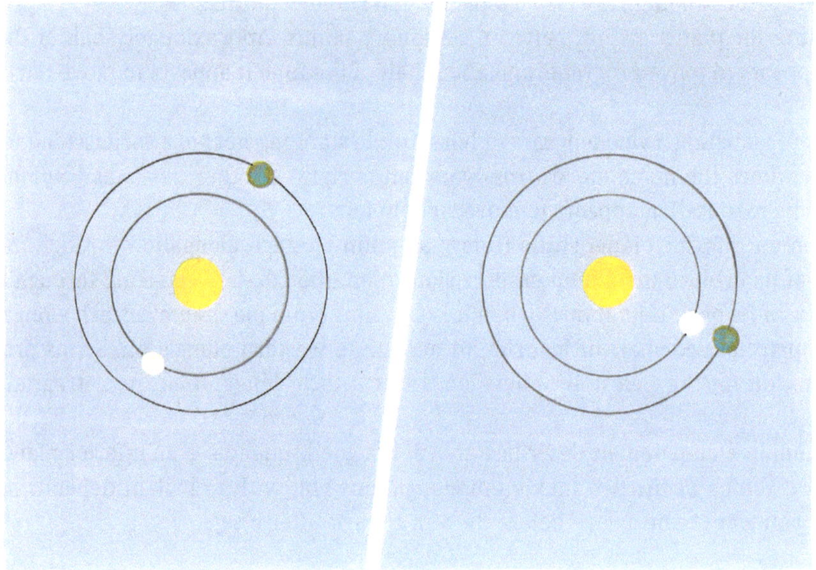

Figure 18.4: Left panel: superior conjunction; right panel: inferior conjunction. Not to scale. Artwork by Christine Tarte.

18.4.4 Venus

Venus cannot be seen in our sky for 47 days. This is a sum equaling that of the interval of invisibility centered upon superior conjunction and that centered upon inferior conjunction. That leaves 251 days for it to appear as your morning star and 286 days for it to appear as your evening star.

Venus goes from being your morning star to being your evening star during superior conjunction. It makes the opposite transition during inferior conjunction.

It takes Venus 220 days to go from superior conjunction to maximum eastern elongation. It then takes another 72 days to move from maximum eastern elongation to inferior conjunction. Similarly, it takes 72 days from inferior conjunction to maximum western elongation. It is another 220 days from maximum western elongation back to superior conjunction.

Based on the above, you might think that an inferior planet like Venus would be brightest around inferior conjunction. However, there is yet another factor that determines the brightness of a planet in our sky. It is the angle of illumination.

Just like the Moon, inferior planets undergo phases. Venus, for example, is a crescent as viewed (through a telescope) from the Earth near inferior conjunction. This geometry is analogous to near New Moon.

Can anybody see the crescent of Venus with the naked eye? It would seem to be impossible; still, there are anecdotal reports of it being accomplished. Intriguing is the fact that there are ancient Mesopotamian personifications of Venus that associate her with a crescent symbol.

If Venus is a crescent, not much of its disk is in sunlight. Its brightness averaged over the area of its full Earth-facing disk is diminished. But at "Full Venus," the planet is farthest from us. It is at superior conjunction and presents to us a complete disk of smaller apparent size than it does at crescent phase. This shrinkage *reduces* the planet's brightness.

These competing factors result in Venus's greatest brightness in our sky occurring when it is a wide crescent near maximum elongation. Venus is most brilliant about 36 days before and after inferior conjunction.

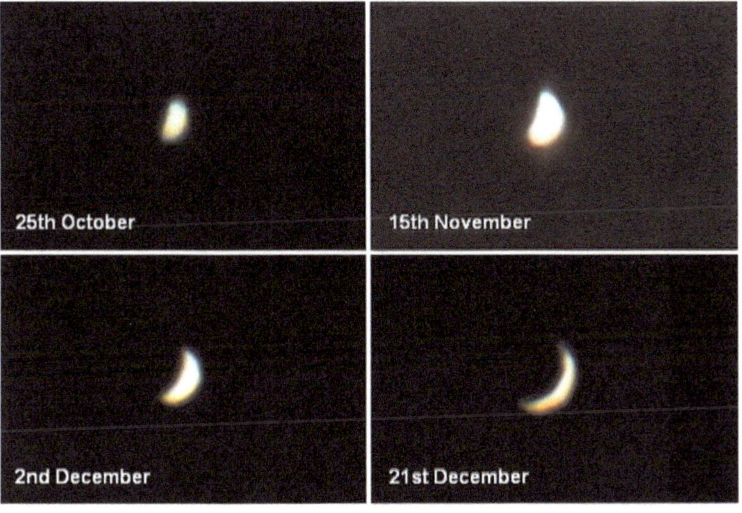

Figure 18.5: The changing apparent size and phase of Venus. Courtesy of Nigel Hoult.

18.4.5 More on planetary brightness

Planets shine only by reflecting sunlight. The brightness of a planet in our sky is a function of its distance from the Sun and its distance from the Earth. It is also a function of the planet's albedo and the angle of illumination. Even though we cannot see the disk of a planet with the naked eye as we can the Moon, it still is true for a planet that the bigger the apparent size of its disk – all else being equal – the brighter the planet in your sky.

Sunlight reflected by Venus is bouncing off high-albedo, white clouds enshrouding the planet's atmosphere. Reflection from cloud particles results in a complicated brightening as a function of angle of illumination.

At its most intense, Venus is 15 times brighter than the brightest star on the Celestial Sphere, Sirius. It can be seen at any air mass – -right to the horizon. In a dark, dark setting, Venus casts shadows! (Jupiter can, too.)

Figure 18.6: A shadow cast by Venus! Venus is more point-like than the Sun (used as a comparison source of light). Its shadow has less penumbra and more umbra than does that cast by the Sun. Courtesy of Bob King.

At any random time, Venus is most likely the brightest planet on your sky dome. Only Jupiter occasionally exceeds it. Diminishing Venus's brightness just slightly is the fact that it is never seen at high altitude/low air mass.

Venus is far away, like all celestial objects, and the Earth's surface is vast compared to you. Even if you are driving a car, at *high speed*, Venus never falls behind you in your sky. It remains in the same direction on the Celestial Sphere. It may do so on the time scale of an hour. Many reports of being chased by a UFO are attributable to bright Venus.

18.4.6 Mercury

When Mercury is visible at all, it is at low altitude where it fights twilight and air mass for recognition. At conjunction it disappears for 35 or 5 days. This leaves only 76 days per synodic year of (poor) visibility.

The light from Mercury reflects off of an atmosphere-less planet. The surface of Mercury is similar to that of the Moon. Its brightness as a function of elongation behaves like the Moon's brightness as a function of phase.

This difference between the two inferior planets results in a simpler brightening behavior for Mercury than Venus. Mercury is brightest several days before and after "Full Mercury" (superior conjunction).

18.5 Superior planets

A planet revolving farther from the Sun than the Earth is a **superior planet**. Its orbital radius is larger than that of the Earth's. There are more of these visible to the naked eye. They are Mars, Jupiter, and Saturn – in that order of distance from the Sun.

With superior planets, you do not get the same geometries that you saw with the inferior planets. The orbit of a superior planet encompasses both the Sun and the orbit of the Earth. Thus, a superior planet is never between you and the Sun. It would have to somehow escape its own orbit to be so!

18.5.1 Conjunction & opposition

Superior planets have only one conjunction. While the planet can be more-or-less behind the Sun as seen from the Earth (elongation[2] equals zero), it never can get in

---

2 Elongation is used here in the same sense as it is for an inferior planet

front of the Sun as seen from the Earth. We do not have to distinguish between superior or inferior conjunction – it is just **conjunction**.

As was one case with regard to inferior planets, conjunction for a superior planet is a time when the planet is as far from the Earth as possible. A superior planet must reach conjunction once every synodic revolution.

Yet, notice that a superior planet also can appear in your sky where no inferior planet can. A superior planet can be 60° away from the Sun, a point called **sextile**; 90° away, a point called **quadrature**; 120° away, **trine**; or 150°, called by the tongue-twister **quinduodecile**. (Astrologers get excited by different planets coincidentally being at different such points at the same time.) In fact, it can have any elongation.

Instead of reaching a second conjunction, every synodic year a superior planet lines up with the Earth and Sun on the opposite side of the Earth from the Sun. This is **opposition**.

At this elongation (180°), the superior planet is as close as it ever gets to the Earth. This was true for inferior planets at inferior conjunction; they were somewhat farther away from the Earth at superior conjunction. However, in the case of a superior planet, the distance from the planet to the Earth at opposition is much, much less than it is at conjunction.

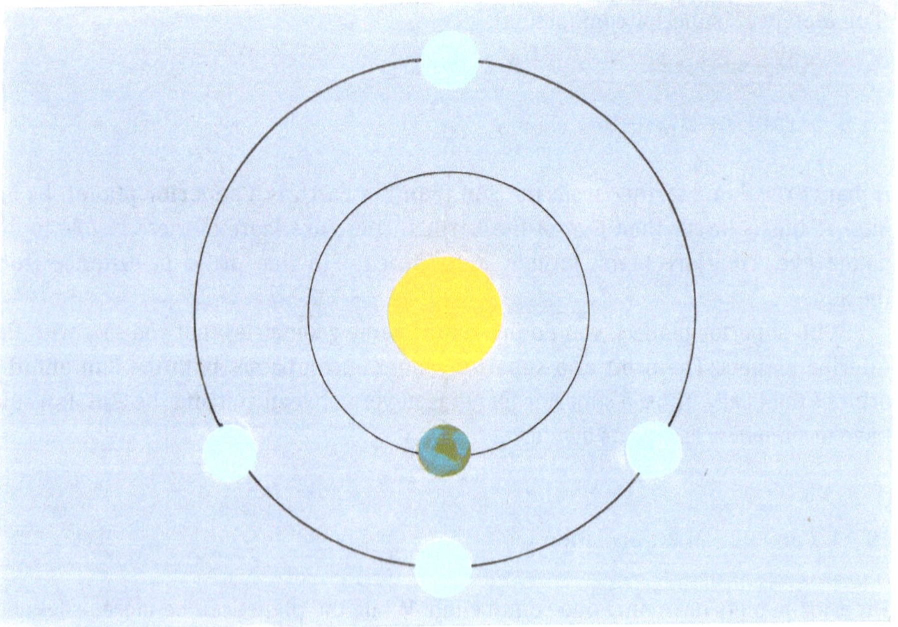

Figure 18.7: Superior Planet at (clockwise) conjunction, western quadrature, opposition, and eastern quadrature. Not to scale. Artwork by Christine Tarte.

Recall that the Full Moon occurs when the Moon is opposite the Sun in your sky and is above your horizon all night. A planet at opposition is similarly visible from sunset to sunrise when at opposition. It culminates at midnight. A superior planet must reach opposition once every synodic revolution, too.

A superior planet is most easily observed at opposition. It is brightest at perihelic opposition, when the planet also happens to be at or near the perihelion of its orbit. It is less bright than its average opposition brightness when it is at or near aphelic opposition and at the aphelion of its orbit.

18.5.2 Retrograde loop

A superior planet appears to travel in a prograde direction most of the time. However, near opposition, it undergoes a brief period of retrograde motion, before returning to prograde. The further the planet is from the Earth when this happens, the briefer the retrograde motion.

Retrograde motion was perplexing to our early ancestors. The retrograde loop, exhibited by superior planets, was particularly unexpected.

The retrograde loop became explainable since Polish cosmologist Nicholas Copernicus's sixteenth-century introduction of a new, watershed theory for the Solar System. In it, the planets, including the Earth, revolve around the Sun. (This is instead of the old Greek theory in which one fewer number of planets revolve about a motionless Earth.)

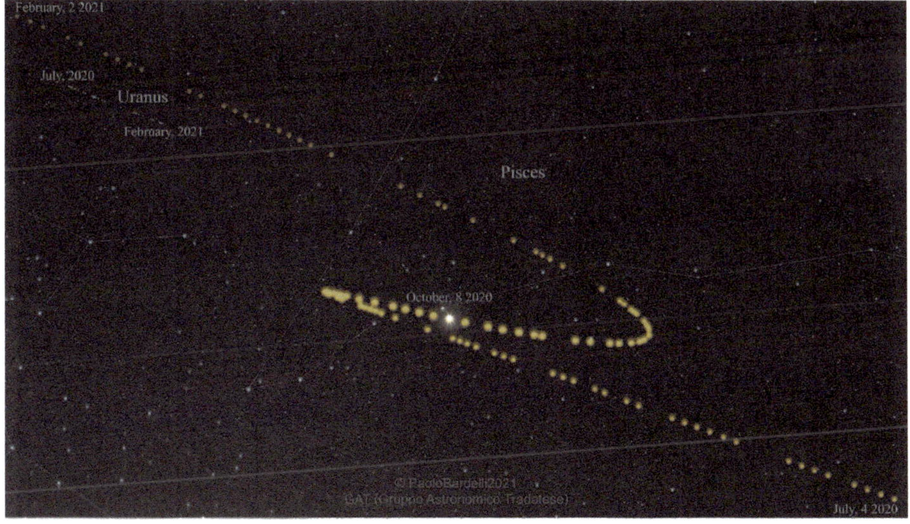

Figure 18.8: Mars executes a retrograde loop in 2020. Courtesy of Paolo Bardelli.

Reiterating, the farther a planet is from the Sun, the more slowly it travels in its orbit about the Sun. This characteristic was formally described by Johannes Kepler; he did so, less than a century after Copernicus, in his famous Laws of Planetary Motion. A superior planet orbits the Sun more slowly than does the Earth. Saturn is slower than Jupiter, which is slower than Mars.

As opposition approaches, the Earth overtakes a superior planet. The superior planet appears, for a time, to be moving *backward* as viewed against a distant background (the stars).

You see the same thing from a car overtaking another car on a curve. The other car (in an outer lane) is seen against a background of much-more-distant trees, mountains, or farms. Before passing, this car is viewed against those features that you both progressively encounter in your direction of travel. But as you start to pass and overtake the other car, it appears again against those background features behind you. Thus, the car being passed looks like it is itself moving backward, until the pass is complete. At this time the passed car resumes forward progress as judged by its encounter with the background ahead of it, as usual.

Notice that it is called it a "loop," not a "zig zag." The "loop" part of retrograde motion comes from the fact that, even if a planet pauses in its eastward or westward motion, it is still traveling a little north or south: Remember that the planes of the planetary orbits are all inclined *slightly* to the ecliptic.

Mars demonstrates the most profound retrograde loop. This is because it is the closest planet to you that is moving in its orbit more slowly than you – on the Earth. Its synodic period is 780 days.

The farther away another superior planet is, the closer its synodic period is to its sidereal period. These planets have successively smaller retrograde loops.

To summarize, the size of a superior planets' retrograde loop is inversely related to its distance from the Sun and, hence, from us. This observation indicates the order of these planets in outward distance.

18.5.3 Other differences between inferior & superior planets

Like an inferior planet, the brightness of a superior planet is governed by: {1} how big its Earth-facing disk appears from the Earth, {2} its average albedo, {3} the angle of illumination, {4} how far the planet is from the source of sunlight, and, of course, {5} how far away the planet is from you (on the Earth). Angle of illumination is now a *minor* effect: A superior planet is always frontlit; it does not exhibit phases the way an inferior planet does. At most, it may appear a bit gibbous if seen through a telescope. The superior planet only does so at a time away from elongation equals 0° or 180°.

Both phase effects and planetary atmospheric effects are minimized at opposition. As it happens, each of the superior planets past Mars has a substantial, relatively high-albedo atmosphere. As sunlight reflects off of such an atmosphere – like Venus but unlike the Moon and Mercury – the change of these planets' brightnesses as a function of elongation is not as drastic as it is for the Earth's satellite or the first planet from the Sun.

All the variables result in the obvious: A superior planet becomes brightest, for a given synodic period, at opposition. This is when Jupiter may take over from Venus, for a while, the role of brightest planet in the sky.

18.5.4 Mars

The closer a superior planet is to the Earth, the more dramatic is its change in distance between opposition and conjunction. Yet not all oppositions are identical.

Each planet, as you know, travels in an elliptical orbit about the Sun. In the case of Mars, the eccentricity of its orbit is high – not in an absolute sense but – by planetary standards. Planetary perihelion and aphelion together are called **apsides**.[3] Mars's orbit is the closest among superior planets to the Earth's. Thus, distance differences between its two apses are a greater percentage difference from the mean than they are for more distant planets. (It is worth pointing out along the way that Jupiter and Saturn are farther from the Earth at *any* time than is any inferior planet at *any* time.)

Whether the Earth or Mars is (or Earth and Mars are) near an apsis can noticeably affect Mars's brightness at opposition. Martian oppositions are best when they occur in August. Very favorable perihelic oppositions of Mars happen every 15 or 17 solar years.

Even then, some favorable oppositions are more favorable than others! So far, we have assumed a mean albedo for each planet. This works well in most cases. However, one hemisphere of Mars has a lower albedo than the other. The reason for this is not entirely understood by planetary geologists.

Mars has an obliquity very similar to that of the Earth. Which hemisphere is pointed toward you at opposition causes Mars's albedo to vary. This affects Mars's brightness in your sky.

At no time does Mars come close enough to the Earth so that you can see its disk with the naked eye, of course. Yet during each favorable appearance in our sky, the internet widely circulates a fable (author anonymous) about the Red Planet. In it, the claim is made that Mars will be closer to the Earth than it has been in some huge number of years. It continues and states that Mars will appear to be exaggerated in size up to the apparent size of the Full Moon!

Such a change in apparent size would be extraordinary. The Earth-Mars distance would have to vary to a *much* greater degree than Kepler's Laws and fundamental parameters for the two planets allow. This myth is, of course, always false.

---

3 or apses

Figure 18.9: The apparent size of Mars at different oppositions. Courtesy of Z. Levay/NASA/ESA.

18.5.5 Jupiter

Jupiter is normally the brightest of the superior planets in our sky. It rivals less frequently appearing Venus. Jupiter is not nearly as close to the Earth as are Venus or Mars. However, it is very large. So, it maintains a high apparent size despite its greater distance from us. As noted before, Jupiter also has a high albedo.

At opposition, Jupiter's diameter as seen in your sky is nearly 60" of arc. Theoretically, then, the human eye/brain should be able to perceive it as a disk. In reality, this never happens.

Orbiting Jupiter are four large satellites, the Galileans. (They are collectively named after Galileo; he first spotted them with his newly devised telescope.) They range in magnitude from 5.6 to 5.0. It would be possible to observe them with the naked eye, dancing around Jupiter at different orbital radii and at different rates, if only they were not lost in the glare of the planet itself.

Some Maori of New Zealand have boasted to westerners that they are able to see the Galilean Satellites by naked eye alone. This would be more believable had the claim first been made before the satellites' telescopic discovery was publicized.

The fourth-century BC/BCE Chinese astronomer Gan De (Warring States Period) represents the first person reliably recorded to have observed at least one Galilean Satellite. He must have been a man with extraordinary vision.

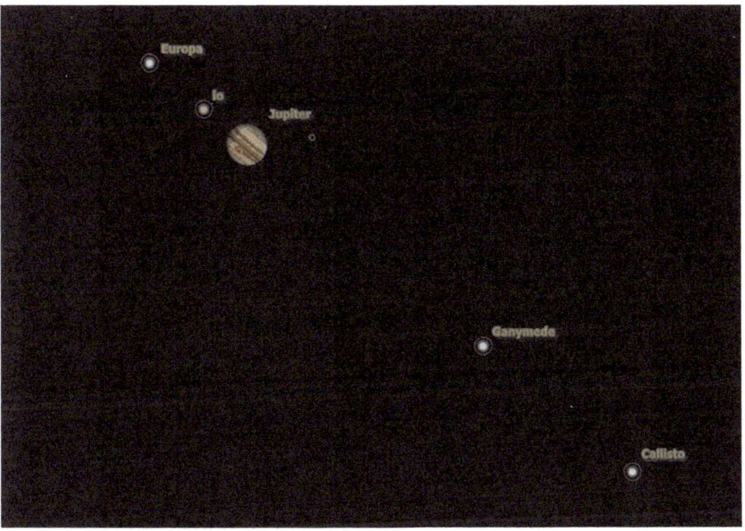

Figure 18.10: Callisto, the Galilean Satellite that orbits furthest from Jupiter, has a maximum angular separation from the planet of 10′ 18″. Ganymede, the next furthest, has a maximum angular separation from the planet of 5′ 51″. Having different orbital periods, you could imagine a time when these satellites are on opposite sides of Jupiter. At such a time their angular separation from each other would be 10′ 18″ + 5′ 51″ = 16′ 09″. This is easily resolved, if it were not for the intervening planet. Courtesy of Tony Netone.

It is well-documented that English cleric William Dawes (who was a respected amateur astronomer) could see all four. Of course, using his telescope he knew exactly where, with respect to Jupiter, to look for them. Nevertheless, his nickname of "Eagle-Eyed" was deserved.

18.5.6 Saturn

The explanation of an unexpected variation in the brightness of Saturn at opposition had to await a telescopic discovery made by Dutch astronomer Christian Huygens (in the seventeenth century): The planet is surrounded by rings. They are very wide but delicately thin.

The rings orbit Saturn's equator. Saturn's obliquity causes the rings to be tilted periodically. Sometimes they present their surfaces obliquely toward you. At other times they are edge-on as seen from the Earth.

The rings of Saturn are made of very shiny ice (or, perhaps, rock covered in ice). They have an especially high albedo. The unresolved planet-plus-rings, as seen by the naked eye, reflects more light toward you every 15 solar years. (This is the period of Saturn divided by two.) Then, the rings are tipped in your direction. The surface area that is planet plus rings is, from your perspective, maximized. In between, again

Figure 18.11: You will never see Saturn and its ring system like this without a good telescope. However, you can (just) see the change in brightness that the unresolved rings cause for Saturn over the course of the planet's year. Courtesy of Richard French, Jeff Cuzzi, Luke Dones, and Jack Lissauer/NASA.

every 15 years, the edge-on rings point in your direction. The rings then contribute effectively no light to your naked-eye appreciation of Saturn.

The real situation is not quite this simple. Also, predicting Saturn's brightness exactly must take into account when the rings create a shadow on Saturn's disk itself, and how much. Thus, they are also responsible for lessening the combined brightness. All else being equal, Saturn is brightest when the system maximizes, first ring light, then planet light.

What about actually seeing the rings themselves? Again, there are anecdotal reports – but only from when the rings are tipped as much toward the Earth as is possible.

18.5.7 Uranus?

"How many planets are visible in the sky to the naked eye?" If you ask this question of someone totally unfamiliar with the sky, you may get the answer, "seven." This would be because the responder naively assumes that all planets can be seen without optical aid. You might even hear, "eight," from someone who – in the moment – forgets that they cannot count the Earth as "in the sky"!

If you ask the same question of someone more astronomically knowledgeable, you will learn, "five." "Five" is the number that this book uses – and for a defensible reason. However, if you ask someone very experienced with observing the naked-eye sky, you could potentially hear, "Six. Sometimes."

Based on magnitude alone, it is possible to observe the farthest-save-one planet Uranus near opposition without optical aid, but barely. It had best be a time at which, and/or location from which, the ecliptic is reasonably high in the sky. Contemporary astronomers Kevin Krisciunas and Brian Skiff have picked out Uranus from the background star field while observing by naked eye on a mountaintop. (Of course, they knew where to look.)

This information is of limited utility. You certainly will not be able to resolve the disk of Uranus. Nor will you be likely to detect the planet's blue-green color. Uranus looks just like one of the myriad, high-magnitude stars that populate the Celestial Sphere.

Even if the star field including Uranus is not crowded, it takes great patience to await detectable motion in this distant planet. Motion on the Celestial Sphere, of course, remains the "gold standard" by which to identify a solar-system object. Detecting this motion probably will be required to verify that you are indeed spotting Uranus.

Notes from cultures around the world
Professional astronomers usually look at known objects in order to learn more about them. A less well-defined program risks wasting limited available time on valuable, high-tech telescopes.

It is no wonder that the discovery of Uranus had to await the eighteenth century and an amateur- (only later) turned-professional astronomer. Englishman William Herschel, a musician and immigrant from Germany, had no obligation to produce any kind of result. He surveyed the skies without specific objective. He used a telescope that he constructed himself. The seventh planet from the Sun was discovered by Herschel in 1781, with the help of his sister Caroline.

The pair came upon an extended object that they initially thought was a comet. (See the next Chapter.) But the characteristics of its orbit gave away its true nature. For the first time in recorded history, the number of known planets changed.

18.6 Keeping track of the planets

The time during which a planet is visible to you, and not near a conjunction, is called an **apparition**. An inferior planet has two apparitions each synodic period between its two conjunctions. A superior planet has only one apparition, centered upon opposition.

At the beginning of an apparition, like a star, a planet can rise heliacally. (A superior planet can rise – and set – acronically; an inferior planet cannot.) How long does it take for such a phenomenon to repeat? The Mesopotamians had the synodic periods of all naked-eye planets worked out by 577 BE/BCE.

18.6.1 Example: Mesoamerica

When you think of the ancient Maya, a landscape of vine-draped ruins may come to mind. But before the jungle claimed them, what are now ruins were once vibrant cities of stone, sheltering a people seemingly obsessed with cycles in the sky. This is known because a precious few surviving Maya books deal with calendrical cycles. The most famous is the *Dresden Codex*.

In the largest Mayan city, Chichén Itzá (eighth century), there is a strange building called the Caracol. It is aligned with Venus!

Figure 18.12: The Caracol. There is nothing else with this architectural style extant within the territory of the former Mayan empire. Courtesy of Rob Shenk.

Why? The Maya felt that the past was prolog. They subscribed to a worldview that saw history as cyclical. They looked for and found recognizable patterns of repetition in nature. (The Maya were well-versed in both the solar year and the synodic month.)

The Maya made a special point to keep track of the morning/evening star, Venus. Not only did they determine its synodic period, but they also noticed that 584 days is commensurate with 365 days (their civil year) once every eight solar years. In other words, they knew about the octaeteris.

Misconception alert

For the Greeks, Venus was Phosphorus (the "morning star") and Hesperus (the "evening star"). This was so until at least 580 BC/BCE.

In about that year, Pythagoras – he is the same guy who convinced other Greek scholars that the Earth is a sphere – wrote a book in which he pointed out that the two lights in the sky must be the same object. Yet even today, people will distinguish between the two.

18.6.2 Importance of planets

A superior planet is of great import to traditional Chinese. You may visit a Chinese restaurant from time to time. Often on the placemat, you can look up when you were born in the Chinese birth-year series. Each year corresponds to an animal (one mythical). These, in turn, supposedly, match with certain astrological character traits. The cycle is 12 years long; a "rat" is born twelve years after another "rat."

Where does the number 12 come from? It is based upon the approximate sidereal period of Jupiter. An interval of 12 years points to the use of 12 birth signs.

However, the Chinese **Year Star**[4] (or counter-Jupiter) is an *imaginary* planet with that period. It supposedly revolves through the ecliptic in a constant retrograde direction. Each year, it occupies one of 12 stations. These are regions of equal angular size around the Celestial Equator.

In the West, it almost goes without saying that the planets play a big role in astrology. A phrase like "when Jupiter aligns with Mars" is so familiar that it has become a cliché. We have previously discussed this subject in another context; it seems needless to delve into it again here.

The Maya Venus calendar, the Chinese Year Star, western astrological practices – none of these have any utilitarian use. But since when are our lives consumed totally with day-to-day prosaic matters? What is more, if you believed that planets such as Venus or Jupiter foretell future events – in an uncertain world, they were practical indeed.

18.7 Days of the week

This topic arguably is tangential. Nonetheless, it probably has occurred to you that the names for the days of the week have something to do with the naked-eye Solar System. Sunday sounds a lot like "Sun day." Monday sounds much like "Moon day."

The seven days of the week correspond to the seven planets and luminaries visible in sky. However, the names of, and personalities assigned to, each day refer to corresponding mythological figures. In English, these are drawn from either Latin or Germanic roots.

---

4 or Tai Sui Star

Yet the order of the planets in distance from the Sun (admittedly, a sequence known without controversy only in the seventeenth century) looks as if it has nothing to with the order of the days of the week! For instance, let us go all the way back to the Hellenic Greeks. One of their geocentric models/theories listed the order of the planets (including the luminaries), outward from the Earth, in this way: The Moon was followed by Mercury. Mercury was followed by Venus. Venus was followed by the Sun. The Sun was followed by Mars. Mars was followed by Jupiter. Jupiter was followed by Saturn.

If this recitation was to have been a list of the minimum distances between each of these bodies and the Earth, the order happens to be the right one. This is true even in the Sun-centered, heliocentric model (theory) you know better represents reality.

So where did the modern week-day ordering come from? If you use the names of the associated planet and not the actual day name, the week looks like: Sun, Moon, Mars (Tyr), Mercury (Odin), Jupiter (Thor), Venus (Frigge), and Saturn. This sequence would appear to be wrong in any model of the Solar System.

The answer is that the order has little to do with any literal model. Astrologers in antiquity decided that every hour of the day was governed by a "planet." If the day began with Saturn, each of the subsequent twenty-three hours of that day would be governed by the next planet or luminary in the Greek order:

Hour 1 = Saturn
Hour 2 = Jupiter
Hour 3 = Mars
Hour 4 = Sun
Hour 5 = Venus
Hour 6 = Mercury
Hour 7 = Moon

At this point astrologers were out of "planets." So, they started over.

Hour 8 = Saturn
Hour 9 = Jupiter
Hour 10 = Mars
Hour 11 = Sun
Hour 12 = Venus
Hour 13 = Mercury
Hour 14 = Moon

And again.

Hour 15 = Saturn
Hour 16 = Jupiter
Hour 17 = Mars

    Hour 18 = Sun
    Hour 19 = Venus
    Hour 20 = Mercury
    Hour 21 = Moon

Now there were not enough hours left in the day to go through the entire pattern.

    Hour 22 = Saturn
    Hour 23 = Jupiter
    Hour 24 = Mars

A new day began. The planet assigned to the first day was the next on the list.

    Hour 1 = Sun

Eventually, each day in its entirety was identified with the object associated with the name of the first hour of that day. Go through the seven bodies a few more times. You arrive at the Moon for the first hour of the third day. And Mars for the first hour of the fourth day, *etc.*

The result of all of this is: Saturday, Sunday, Monday, Tuesday, Wednesday, Thursday, and Friday. The powerful Christian Church required that the week begin with Sunday, though there are other traditions.

Notice that the concept of the week *itself* is based on no celestial period. This unit of time – unlike every other commonplace such unit – is uniquely nonastronomical.

Some sort of ancient market cycle has been suggested. Still, the origin of the week is unknown with any degree of certainty. Of course, the *Bible* used by major religions references seven days. However, it does not explicitly define this time span as the week.

Some societies have used an eight-day week. Nine-day weeks once were popular in Asia. There have been those who experimented with a ten-day week. Some peoples had or have no use for a week at all. Thus, this chapter ends with an enigma.

18.8 Chapter summary

Five easily seen naked-eye planets move on the Celestial Sphere. They do so because they, plus the Earth, orbit the Sun. Each has its own synodic period and brightness variation behavior. Different planets have different periods of visibility called their apparition.

Inferior planets have smaller orbital radii than does the Earth. For this reason, their elongation range is small, and they are difficult to see near inferior and superior conjunction.

Superior planets have greater orbital radii than does the Earth. They may be at any elongation, several of which are named. Superior planets have only one conjunction and can appear at opposition. At the time of opposition, a superior planet undergoes a retrograde loop.

The planets have been and are important to diverse civilizations. The Mayans had special interest in Venus; a counter-Jupiter is the Chinese Year Star. The days of the modern week are named after the Sun, Moon, and naked-eye planets.

18.9 Chapter review

1. How can you tell a planet apart from a star?
2. What does the sidereal period of a planet mean? What does the synodic period of a planet mean?
3. What direction do planets appear to move (when prograde) in the night sky? Why is it the case?
4. Why do planets sometimes appear to move retrograde?
5. Why is Venus only visible near twilight? Why can superior planets be seen at any time of night?
6. Neither the orbit of the Earth and nor that of the inferior planets are very eccentric. Ignoring changes in distance, what causes an inferior planet's brightness, as seen from the Earth, to vary?
7. What factors causes a superior planet's brightness, as seen from the Earth, to vary? Which one of these factors often is the most significant?
8. Match the planetary configuration in your sky on the left with the view as would be seen from above the Solar System on the right.

inferior conjunction	superior planet on the opposite side of the Sun from the Earth
opposition conjunction	inferior planet on the opposite side of the Sun from the Earth
superior conjunction	inferior planet between the Earth and the Sun
	the Earth between the Sun and a superior planet

19 Your Solar System

It is time to discuss sights that do not involve just one solar system body. In a way, our discussion of eclipses began with conversation of this subject. However, first, we will consider the solar system itself, no longer limiting ourselves to just the planetary system.

19.1 Chapter learning outcomes

After reading this chapter, you will be able to
- sketch the solar disk, including visible features
- diagram different alignments involving the Moon and planets
- discriminate between temporary sky phenomena such as aurorae, meteors, and comets.

19.2 Solar disk

The solar system is named after the body without which it would not exist. The formation of objects such as planets is thought to be a by-product of star formation. The Sun is "your" star. It is the largest and most massive object in the solar system by far, and everything else in the solar system orbits it.

19.2.1 Limb darkening

Notwithstanding, there is little to see upon the giant Sun *itself* with the naked eye. (For cautions regarding observation of the Sun, this is a good time to review Chapter 2.)

An incandescent gas, the Sun, is a bit darker, and redder, on its limb compared to its center. This is because limb light emanates mostly from the Sun's outer, cooler layers. This inhomogeneity is called **limb darkening**. It is so subtle as to be very difficult to make out with the naked eye.

https://doi.org/10.1515/9783111441245-019

Figure 19.1: Limb darkening. This image has been processed to enhance color differences. Courtesy of Brocken Inaglory.

19.2.2 Sunspots

In contrast to the subtlety of limb darkening, you *may* see a very dark, discrete **sunspot** or sunspots on the otherwise featureless, yellowish photosphere. These are temporary, cooler regions on the solar disk. They appear black to the human eye in response to the overwhelming brightness of the photosphere. If you could somehow magically remove a sunspot from the Sun and see it by itself against the true blackness of night sky, it would look plenty bright.

Sunspots are produced by the Sun's magnetic field. They start out small; most of them die out before achieving any appreciable size. But occasionally there is an eventual big one.

Or "ones." Sunspots typically appear in groups.

The biggest sunspots are several arc minutes across as seen from the Earth. Even sunspots that are just visible to the naked eye may go unnoticed, of course, unless somebody is specifically looking for them. Nonetheless, the Chinese documented pre-telescopic sunspots regularly, starting in at least 165 BC/BCE.

Between first and second solar eclipse contacts, you *might* be able to watch a sunspot or sunspots disappear as the Moon covers it/them. In a total solar eclipse, they reveal themselves again between third and fourth contact as the Moon exposes the same sunspots once again to view.

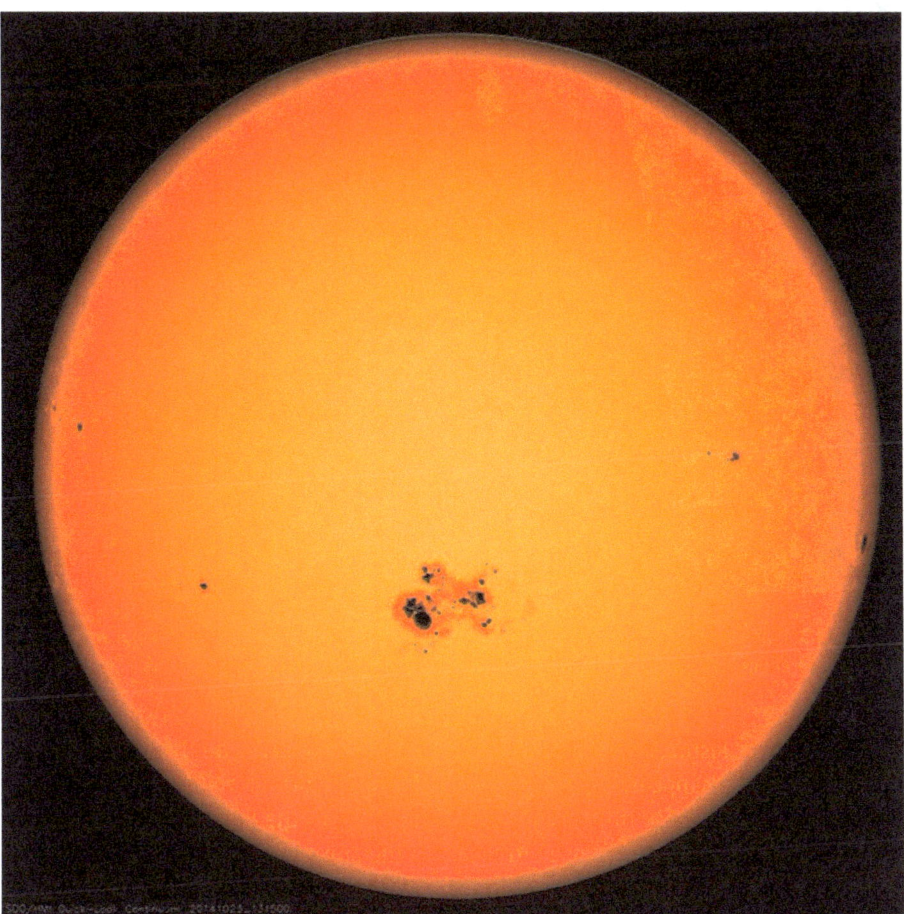

Figure 19.2: This naked-eye sunspot of 23 October 2014 was the largest of the 2008–2014 sunspot cycle (See the section below.) Courtesy of NASA.

Individual spots last from a few hours to a few months. There are exceptions, though.

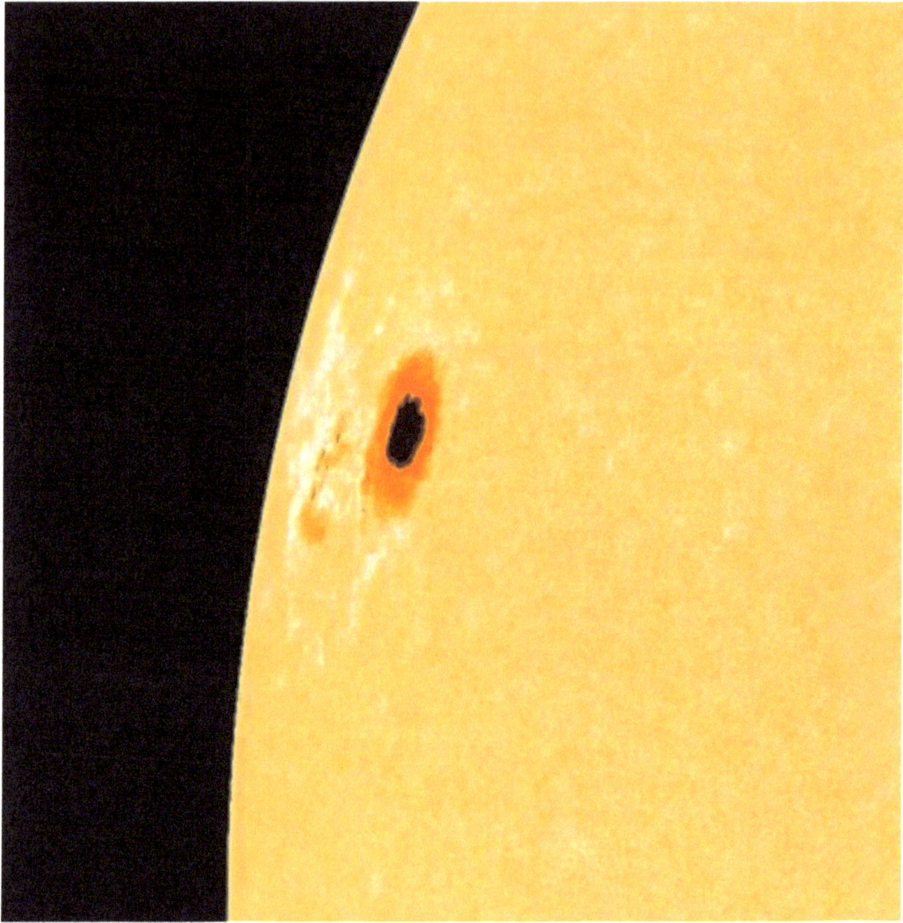

Figure 19.3: If not for its foreshortening due to the spherical nature of the Sun, this sunspot would appear several times larger than would the Earth at the same distance. Courtesy of NASA.

A larger, longer lasting sunspot might be watched as it appears on the eastern limb of the Sun until it disappears over the western limb. Its normal, roughly circular shape is distorted at the limb, thus proving the sphericity in the Sun – if such proof was needed. Indeed, sunspots are used to measure the Sun's latitudinally dependent rotation period, which averages about 28 days. Yet if the projection of a sunspot is made smaller, this makes observing it with the naked eye even more difficult.

19.2.3 Sunspot cycle

In the nineteenth century, a German pharmacist and amateur astronomer named Heinrich Schwabe took to counting the number of sunspots he could see on the Sun at a given time. (He used a telescope.) He did this for many years. Schwabe found that, on average, there are a lot more spots in some civil years than in others.

Furthermore, this rise and fall in the number of spots turns out to be quasi-periodic. There are a maximum number of sunspots every 11 civil years (approximately). Likewise, in-between, there are a minimum number of sunspots every 11 civil years (approximately). This is the **sunspot cycle**.

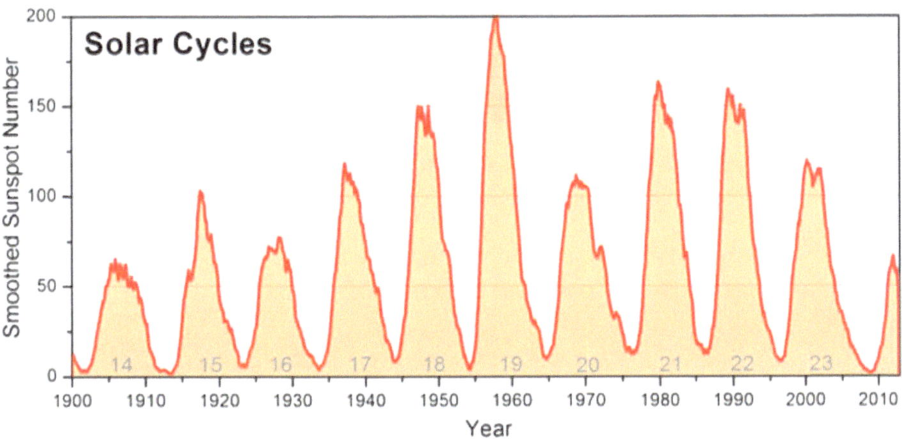

Figure 19.4: The sunspot cycle. Courtesy of NOAA.

At sunspot maximum, there may be a hundred times as many spots as there are at sunspot minimum. As you might expect, your odds of catching a naked-eye sunspot are greatest at sunspot maximum.

On the other hand, at minimum, no one might see a spot regardless of how much technology is used to help. During a particularly long sunspot minimum in 2007–2009, even professional astronomers found hardly any sunspots for over a civil year.

Where to look for sunspots? (Use a proper filter, of course; please review "Tools of the naked-eye observer" in Chapter 2.) The solar equator is very close to mid-disk. Early in the sunspot cycle, spots appear in the Sun's high latitudes. Later in the cycle, look at lower latitudes.

In the likely configuration of a pair, one sunspot corresponds to a magnetic north pole and the other a magnetic south pole. Theoretically, you can see evidence for the sunspots' magnetism for yourself – admittedly, with a great deal of luck: During a total eclipse of the Sun, you might observe a naked-eye prominence in which its curved nature is discernible. The prominence is caused by gases cooling and following the Sun's

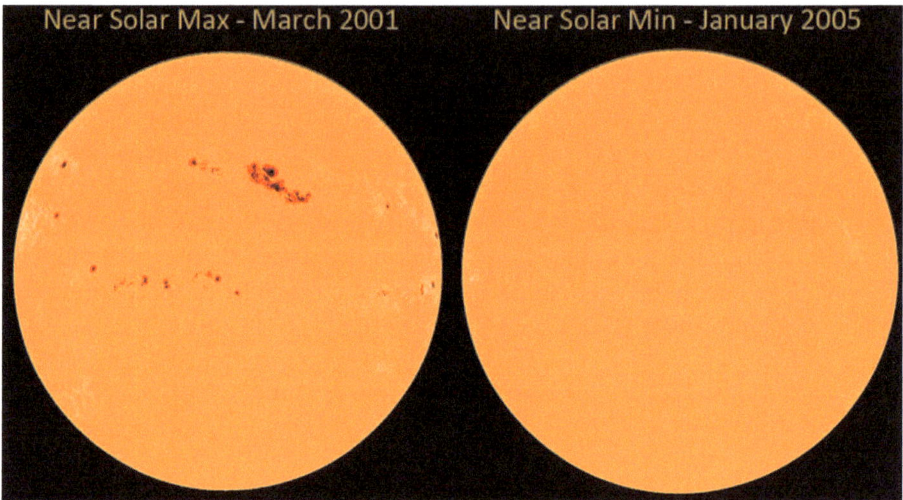

Figure 19.5: At the time of sunspot cycle maximum (left) and at the time of sunspot cycle minimum (right). Courtesy of NASA.

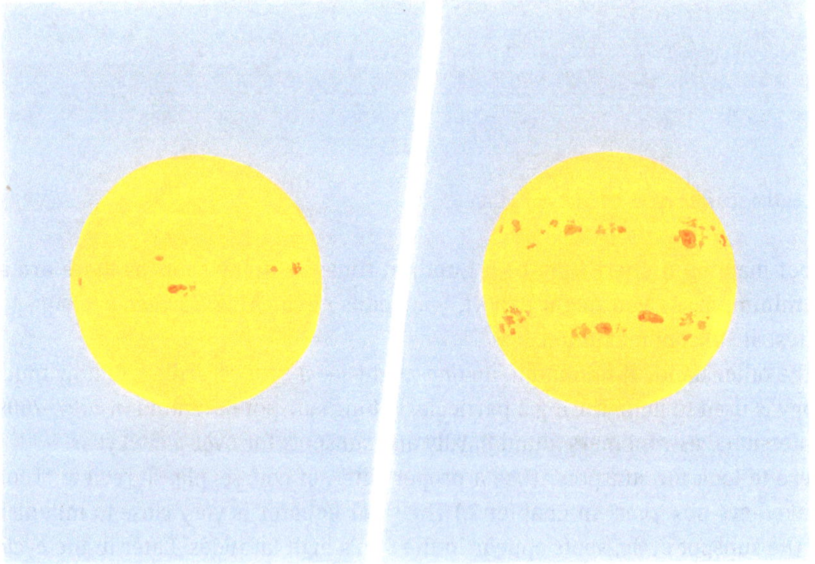

Figure 19.6: Sunspot appearances migrate poleward during the sunspot cycle (latitudinal variation exaggerated). Artwork by Christine Tarte.

looped magnetic field lines. You might even notice that, extrapolating the prominence in the direction of the unseen photosphere, it intersects at the location where you know there to be a naked-eye sunspot group. Admittedly, there are lots of "if"s!

19.2.4 Aurorae

You now come to the tiniest astronomical objects in this book. They constitute a continuous stream of subatomic-sized particles emanating from the Sun. This radiation is harmful to life.

Fortunately, the Earth's magnetic field protects you from most of it. On occasion, though, the Sun releases a particularly strong beam of such particles. If this beam reaches the Earth, the magnetic field may be overwhelmed. This is particularly true in the Far North and South, near the planet's magnetic poles.

Yet the Earth has a second line of defense. It is the atmosphere. High up in this blanket of air, molecules of gas absorb the dangerous radiation and release its energy as harmless light.

You can see the atmosphere do battle on your behalf. Appearing in your sky is the **aurora borealis** or (in the southern hemisphere) the **aurora australis**, often appearing as undulating sheets, "curtains," and streamers of colored light. You may know this phenomenon as the "northern lights"; antipodeans might call it the "southern lights."

Green is the most common auroral color. Violet, blue, and red are also possible.

The green glow is produced by offending radiation interacting with atmospheric oxygen roughly 200 km above you. Red is the result of this interaction higher up, where the atmosphere is even more rarified. Blue and purple are seen when solar particulate radiation reaches a much lower elevation and interacts with atmospheric nitrogen.

The mid-latitudes are not well situated to see the aurora. However, especially near sunspot maximum, it might indeed reach this far equatorward.

The aurorae seen there in May 2024 may have been one of the most intense displays in recorded history. The solar activity that created it buffeted the Earth-orbiting satellites; the Global Positioning System ("GPS") did not always work. High-frequency radio communication was interrupted. And some generator plant operators had to constantly regulate their output due to extra current induced in cables by "electrical storms."

Figure 19.7: The aurora borealis as seen at Winnipeg, Canada. Courtesy of Trang Pham.

19.3 Planetary conjunctions

So far, we have considered what you can expect to see in the sky over a relatively short period of time, or at least is easily predictable. Some celestial phenomena do fit well into such patterns of visibility.

Of course, the number of planets visible on your sky dome at any one time varies. Occasionally, by happenstance, this number is high. When multiple planets stretch across a single sky, it is very easy to map out the path of the ecliptic.

Planets may also appear near each other in the sky. They come into conjunction with the Sun; there is no reason why they cannot appear to approach each other, too. Any time one planet is near another, in your sky, is also commonly called a conjunction.

"Near" is vague. *Technically*, a conjunction means that the two planets cross the celestial meridian at the same time – that is, they are angularly close north and south of each other (share the same right ascension).[1]

For centuries, people have tried to equate the so-called Star of Bethlehem with a historical conjunction of planets – often in the symbolically Christian constellation of

---

1 Some say that the proper definition for a conjunction is that the two objects share the same elliptical coordinate.

Figure 19.8: A planetary conjunction. The two planets were just a tenth of a degree apart in 2020. Courtesy of Giuseppe Donatiello.

Pisces. Today, most scholars agree that the conjunction of 7 BC/BCE, the most likely candidate, was too early to signal the imminent birth of Jesus. An objective reading of the *Gospel of Matthew* implies a miraculous occurrence. In that case, there is little sense in searching for a historical celestial event.

The grouping of all three naked-eye superior planets is given a special name, a **grand conjunction.**[2] Note that the planets being in alignment (forming a literal line segment) is not a requirement of the definition. Grand conjunctions occur about every 20 civil years or so.

It is the periods of Jupiter and Saturn that govern the frequency of grand conjunctions. These planets move so slowly on the celestial sphere, compared to Mars, that Mars always can be counted upon to catch up with them sometime during a Jupiter-Saturn conjunction. Depending upon how one interprets a "conjunction," the media-hyped grand conjunction of 2020 was the closest since 1226!

Notes from long ago
Harvard University's preeminent historian of astronomy, Owen Gingerich, reminded us about Nicholas Copernicus. Copernicus watched Saturn, Jupiter, and Mars approach each other in grand conjunction during 1503/1504. Gingerich speculated that doing so might have caused Copernicus to wonder whether there might be a better way to model the planetary system than the Earth-centered view of his predecessors.

---

2 Or great conjunction

More generally, any three or more planets, in close proximity on the celestial sphere, are called **massing**. A massing may be joined, temporarily, by the Moon. Massings of all five easily naked-eye planets occur roughly every 400 years.

In 185 BC/BCE, the Mesopotamians recorded such a massing. This will happen next in the 2040s.

Figure 19.9: Close to the record. Here, *four* planets near each other in the sky. They are, from left to right, Jupiter, Venus, and Mercury (all top, left) with Mars (below). The line segment (further below, right) was created by a landing aircraft in this long-exposure image. Courtesy of Tatiana Gerus.

The words "close" and "near" are, of course, purposefully imprecise. What if you do not insist that a massing requires that the planets involved share some great circle in your sky? Then, how far away from each other the planets involved in a massing become ultimately is "in the eye of the beholder."

There is a limit to proximity, though. In the case of all naked-eye planets massed, it is 18 min of arc. That is, the orbits of each are such that in no circumstance can they ever appear within the same 18-min circle in your sky. There is never a time when they are not resolvable as five distinct planets.

It *is* possible that the *two* bodies involved in a conjunction appear so close to each other that you cannot resolve one from the other. Two celestial objects at their closest in your sky before they begin to separate is an **appulse**; here, the appulse results in the two objects appearing as one.

19.4 Occultations

The larger the apparent size of a closer celestial object, compared to a more distant one, the more likely it is that a conjunction becomes something else. In an **occultation**, one object passes in front of the other. View of the more distant object is temporarily hidden. A total solar eclipse could be called a special sort of occultation. However, the Sun itself produces too much glare to watch *it* occult anything.

Once more, stars have no apparent size. Our Moon, with its relatively large apparent size, occults a background star often. Such an event is more striking, though, if the Moon occults a very bright star.

Remember that the Moon's path requires such an event to occur near the ecliptic. The number of, say, first-magnitude stars is a small fraction of all those that can be seen with the naked eye. Therefore, the occultation by the Moon of one of these stars is infrequent. Candidates include Regulus [Alpha Leonis], Spica [Alpha Virginis], Antares [Alpha Scorpii], and Aldebaran [Alpha Tauri].

No matter the star, but depending upon the lunar phase, the star may be approached and covered by the illuminated limb of the Moon. Depending on the chord across the lunar disk that represents the path of occultation, it may again be revealed by the illuminated limb. But more likely, the star then reappears on the unilluminated limb. The same can be said of a star that is occulted by the unilluminated limb of the Moon.

When the Moon covers another object such as a star, the time it takes to do so is that of **immersion.** When it uncovers that body, the time it takes to do so is called **emersion.**[3] As the Moon has no atmosphere, immersion and emersion are sudden and brief. Most would agree that an immersion or emersion at the Moon's dark limb is more dramatic than one at its bright limb. (Immersion and emersion are generalizable to any occultation scenario.)

The occultation of a *planet* by the Moon is certainly rarer. (Notwithstanding, we know that none other than Aristotle watched Mars disappear behind our satellite on 4 April 357 BC/BCE.) While it does not make much difference for a superior planet, an inferior planet's phase at the time of occultation does. It yields the possibility of an immersion or emersion involving the dark limb of the planet, into/from the dark limb of the Moon. If you decide to take it upon yourself to time the exact instant of immersion or emersion, the "black-over-black" scenario may make this difficult.

---

3 Or ingress and egress

Moreover, because a planet has an apparent size – although one not obvious to the naked eye – the vanishing and re-appearance of the occulted planet are not instantaneous. If the planet involved is one with an atmosphere, it may even be possible to detect a gradual planetary dimming and re-brightening at (in particular, the dark) lunar limb. For a planet, a partial occultation is even possible as well.

Figure 19.10: Moon occults Venus. Courtesy of Bob King.

The occultation of a star, especially a bright star, by a planet is very rare. The last such occurrence was Regulus (Alpha Leonis; magnitude = 1) by Venus in 1959. It will happen again in 2044.

The most serendipitous of occultations? That would be the naked-eye observation of a star occulted by one of the large *satellites* in the outer solar system – even though the satellite is itself unresolvable. It is theoretically possible to make this observation in those all-too-infrequent optimal conditions. It has not happened in practice.

An occultation of a planet *by* a planet is extremely rare. Apparent size, even though it is indeterminable to the naked eye, is a limiting factor. So is whether the larger appearing planet can pass between you and the smaller appearing planet.

The possibilities reduce to Saturn by Mars, Saturn by Jupiter, and any naked-eye planet by Venus. However, apparent size changes with distance, so sometimes Mercury can occult Venus and even Mars, too. Counting partial occultations increases the number of possible combinations.

Grazing occultations: Perhaps a true occultation does not take place. Say, for instance, that the Moon merely grazes another celestial object. This sight is also interesting to watch. The nearly occulted (again, for example) star may even wink in and out several times as its light alternately passes through lunar valleys and is blocked by lunar mountains, on the rough lunar limb.

One more variation on the occultation theme is worth mentioning. It is possible that the Moon occults a *cluster* of stars (i.e., the Pleiades). In such a case, you can see an occultation of multiple stars more or less all at once.

19.5 Transits

When a celestial object of smaller apparent size passes in front of or near enough to partially block the sight of an object of larger apparent size, it is called a **transit**. An annular eclipse – *observed, as always, with proper eye protection* – technically could be called a **transit** of the Sun by the Moon.

The most famous transit is that of the Sun by Venus. *Using eye protection,* the black dot of the planet can just be seen passing over the solar disk.

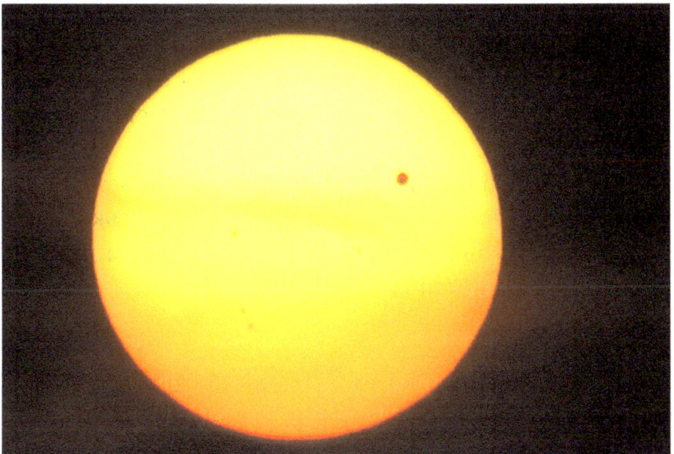

Figure 19.11: Twentieth-century transits of Venus. Courtesy of NASA.

Transits of Venus occur in a pattern: two transits 8 solar years apart, with each pair separated in time by 128 solar years. Transits of the Sun by Mercury are more common, but Mercury's apparent size is too small; you cannot see such a transit with the naked eye.

The most recent pair of Venus transits took place in the years 2004 and 2012. Venus transits are always in December or June. These times are when a ray from the Sun through a node of Venus's orbit points in the direction of the Earth.

Because it is comparatively close, Venus's transit silhouette is seen at different latitudes on the Sun depending upon where you are on the daytime Earth. This geometrical fact once was used in an attempt to determine the distance between the Earth and the Sun.

Unfortunately for the experiment described above, Venus's contact times must be measured exactly. The moment at which a transit of Venus begins and the moment at which it ends are surprisingly difficult to determine even with a telescope.

The extreme contrast between the larger Sun and the smaller self-shadowed planet causes an optical phenomenon called the **black drop effect**. The silhouette of Venus becomes elongated. The photosphere seems to "reach out" and engulf Venus. Dark Venus appears to bleed across the solar limb. It is as if for a few moments both are fluid.

The black drop effect is an optical one that occurs when light passes through a tiny gap. It is unavoidable and the bane of those making measurements with a telescope.

Not everyone can see something as small as the outline of Venus on the disk of the Sun at all. Many reports of naked-eye transits of Venus are likely reports of sunspots instead. The great Islamic scholars Al-Kindi, Ibn Rushd, and Ibn Bajjah were all thus fooled. So was the founder of modern planetary astronomy, Johannes Kepler.

When we read in history of such an event lasting for days, it cannot be caused by something revolving closely about the Sun; that is too long. It must be due to a phenomenon on the disk of the Sun itself. A transit of Venus lasts, at most, 7 h and 50 min.

With no believable reports of a Venus solar transit from before the telescopic age, you are led – like many time-dependent phenomena described in this book – to this conclusion: if you do not know when and exactly where to look ahead of time, you will not see Venus in transit across the disk of the Sun.

> **Notes from cultures around the world**
> The Moon and planets routinely cross the Milky Way or one of the apparent rifts within it. While not a transit in the formal sense, such an event is of interest to, in particular, indigenous people of the southern hemisphere.

19.6 Meteors

Meteors require the existence of an astronomical object but also the presence of the Earth's thick atmosphere. They are traditionally associated with naked-eye astronomy.

19.6.1 Cause of meteors

Orbiting the Sun are many rocky/metallic bodies, ones much, much smaller than any planet. They are to be found between and beyond the planets' orbits. These are the objects that cause craters when they strike a surface such as that of the Moon. Their tiny apparent size makes them normally far too small to be visible with the naked eye.

But **meteoroids** do routinely hit the Earth. You rarely see a resulting crater, though. Most are destroyed by the extreme frictional heat produced by tearing through our atmosphere on their way to the ground. You can see their fiery demise in your night sky.

Meteors most likely appear as random, brief, white streaks on your sky dome. Some consider catching sight of one to be a sign of "good luck."

Where did that come from?
Meteors are also called "shooting stars," or even less appropriately, "falling stars." These nicknames are unfortunate because meteors have nothing to do with stars. The Latin prefix "meteor-" means "high up" and reflects an ancient uncertainty about whether or not they belong to the celestial realm.

If you want to see meteors on any given night, wait until the wee hours. At this time your location on the Earth is facing the same direction as the Earth's direction of travel about the Sun. You will be more likely to see the Earth's interception of a meteoroid.

The typical meteoroid discussed so far weighs less than a gram and is the size of a dust grain! Occasionally, something more massive reaches our upper atmosphere. The result is a meteor that is the brightness of Venus or brighter. It may also exhibit some combination of lasting for several seconds, varying in intensity while doing so, exhibiting the colors of hot gases, and leaving a trail of smoke behind. Its path even

Figure 19.12: 2023 fireball in Tucson, Arizona, USA. Meteors may enter the Earth's atmosphere traveling at 100,000 km/h. The atmosphere slows them to roughly 1,000 km/h. The result is the release of a tremendous amount of energy. Courtesy of Eliot Herman.

may be seen to change, resembling that of a baseball pitcher's curveball. Some call these infrequent visitors "fireballs." They even may be visible in daytime.

Thousands of fireballs occur daily. However, statistically, most must happen over the empty ocean or in the daytime (during which their impressive nature is diminished).

19.6.2 Bolides

You might expect at least parts of larger meteoroids to reach the ground. Sometimes they do. However, the structural forces within the parent body may be weak. The energy of the incoming mass may be great enough to cause it to explode when it hits – not the ground, but – the thicker layers of the atmosphere. It is called a **bolide**.[4] The result can be visually dramatic. A pre-colonization rock art panel in South Africa shows a dance and sacrifice, as a bolide-like object streaks above the participants and breaks in two.

On 2 February 2016, a bolide with an impressive "wow factor" occurred over the South Atlantic, but nobody was there to see it! It was detected by NASA instruments. On the night of 10 December 2017, a probable bolide *was* seen by hundreds, as it lit up Canadian skies from British Columbia to Saskatchewan provinces.

Figure 19.13: The Chelyabinsk bolide. A small remaining portion of the meteoroid survived to the ground and was retrieved. Courtesy of M. Ahmetvaleev/NASA.

---

4 Some astronomers do not make this distinction between fireballs and bolides.

Then there was 15 February 2013. A (daytime) meteoroid, the size of a *house*, fell over the city of Chelyabinsk, Russia. The light from its fiery ablation was as bright as the Moon.

The resulting "superbolide" event unleashed the energy equivalent of a nuclear bomb. Sixteen hundred people sustained injuries from flying glass, smashed by the atmospheric shock wave generated.

The behavior of both typical meteors and bolides is unpredictable. In our era, it is even possible that "meteors" might be the result of de-orbiting objects originating in the space program. Their cause is manufactured celestial junk!

If the remains of a meteoroid stay intact and reach the ground, it is renamed a **meteorite** (or meteorites). It becomes a rock/or rocks and is/are the province of geologists. The subject becomes no longer germane to the subject of this textbook.

19.6.3 Meteor showers

Random meteors are said to be sporadic. On average, 0–3 are to be seen per hour.

In addition, sometimes the Earth intercepts a whole swarm of meteoroids – the smaller-mass and less-energetic variety, anyway. The average number of meteors you can see in the sky on a given night becomes higher than normal. You experience a **meteor shower**.

Figure 19.14: A long photographic exposure of a meteor shower makes the meteor flux look greater than it is. Notice that the radiant happens to be located within the Milky Way. Courtesy of Preston Dyches/ NASA.

Most clusters of meteoroids are thought to be the remains of comets (see below). They may fill most of the orbit belonging to the comet. During a meteor shower, the Earth is crossing the orbit of that ancient object.

The Earth arrives at the same place in its orbit at the same time each year. Say that this place is where the swarm is. For this reason, meteor showers are predictable. They occur annually.

Meteor showers are optimistically named. The frequency of meteors during these days is indeed greater, but usually not vastly greater, than the normal background of sporadic meteors on any night. The Perseid shower each August produces around 100 meteors per hour, but that is on the high side for showers.

Nevertheless, there have been spectacular exceptions. If the Earth crosses a shower meteoroid swarm where it happens to be particularly dense, for a short while meteors may storm downward at an almost uncountable rate.

There are patterns to such meteor storms. For example, the November Leonid meteor shower is usually a typical one but may become very active at an interval of every 33 years. Another meteor shower, more constant and a bit less dramatic, is still capable of putting on a good show; it is December's Geminids.

> **Tools of the naked-eye observer**
>
> A *Graphical Timetable* is a visual representation of an astronomical almanac for a given year. It is a graph, traditionally drawn with date on the vertical axis and time of night on the horizontal axis. The chart is hourglass-shaped due to the varying length of nighttime throughout the year. A curve representing the end of twilight might be included.
>
> Normally plotted on a graphical timetable are planet and bright-star rises, planet and bright star sets, and auspicious culminations. Moon phases and peak dates for major meteor showers may be included. So may be the date/hour for maximum brightness of a given planet.

Meteors in a shower all seem to come from the same point on the celestial sphere. Each streak diverges from this point; it is the shower **radiant**.

The effect is the same if one watches a row of cars, side by side in multiple lanes, traveling toward you on the motorway.[5] (This is not recommended!) On the horizon, the cars all seem to occupy a single point, but as they get closer, they appear to move outward from this point and turn into a line segment.

A radiant indicates that meteoroids originate at a great distance beyond the Earth's atmosphere. Meteor showers are named for the constellation in which their radiant is located. (Shower-producing meteoroids seldom result in a meteorite.)

---

5 Freeway

Figure 19.15: The 1833 meteor storm over New York, USA. Courtesy of E. Weiss.

19.7 Asteroids

A potential source of random meteoroids is collisions in the asteroid belt, which fills an orbit outside of Mars's. The rocky **asteroids** represent leftover building blocks of planets that may have been prevented from doing so by the gravity of nearby Jupiter. Their population is innumerable.

The asteroid Vesta (discovered by Heinrich Olbers on 29 March 1807) is the one inhabitant of the asteroid belt, out of millions at the very least, that is unarguably visible to the naked eye. (Claims for asteroid Ceres[6] have been made.) It can reach magnitude 5 at opposition but is more likely to be at the very limit of naked-eye visibility. Vesta is neither the closest asteroid nor the largest. The major factor that makes it visible, just barely and not all the time, is an atypical albedo.

Asteroid albedos tend to be low, like the Moon. Vesta is an exception; it is shinier than most. (After Vesta, asteroid size and albedo fall off rapidly.)

Not every asteroid inhabits the asteroid belt. There is a subset with eccentric orbits that carry them across the orbits of planets, including the Earth. This is the setup to a dangerous, albeit extremely improbable, collision.

More benignly, it is possible that a nearby asteroid – almost certainly not on course to an impact! – becomes visible to the naked eye. There may be a very brief window of opportunity, during which the asteroid is close enough to see. There may be little warning as it is thought that there are many, many smaller asteroids that have yet to be discovered.

Notes from yesterday and today

Large asteroid impacts with the Earth take place on a geological time scale. However, every 50 to 100 years, on average, the Earth does encounter at least a *small* asteroid.

The result is a crater, just like on the Moon. Eventually erosion and plate tectonics on the Earth obliterate such an impact feature. However, there remain some craters that you can visit. They have lasted due to their age, size, and/or location.

These include the 1,300-m-diameter Barringer Crater in the United States ("only" 50,000 years old). The Lonar Crater, of similar age, is 1,800 m in diameter and can be found in India. An island off Estonia alone has nine separate impact craters, thousands of years in age. Australia's Wolf Creek Crater is older (perhaps a couple of hundred thousand years), but smaller in diameter (900 m). Barely yet distinguishable is the Vredefort Crater in South Africa, which is 200–300 *km* in diameter and 2 billion years old!

19.8 Interplanetary dust

The inner solar system is dusty. These particles cannot be seen individually without a microscope! But in quantity they make their presence known.

Under optimum dark conditions, it is possible to see a pyramid of faint, white, light towering into the sky away from the just-set or the soon-to-rise Sun. (Many mistake it for the first or last vestiges of twilight.) This **zodiacal light**[7] follows the ecliptic and is as much as 120° at its widest.

---

6 The largest asteroid, it is also classified as a dwarf planet.
7 In the morning, also called the "false dawn"

At the same time, there may be slight, white, sky brightening 180° from the Sun, where it is then called the **gegenschein.**[8] The gegenschein is perhaps 10° in width. Look for the zodiacal light and gegenschein in spring post-dusk and fall pre-dawn.

The zodiacal light is **interplanetary dust** scattering the light of the Sun. The gegenschein is caused by sunlight reflected by this same dust. The zodiacal light was first recorded by Nicephoras in 410.

Figure 19.16: Zodiacal light (left) and gegenschein (right). Courtesy of Y. Beletsky/ESO.

Interplanetary dust that reaches the Earth lofts down invisibly. The low mass compared to surface area, of each piece, allows it to semi-float in the atmosphere. Still, if you find bits of metal-enriched "dirt" in your eave troughs, you may wonder from where it came. Interplanetary dust falling upon your rooftop and concentrated by rain is indeed a possible source.

19.9 Comets

Beyond most of the planets is the home of **comets**. While asteroids are primarily made of out rock and some metal, comets are much-smaller-than-planet-size lumps consisting of mostly ice with some small rocky particles. Notwithstanding, a comet can be one of the loveliest of celestial objects to behold with the naked eye.

---

8 German for "counterglow"

Unfortunately, because their appearance is almost always unpredictable, comets gained the unearned reputation as a sign of impending disaster (but see below). In fact, "disaster" means "bad star."

Just one example: The king of the Inca Empire died coincidentally just after the appearance of a comet in the sky. So, when shortly thereafter *another* comet just happened to show up as the Spanish invaded, the Inca were too demoralized to effectively fight off their conquerors.

Figure 19.17: The Norman invasion of England is commemorated in the Bayeux Tapestry. In this panel, English King Harold gets the bad news. Notice the comet shown menacingly looming overhead.

19.9.1 Comet behavior in your sky

Comets have very large and eccentric orbits about the Sun. As you know, Sun-orbiting objects travel more slowly at, and thus spend more time near, aphelion. Most of the time, potential comets are too far away to see without a telescope or are not detectable at all. Occasionally, though, their brief encounter with the Sun at perihelion also brings them nearer to you – residing as you do in the inner solar system.

Here, the Sun's warmth vaporizes volatile ice. The small, solid comet **nucleus** becomes surrounded by a many-powers-of-ten bigger envelope of bright gas and some dust called the **coma**. The coma may even fluoresce in the now stronger sunlight reaching the comet. It is at this point, when the "head" of the comet (nucleus plus coma) can be larger than the Earth, that your naked eye may be able to see it as a fuzzy disk.

Even closer to the Sun, electrically charged particles, emanating from our star, blow gas in the comet's coma outward (away from the Sun). The resulting tail is the comet's "trademark" feature. A comet with tail is even easier to spot. There is nothing else like it in your sky.

A **comet tail**, while made of very thin gas and dust, can be the longest object in the solar system! It may also, for a while, be the apparently largest object you see on the celestial sphere.

As the comet approaches, the tail trails; as the comet recedes, the tail leads. It behaves in the same manner as a ship's flag, blowing in a terrestrial wind and pointing in the direction opposite that from which the wind comes – regardless of the direction in which the ship is sailing.

Upon departing for the outer region of the solar system again, a comet's tail once more shortens to just the coma, which itself eventually fades to invisibility. The comet returns to being just a small, frozen nucleus and effective invisibility. The naked-eye comet apparition lasts for days, weeks, or months.

Misconception alert

Some people think that comets pass through the sky on a timescale of minutes or hours instead of days or weeks. Perhaps their tails give the impression of great speed. Alternately, it can be the result of confusion between meteors and comets.

Notice that a meteor is not so much a thing as a visible phenomenon caused by a thing. A meteor is a flash of light in your sky.

Like lightning, a meteor is not a single solid object; it is what we sense, produced by a physical phenomenon. A meteor is *caused* by a solid object (the heated meteoroid). No such distinction applies to comets.

19.9.2 Comet tails

A comet may exhibit more than one tail. Once liberated by vaporization of ice surrounding it, a dust component separates from the gas component. The resulting dust tail follows the more curved path of the comet, while the gas tail, made of much lower-mass gas molecules, appears radially straight with respect to the Sun

The gas tail itself may separate into components, depending on the prevalence of different molecules and ions within it. Different tails might be discernible by differences in color.

Figure 19.18: Comet Hale-Bopp (1997) with its two tails. Courtesy of Fred Espinak/NASA.

A comet may exhibit a fainter, shorter, **anti-tail.** An anti-tail appears to radiate from the head of the comet in the direction opposite that of the tail.

This is a projection effect. You are indeed seeing comet stuff, but you are seeing it behind the comet, where it bends along the comet's orbit and appears to be going the "wrong" way. Comet debris left in its orbit becomes visible when the Earth happens to pass through the comet's orbital plane. An anti-tail is a rare suggestion of depth on the celestial sphere.

Back to the comet's proper tail, sometimes it will disconnect from the rest of the comet. Catching such a disconnection requires very frequent observation of the comet. But soon the coma likely "grows" a new tail. You are more apt to see the resulting apparent gap in the comet's tail, between the old and new.

Figure 19.19: Comet PanStarrs (2013) exhibits an anti-tail. Courtesy of John MacLean.

Figure 19.20: Comet Leonard (2021) undergoes a tail disconnection. Courtesy of Gerald Rhemann.

19.9.3 Long- and short-period comets

There are few comets that come to naked-eye visibility and also have their orbit computed from multiple observations so that their arrival in your sky is predictable. Most bright comets have very distant aphelia and extremely lengthy periods.

Long-period comets often have periods so long that the comet has not been seen before in recorded history. You must await the announcement of the comet's (usually telescopic) discovery before searching for it. This is because the known short-period comets (those orbiting the Sun in 200 years or less) have made many trips around the

Sun, each time deteriorating by losing the material. Once it leaves the comet, the tail matter is gone for good. These comets tend to be dimmer.

Incidentally, long-period comets may appear anywhere on the celestial sphere. They are not restricted to the ecliptic. This includes declinations that may not be visible to you.

The most famous short period comet is Comet Halley. It is the only short-period comet that is consistently visible to the naked eye.

Halley visits the inner solar system every 76 years and has been recorded as having done so on almost every appearance since 2004 BC/BCE (throughout most of history, though, it was assumed to be a different comet each time). Its next apparition is in 2061.

Figure 19.21: Comet Halley (1986). William Liller/NASA.

Comets are traditionally named after their discoverers. (Halley's Comet is an exception.) A recent comet, named NEOWISE, was named after the orbiting observatory that was *used* to discover it! NEOWISE reached satisfying the naked-eye visibility in 2020.

19.9.4 Great comets

Long-period comets are the ones least eroded because they have made fewer passages near the Sun; they are often brighter than an average short-period comet. Truth be told, though, it is difficult to predict how bright a particular comet will become in your sky. Some of the variables involved are well-known, others not so well. Comet brightness is a function of distance from the Earth, distance from the Sun, other orbital characteristics, placement in your sky, area of the comet nucleus, how many trips near the Sun the comet has made, the structure of the comet nucleus, and chemical composition.

A truly **Great Comet**[9] is one that is easily spotted even in a sky that is not entirely dark. This is admittedly subjective; think of a bright center of a comet's head that at

Figure 19.22: Great Comet McNaught. Courtesy of Sebastian Deiries/ESO.

---

9 or Grand Comet

least reaches a negative apparent magnitude. Based upon this definition, a Great Comet is overdue: Statistically, one shows up every (very roughly) 10 years. The last graced the skies of the *southern* hemisphere in 2006. (It was lovely.)

In 2011, it was not too difficult to make out Comet Lovejoy with the naked eye. However, it fell short of Great Comet status.

The last bona fide Great Comet for nearly all to see was Comet Hale-Bopp in 1996. The comet was bright because of the comparatively large size of its frozen nucleus. This provided plenty of reflective source material for its coma and tail.

Coincidentally, the year before Hale-Bopp, Comet Hyakutake became a Great Comet. In this case, the nucleus of the comet was not large, but the comet was closer. It stretched across almost an entire sky dome.

Comet NEAT (2004) put on a pretty good show, too. However, it missed becoming a Great Comet because it was at its brightest when very near the Sun.

Indeed, some comets break up into smaller comets at their closest encounter with the Sun. Others do not survive in proximity to our star and are never seen again.

19.10 Chapter summary

The Sun is the predominant object in the solar system. Yet *safe* solar observing with the naked-eye yields little; an exception is the occasional large sunspot.

The likelihood of you seeing a sunspot increases with the total number of spots at the maximum of the 11-year sunspot cycle. This is also when aurorae are most likely visible.

Solar system objects appear to interact. You can see conjunctions, occultations, and (more rarely) transits.

With one exception, you will not see asteroids. However, asteroids are one source of meteoroids, which you can see when they self-destruct in our atmosphere as meteors. The best opportunity to see meteors is during an annual meteor shower.

The appearance of a naked-eye comet remains difficult to predict. When one does show up, though, it can be striking due to its distinctive coma and tail.

Perhaps surprisingly, it is possible to see one of the solar system's smallest components. Interplanetary dust appears faintly in your dark sky as the gegenschein and zodiacal light.

19.11 Chapter review

1. What are sunspots? How are sunspots related to the aurora borealis?
2. Given that there were a lot of aurorae seen at even lower latitudes in 2024, when might you expect to see that occur again? Why?

3. Match the apparent interaction on the left with the example on the right.

    conjunction   A planet passes in front of your view of the Sun.

    occultation   The Moon passes so near a star or planet that they cannot be resolved.

    transit   Two planets are near each other in your sky.

    apulse   The Moon covers your view of a star or planet.

4. Describe a grand conjunction.
5. Why is a transit of Venus a rare event?
6. If you want to see a meteor, when would you plan to observe (e.g., time of year and time of day)?
7. What is the difference between a meteor, meteoroid, meteorite, and a bolide?
8. What are asteroids? Why is the asteroid Vesta the most likely to be seen with the naked eye, even though it is not the closest or largest?
9. What is interplanetary dust? What evidence do we have that this dust reaches the Earth's surface?
10. What makes a Great Comet great?

20 Your naked-eye universe

Let us look at the challenges. These celestial objects and events hover at the border-line of naked-eye detection or are so rare that there is no guarantee that they will be visible during your lifetime.

Odds of seeing the former can be increased with practice. Odds of seeing the latter depend upon serendipity – but the "payoff" justifies hope.

So far, this book has dwelt extensively on what you can see, with the naked eye, of the solar system. This is largely because the scale of the galaxy takes a tremendous jump when your upward gaze goes toward that which lies beyond the Sun, Moon, planets, and occasional comet.

Because of the distances involved in interstellar space, brightness becomes a huge factor in visibility. Of course, you can see individual stars. Somewhat amazingly, you can see unusual properties of some of these stars as well.

Tools of the naked-eye observer

A physical model of the celestial sphere is a *celestial globe*. Star and constellation names, plus coordinates "lines," are engraved on such a globe so as to be read from the *inside*.

Many celestial globes are semi-transparent and include a small model of the Earth (terrestrial globe) inside. Below you will read why such a representation can never be to scale.

20.1 Chapter learning outcomes

After reading this chapter, you will be able to
- contrast binary stars, variable stars, novae, and supernovae in terms of brightness
- compare deep-sky objects such as different types of star clusters, nebulae, and galaxies
- evaluate different potential solutions to Olbers's paradox

20.2 Binary stars

In Chapter 2, you were introduced to optical double stars used as a test of resolution. So too are other pairings that are true binary stars. **Binary stars** orbit each other (e.g., Alcor [g Ursae Majoris] and Mizar [Zeta Ursae Majoris], which may be loosely related).

Optical doubles and binary stars are interesting to observe in their own right, especially two stars of different brightness and color in the same small portion of your field of view. Choice examples include Algiedi [Alpha Capricorni], Xamidimura [Mu

https://doi.org/10.1515/9783111441245-020

Scorpii], Epsilon Lyrae, Omicron Cygni, and Zeta Corvi. The point is that they are sprinkled all over the celestial sphere. (Naked eye *triple* stars exist, too!)

Binary stars are described not only by the brightness and color of their components, but by their orientation on the celestial sphere. This last descriptor is position angle. Imagine a ray beginning at the brighter member of the stellar pair and passing through the dimmer member. The angle between this ray and the direction of the celestial pole from its origin (measured clockwise) defines the binary's position angle.

20.3 Variable stars

Up to this point, you have been presented with stars that in themselves are unchanging. This is normally true over a human lifetime, if not the lifetime of human civilization.

However, exceptions exist. There are stars that *change* in brightness, **variable stars**. Few do so to the degree that the naked eye can perceive any modulation, though.

A famous counterexample is Algol [Beta Persei]. After several days shining at a constant magnitude, Algol appears to drop in brightness for approximately 10 hours. The change in light received is great enough that you can see the difference with your eyes alone.

Algol itself is not "at fault." Algol is one component of a binary star. That there are actually two stars present is irresolvable to the unaided eye. You mostly see the light of brighter Algol, and (not knowing otherwise) would assume that it shines by itself.

The unseen second star is less massive than Algol and does most of the moving within the pair. Coincidentally, the orbital plane of this dimmer star *about* Algol intersects your line of sight *to* Algol. Once per revolution, the junior star disappears behind Algol. You do not see the sum of the light from two stars, you see the light of only one. Their combined brightness lessens.

Half a revolution later, the smaller partner to Algol travels across the face of Algol. It blocks out some of the latter's light. As the surface brightness of the companion is less than the surface brightness of Algol, again, the sum of the brightness contributed by the pair decreases. There is another dip in the magnitude of what most construe to be merely Algol. Together with its lurking second, Algol is an **eclipsing *binary***.

Algol was once upon a time called the "Demon Star." This might reflect an abhorrence to a clearly fluctuating star in the celestial realm, which was supposed to be immutable.

Sheliak [Beta Lyrae] is an eclipsing binary as well. The effect on this star is also a variation in brightness – one of lesser amplitude but, nonetheless, one that is still detectable to the naked eye.

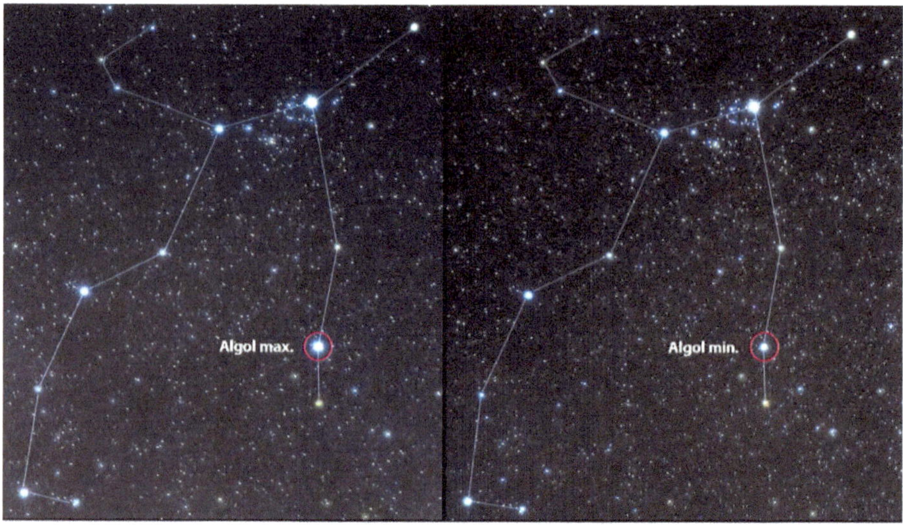

Figure 20.1: Algol. In digital images, the more light the collector receives, the more it spreads into adjacent pixels. This is why brighter stars look bigger in photographs. Courtesy of Bob King.

The star Delta Cephei is a more common kind of variable star. Here, a single star pulsates. It fluctuates in size: bigger, smaller, bigger, smaller. In the process, its luminosity goes up and down. Delta Cephei swells and shrinks and swells again, brighter to dimmer to brighter, every 5 days.

A lot of pulsating stars tend to have periods of variability that, on average, are longer than those of many eclipsing variables. If most stars were to expand and contract too rapidly, they would fly apart!

Delta Cephei is actually an unusual pulsating star. It and others in its class exhibit an interesting property: Their average luminosity is proportional to their period of variation. Just by timing its fluctuation in magnitude, you can conclude that – even though Delta Cephei is not particularly bright in your sky – nonetheless, its luminosity is great. This star produces 2,000 times more light than the Sun!

The most well-known variable star, because of its high average brightness and funny sounding name (to many of us, at least), is Betelgeuse [Alpha Orionis]. It is considered to be the first magnitude star. However, upon close inspection, Betelgeuse's magnitude varies. It does so with modest amplitude and an irregular period lasting *years*. Not everybody who tries notices Betelgeuse's variability.

For variable stars, the Bayer identification changes a little bit. Instead of the Greek, Latin letters are used in front of the genitive case for the constellation in which the star is found.

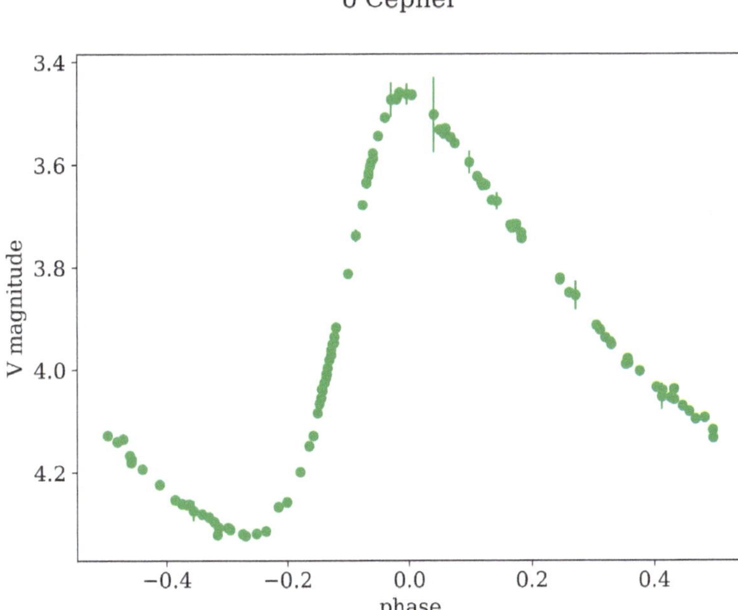

Figure 20.2: A light curve is a graph of magnitude versus time. This is a typical light curve for Delta Cephei. The units on the horizontal axis are days. Courtesy of Warrick Ball.

Figure 20.3: In 2020, Betelgeuse (red; upper left) was as dim as it has even been seen. (Right panel) It has since brightened. Courtesy of Brian Uttum.

20.4 Novae and supernovae

Occasionally, a star will brighten markedly and unexpectedly. Material falls upon a star and an explosion takes place. This sudden stellar eruption is called a **nova**. In June 2025, two naked-eye novae were visible at once, for only the second time in recorded history.

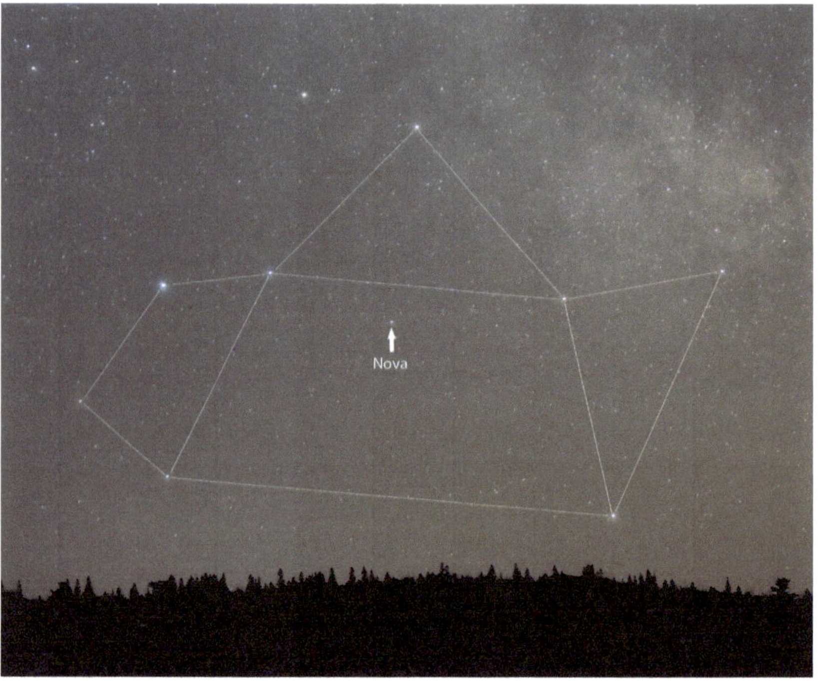

Figure 20.4: This nova took place in the constellation of Sagittarius in 2015. Over the course of several months, it peaked at magnitude 4.3. Courtesy of Bob King.

A nova is interesting. A **supernova** is spectacular.

At the risk of becoming too astrophysical, the energy that stars generate (through nuclear reactions) results in an outward pressure that balances the inward pull of gravity produced by the star's own mass. Once a star has run out of nuclear fuel, reactions cease, and this outward pressure disappears. Self-gravity causes the star to collapse upon itself.

If the star is much more massive than average, the result is a terrific explosion. The star blows up. This is a supernova.

A supernova is much rarer than a nova. Novae may repeat; a supernova is a blazing, one-time phenomenon.

A nearby supernova, over the course of a few hours, may become brighter than all other stars on the celestial sphere put together! There are reports of supernovae by the light of which one could *read* at night.

Figure 20.5: The supernova of 1054. On the left is Holy Roman Emperor Henry III. From ~1450.

Where did that come from?

"Nova" is Latin for "new." Also Latin, the prefix "super-" means "beyond."

Novae and supernovae are not literally new but may be stars previously too faint to see with the naked eye now elevated in brightness so that they become visible. Eventually the phenomenon fades and disappears to the unaided eye.

The earliest record of a naked-eye supernova depiction *might* just be a **petroglyph** found in the 1990s within Chaco Canyon. A petroglyph is an example of rock art in which the image is produced by scratching away patina that has built up on an exposed rockface, revealing a lighter mineral below.

The petroglyph of interest consists of what could be a star image next to what looks very much like the crescent Moon. The Moon in this phase would indeed have approached the supernova of 1054 in the Chaco sky. (We learn about this event from reports made by Chinese astronomers.) Moreover, we know that the Canyon was inhabited in 1054.

Figure 20.6: Chaco Canyon petroglyph purported to represent the supernova of 1054. Courtesy of Bob King.

Critics point out that rock art cannot be dated with any precision. The "supernova" could just as well be something like the planet Venus. Venus and the crescent Moon would have appeared in the sky near each other many times over the age of the Chaco-based civilization.

Notes from long ago
Only seven nearby supernovae have been reliably recorded in all of human history: in the years 185, 393, 1006, 1054, 1181, 1572, and 1604. For the first five cases, we are indebted to Chinese astronomers. They gave us enough information that each of these supernovae, along with a pair from the sixteenth and seventeenth centuries, can now be identified with a *supernova remnant* (an expanding cloud of gas) visible through modern telescopes.

A moderately bright, naked-eye supernova was visible in 1987A. (The "A" means that it was the first supernova of any kind discovered that civil year.) 1987A was not dazzling because it did not occur anywhere in our own Milky Way Galaxy. It appeared in the next galaxy nearest to us. It was much farther away than those previously recorded. Still, the event was a Big Deal.

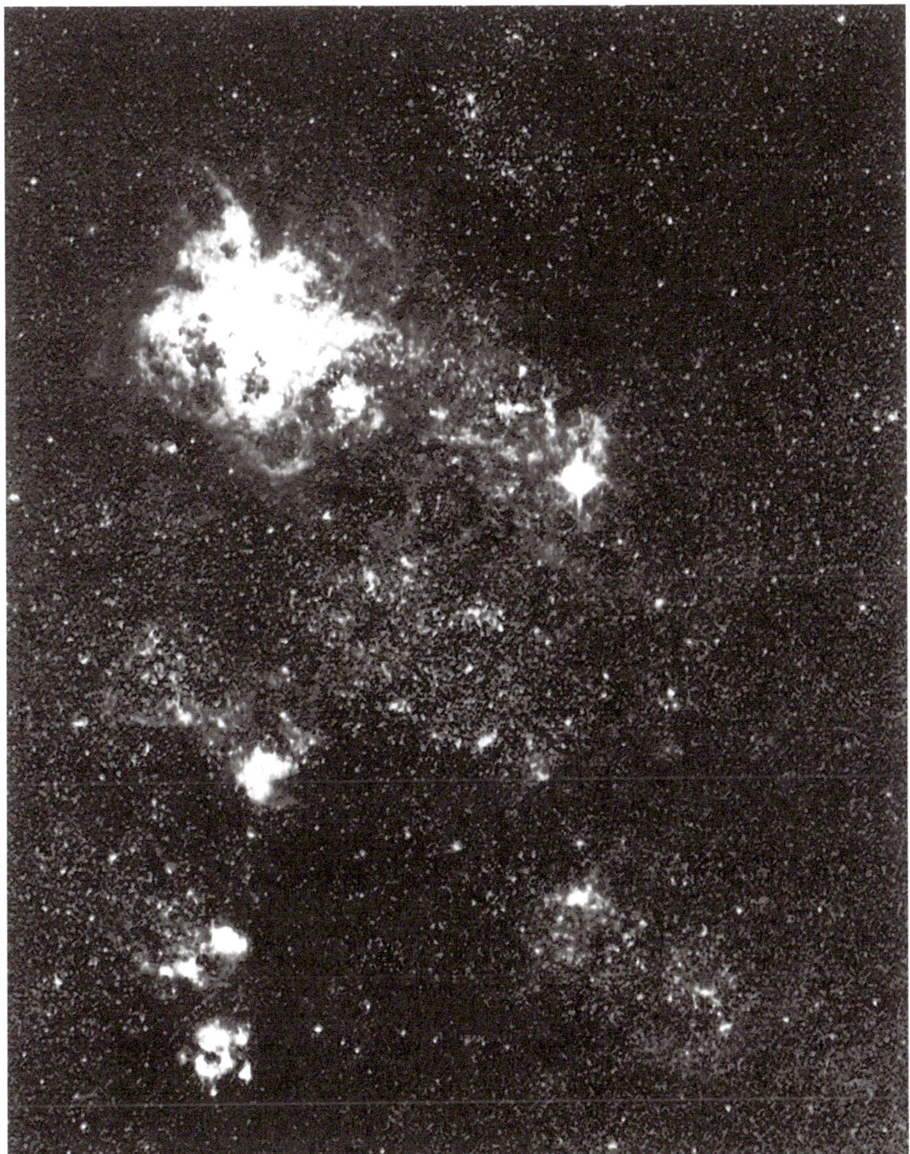

Figure 20.7: Supernova 1987A (top, right) was visible to the naked eye for the better part of a year. It peaked in brightness at magnitude 3.0. Courtesy of ESO.

Back in 1885, a supernova in the more-distant Great Andromeda Galaxy shown as a distinct star amid this faint, unresolved mass of stars. Still more distant supernovae, in other galaxies, may become visible to the naked eye, whereas their host galaxies are not. You then see the same sort of sight as when an invisible star goes nova and becomes visible.

> **Notes from cultures around the world**
>
> Sunspots. Comets. Novae. Change. Most reports from antiquity come from the East (i.e., China). Why? These phenomena appear in every sky.
>
> In the West, the tradition was long held that the sky was the home of perfection. Of immutability. What if something is observed that produces cognitive dissonance with belief? Remember the "demon" Algol?
>
> One solution, which restores mental equilibrium, is to ignore. And certainly not to record in writing.

20.5 Deep-sky objects

Objects beyond the solar system, yet with apparent angular extent, are called **deep-sky objects**. These will include open star clusters, globular clusters, nebulae,[1] and – beyond our galaxy – other galaxies.

Magnitudes are designed for point sources of light (i.e., stars). For this reason, it is more difficult to assign an apparent magnitude to extended objects.

20.5.1 Open star clusters and nebulae

Open star clusters were introduced earlier, in the context of star grouping often mis-identified as constellations or asterisms. There are naked-eye open clusters, though, clearly distinct from the constellation borders inside of which they technically fall.

An example is the Double Cluster in Perseus, *two* clusters side by side. Both contain a seemingly uncountable number of faint stars.

Another example of a naked-eye open cluster is the Praesepe[2] (or Beehive Cluster) in the constellation Cancer. This open cluster is of literary interest inasmuch as it is mentioned in the Hellenistic Greek poet Aratus's famous poem, *Phenomena* [c. 315 BC/BCE].

The center of the Big Dipper asterism is actually an open star cluster. Because it is the nearest such cluster to the Earth, it appears much more spread out than other open star clusters. Indeed, it is not readily recognizable as one.

---

1 "Nebulae" is plural, "nebula" is singular.
2 or Nephelion, meaning "little cloud" in Latin

Figure 20.8: Double Cluster in Perseus, which individually go by the misnomers h Persei and χ Persei. Courtesy of Jonas Thomén.

Open star clusters are scientifically interesting because they are newly created stars. It is thought that they formed together out of a cloud of interstellar dust and gas called a **nebula**.[3] They will eventually disperse.

Dark nebulae came up in Chapter 4's discussion of the Milky Way. Ordinarily, it would be impossible to see a black cloud against a starless sky. We miss many this way.

It is possible to see a *bright* nebula, though. It glows by virtue of the light produced by stars that formed within it. Alternately, leftover dust scatters starlight and makes the nebula visible. Yet the open star clusters themselves may be difficult to pick out.

20.5.2 Globular star clusters

Very different from open clusters are **globular clusters**. These are old stars, so close together that gravity maintains the cluster for the entire lives of the stars within. (They are not "open.")

---

3 Latin for "mist," "vapor," "fog," "smoke," or "exhalation"

Figure 20.9: The stars in open clusters tend to be about the same physical distance from each other. The Praesepe stars look far apart, compared to those of most open star clusters, because this grouping is comparatively near to us. Courtesy of Alan Tough.

The best example, based upon visibility, is Omega Centauri. It contains millions of stars!

The stars in this globular cluster are so densely packed (hence the name), and so far away, that – except from around the periphery – individual stars cannot be resolved. This cluster looks like a faint disk or (with imagination) a sphere, only half a degree across. The Omega cluster is best seen from the southern hemisphere.

Misconception alert

With the star-like name, Omega Centauri, it is easy to conclude that this object is some unusual type of individual star. That is exactly how it got its Bayer designation.

It was not until 1687 when Edmund Halley (of comet fame) had a "lightbulb" moment and realized that a star should not be fuzzy. However, by this time, and without knowledge of what the non-stellar object actually was, it was decided that changing its name would only result in confusion.

There are smaller, but better placed, such clusters for observers in the northern hemisphere. The Great Hercules Cluster is the best example. However, even though it is the "Great" globular cluster, excellent observing conditions are required to make out its angular extent with the naked eye. It is easily mistaken for an ordinary star. The

Figure 20.10: The few stars that shine through the Coalsack Nebula (or Black Magellanic Cloud) are reddened, giving the whole nebula a reddish appearance just short of absolute black. Courtesy of S. Brunier/ESO.

same goes for the Ptolemy (Globular) Cluster in Scorpius and the – straightforwardly named! – Sagittarius (Globular) Cluster.

20.5.3 Galaxies

What about the naked-eye view of the truly distant Universe? Galaxies of stars, home to all the kinds of objects described thus far, are the building blocks of the visible universe.

Somewhat surprisingly, the unresolved stars of your home Milky Way Galaxy are not the most distant you can see without optical aid. From a dark location, on a moonless night, look in the direction of the constellation Andromeda. You will see a tiny smudge of light, not point-like, as you would expect from a single star.

Figure 20.11: The Orion Nebula is equivalent in brightness per unit area to a magnitude 4 star. Courtesy of S. Brunier/ESO.

This is, in fact, another galaxy of stars, *beyond* the Milky Way. It is the Great Andromeda Galaxy. It contains even more stars than the Milky Way Galaxy and is the largest number of stars (though unresolved) that you can see in a single view. The light from hundreds of billions of stars combine to make the Andromeda Galaxy visible to most everybody's naked eye. Notwithstanding, it was not mentioned in writing until the Persian astronomer Abd al-Rahman al-Sufi did so in 964.

Observers in the southern hemisphere can easily see two *satellite* galaxies of the Milky Way Galaxy because they are close by – on the scale of galaxies. They are the amorphous (and misnamed) Large and Small Magellanic Clouds. (The former is home to the 1987 supernova.) For many antipodeans, they are circumpolar.

These "star clouds" do not appear to be radially symmetric disk galaxies like Andromeda and the Milky Way, and they contain a great deal fewer stars than either. The Large and Small Magellanic Clouds only happen to be situated near each other in the sky; the first adjectives in their names refer to their apparent sizes.

Figure 20.12: Omega Centauri. Courtesy of ESA.

The Africans who live near the lakes that feed the Nile River used to use the Magellanic Clouds for calendrical purpose. During their dry season, the Clouds line up north-south in the evening sky. During their wet season, they line up east-west.

The Andromeda Galaxy, both Magellanic Clouds, and our own home Milky Way Galaxy, are all part of a small cluster of galaxies named the Local Group. So is the less conspicuous Triangulum Galaxy.

Like those other nearby galaxies, beyond the Milky Way but within the Local Group, the Triangulum Galaxy makes it onto our naked-eye visibility list. Andromeda can be spotted because it benefits from the luminosity of a huge number of stars. The Magellanic Clouds do because they are in orbit about, and never stray far from, our Milky Way Galaxy.

Neither of these circumstances hold true for the Triangulum Galaxy. In fact, this galaxy in the constellation of Triangulum is rather insignificant as galaxies go. Still, it is *this* whisp of light in the sky that is the most distant object, with even a hint of apparent shape, that can be seen by acute naked eyes.

20.5.4 The scope of naked-eye astronomy is enormous

What about *beyond* the Local Group? Here we join the quest for a "holy grail" of naked-eye astronomy. You see, at most, the only thing that you and a galaxy like this would share is that you and it are both part of the same – not cluster of galaxies,

Figure 20.13: The Andromeda Galaxy is thought to have a spiral shape. Ignoring brightness and resolution, Andromeda happens to present itself to us edge on. Unfortunately, under no circumstance can the spiral pattern in its perpendicular disk be made out with the naked eye. Courtesy of S. Brunier/ESO.

but – *super*cluster of galaxies. Superclusters are the largest structures in the universe (as far as we can tell).

The galaxy Centaurus A is an example of such a far-off galaxy. It is situated in another cluster of galaxies separate from the Local Group. It is fantastically luminous. Centaurus A generates more light than can be accounted for by its (admittedly prodigious) number of stars.

Centaurus A's luminance is inflated by the presence of a tremendous mass in its middle. While itself invisible – even the best telescope fails to make this component appear – an exceptional amount of material nevertheless falls into it. That plummeting matter emits an incredible amount of electromagnetic energy as it does so, including visible light.

What is the mysterious central mass at the heart of Centaurus A? If you were to see Centaurus A, you would be looking toward a supermassive black hole!

Figure 20.14: The Magellanic Clouds as seen in Australia. Courtesy of Ed Dunens.

Because of its ultra-high luminosity, even though Centaurus A is the most distant of any object we have discussed, there have been rumors of its naked-eye visibility (again, far to the South). A credible claim for this dimmer-than sixth-magnitude object is made by contemporary astronomer Stephen O'Meara.

Who will be next? Will it be you?

Notes from yesterday and today

What is the faintest astronomical object (or objects) visible to the naked eye? There are two good candidates and, surprisingly, they are not deep-sky objects.

Polish astronomer Kazimierz Kordylewski noted that the Earth's and Moon's gravity should balance each other in such a way that material 60° ahead of or trailing the Moon in its orbit will be trapped there, at least temporarily. We know of such material; it is interplanetary dust. Beginning in the 1950s, Kordylewski looked for marginally denser-than-average clouds of dust at these positions.

In a lovely illustration of how theories work, Kordylewski eventually found the ever-so-slight brightenings, for which he was looking, moving against the stellar background at the pace of the Moon. He used a telescope of course.

However, other astronomers argued for a long time about their existence and whether a sight-gifted individual could see the eponymous *Kordylewski Clouds*[4] under ideal conditions. The matter was not decided in the affirmative until 2018!

---

4 or Ghost Moons

20.6 The Universe: is your ability to see the sky pre-ordained?

Probably the most obvious thing about the sky is that, when the Sun sets, it turns black. There are individual sources of light within it. But even in a clear night's sky, most of what you see, or (depending upon how you think about it) do *not* see, is utter blackness. To state the obvious, your view of the universe, the sky, appears to be bright objects "painted on a dark canvass," rather than dark objects "painted on a light canvas."

20.6.1 Olbers's paradox

"Why is the night sky black?" It seems like a very simple question. Yet it even has a name, **Olbers's paradox.**

Your first answer to this innocuous query (at first glance) reasonably might be, "Why not?" The black is merely the result of emptiness, right? If a particular direction in the sky is dark, that is, because there is nothing in that direction to produce light.

It is true that, throughout history, most people have not bothered to ask why the night sky is black. It just is. This blackness serves as a backdrop on which we see the discrete celestial objects that interest us.

Yet recently the question posed at the beginning of this section has become more interesting.

Our blackness theory only works if the universe is finite. The traditional example is standing in a woodland. If it is a copse, in some directions your view runs into a tree. But in other directions you see an ambiguous view coming from spaces between the trees.

If you are in the midst of a *forest* – very big, or of imaginary infinite extent! – in every direction in which you look, your view runs into a tree. That tree may be far or near, but, notwithstanding, it and all the others prevent seeing beyond. All you see is trees.

In this analogy, the trees are sources of light. Most typically in your sky, these are stars.

It is understandable that in a small, finite universe, you will see between the stars into – well – nothing. It does not get much darker than that.

Here is the problem: We now know that the Universe is terrifically large. In fact, there is no evidence that it has any boundary ("end" or "edge") at all. It may go on forever!

This means that in any direction you look, you should see the light of some star populating the Universe. The result should be the opposite of what you witness in reality. The sky should be ablaze with starlight. It should be closer to white than black. Indeed, every bit of it should be, on average, the brightness of the *Sun*.

Figure 20.15: If the trees end, you see gaps between them (left). If they do not, every line of sight intersects a tree (right). Photographs by the author.

This logical dilemma is named after the nineteenth-century astronomer, Heinrich Olbers, who supposedly first posited it. It is a historical curiosity that, in astronomy, many things are named after the wrong person. It turns out that Olbers was not the first person to worry about the dark of night. Regardless of what you call it, how do you escape Olbers's paradox?

20.6.2 Do more objects in the Universe emit or absorb light?

Maybe there are far more objects in the Universe, which not only do not produce light but absorb it, than there are stars. This material need not be exotic nor discrete; it could be dust. If you put enough such material into the universe, it obscures distant stars. You only see nearby ones.

The result is insufficient visible stars to turn the sky bright. You see what you really do see: a certain number of stars against a black background (made up in large part by dark dust).

Here is the catch, why this possible solution to Olbers's paradox fails. For the above to work, it is necessary to "pour" so much dust into the universe that you cannot see your own Sun.

You *can* see the Sun. Observable fact trumps theory. You need some other explanation for Olbers's paradox.

Maybe it is the other way around. Maybe there are just not enough stars.

Astronomers have ways of counting the stars. There are plenty. That is not the solution to Olbers's paradox, either.

Of course, the stars could be arranged in just such a way so as to produce "gaps," but why? That argument relies upon some very special pleading.

20.6.3 Olbers's paradox points toward a description of the entire universe!

This next potential key to the puzzle comes from cutting-edge astrophysics. Astronomers have discovered that the Universe is *expanding.*

While the number of objects in the Universe stays more or less the same, the volume they occupy grows with time. A result is that light from distant objects like stars gets altered; it becomes radiated heat instead of visible light.

You cannot see radiated heat. Thus, the light from these stars does not contribute to the brightness of your sky.

This answer to Olbers's paradox is partially correct. The trouble is that the heat nevertheless reaches us. The sky is not hot. There still must be more going on in order to "fix" the paradox.

What appears to be the ultimate answer to Olbers's paradox is rather amazing. The Universe has not been around forever. The expansion/creation started 14 billion years ago.

Light travels at a finite speed. Light from stars far away has not reached us yet. They have not yet contributed to an imagined whiteout.

If nothing – including, of course, stars – existed 14 billion years ago (time equals zero), you cannot hope to see anything that corresponds to a distance equivalent to a light travel time of, say, 14 billion years. Its direction would appear vacant (black) to you.

You started with something that seems very obvious. It gets dark at night. You end with observational evidence in support of one of the major scientific discoveries of the twentieth century: the Big Bang theory for the origin of the universe!

20.7 Chapter summary

Stars deserve another look. There are stars that revolve about each other. There are stars that change in brightness. There are stars that erupt (novae), and there are stars that blow up (supernovae). A nearby supernova would be an extraordinary sight, but no one has seen one for over 400 years.

Stars also group. Open clusters contrast with globular clusters. The largest stellar structure visible to the naked eye is a galaxy.

There is also material between the stars. In some cases, it is visible to the naked eye as a nebula.

The sky is so busy that it is surprising that not every bit of it is occupied by a source of light. This is Olbers's paradox. At first, it sounds trivial; on further consideration, it is evidence in favor of our modern cosmology. Thus, we end where we began, in search of our place in the Universe.

20.8 Chapter review

1. It is easier to see that Delta Scuti is a variable star than it is to see that Betelgeuse is a variable star. Why?
2. Both a pulsating variable star and an eclipsing binary star appear as a single star in your sky. What is the difference between the two?
3. Why do most accounts of changes in the night sky from antiquity come from eastern civilizations? Given this, why do you suppose Algol was known in the West as the "Demon Star?"
4. What is the difference between a nova and supernova? If a nova and supernova are at the same distance from the Earth, why is the supernova so much brighter than the nova?
5. Match the deep sky object on the left with its description on the right.

Open star cluster	Collection of perhaps billions of stars; the biggest example can be seen in the direction of the constellation Andromeda
Nebula	Compact group of hundreds of thousands of stars
Globular star cluster	Cloud of dust and gas between the stars
Galaxy	Loose collection of young stars

6. What is a nebula? How does this relate to a supernova?
7. What makes the Big Dipper part of an open star cluster?
8. Why are the Magellanic Clouds and the Andromeda Galaxy easier to see with the naked eye than the Triangulum Galaxy?
9. What is Olbers's paradox?
10. What are possible solutions to Olbers's paradox? Evaluate these possible solutions and explain why it is mostly dark at night.

Afterword

This has been a book of the possible. The intent was to explore the phenomena you might find yourself part of, and celestial objects you might see, with the naked eye. Whether you actually do or not is almost beside the point. It is just as you may never visit all the places described while reading a geography text; in both cases, the author hopes that you are nonetheless richer for the experience.

This book extolled the virtues of naked-eye astronomy. However, that is not to claim that astronomy practiced in this way is exclusively virtuous. There is no question that simple optical aid, such as opera glasses or generic binoculars, will gather more light and provide greater magnification than your eyes alone.

Many books are written entirely on the topic of using a telescope, and this is as it should be. Here, the philosophy has been that the instrument used to observe should not risk overshadowing the subject of observation. But that risk is only for the beginning skywatcher or armchair astronomer.

Thank you for reading this book first. Where you travel to now in your mind's eye is up to you.

https://doi.org/10.1515/9783111441245-021

Appendices

I The most dependable of clocks

Water clocks dry up. Sandglasses run out. Spring or weight-driven clocks are over- and underwound, and electrical clocks suffer from power outages.

Only a *sundial* cannot be stopped. It was invented by the Chinese at least 2,000 years ago.

A *gnomon* is at the heart of every sundial. The gnomon is home to a shadow-casting edge called a *style*. It is placed perpendicular to (or, less often, another orientation with respect to) a disk below it.

That disk is the *dial* and is most often marked with units of hours. As the Sun appears to move on your sky dome, the style's shadow moves as projected onto the dial. This is marked with hours of the day. However, such a sundial may not produce hours of equal duration nor angular separation.

Tilting the gnomon toward the celestial pole, so that the style's angle with respect to the horizon equals the sundial's latitude, mostly solves this problem. The hours from sunrise to sunset are of equal duration. The marks on the dial representing these hours are angularly symmetric.

Figure A.1: Morehead Planetarium Sundial, University of North Carolina. Courtesy of William Yeung.

There are more satisfactory versions of the sundial. For instance, aligning the gnomon parallel to the axis mundi introduces better proportionality.

https://doi.org/10.1515/9783111441245-022

There are sophisticated dials that are not flat but, instead, a segment of a hollow cylinder. The style is a wire through the axis of this cylinder. The angular intervals between hours, marked radially on the inside of the cylinder, are of equal duration. They are also of equal angular separation.

Of course, a sundial keeps track of real Sun time. The time it shows must be corrected using the equation of time.

The sundial was invented in the northern hemisphere. In this hemisphere, the Sun works its way across your southern sky during the course of the solar day. Which way will the shadow of the sundial's style appear to turn, from morning to afternoon? Clockwise.

The hands of our mechanical clocks could just as well turn in the opposite direction. They would function as well. The direction we identify as clockwise (the same direction in which a person screws on a jar lid) was chosen because that was the direction of the most familiar of circular motion associated with timekeeping, that of a sundial. Any other clock was *expected* to move in the same direction as sundial "moves."

With no moving parts, the sundial remains the most dependable of clocks. Unless it is cloudy . . .

II Selected heliacal star sets and rises

Dates are approximate.

Gamma Pegasi	13 March	6 April
Alpha Piscium	29 March	27 May
Alpha Arietis	12 April	9 May
Alpha Ceti	12 April	11 June
Êta Tauri	1 May	9 June
Alpha Tauri	10 May	20 June
Beta Orionis	7 May	14 July
Gamma Orionis	13May	10 July
Beta Tauri	27 May	26 June
Epsilon Orionis	13 May	16 July
Alpha Orionis	23 May	14 Julv
Alpha Canis Majoris	21 May	7 August
Alpha Geminorum	28 June	18 July
Alpha Canis Minoris	13 June	5 August
Beta Geminorum	30 June	24 July
Alpha Leonis	25 July	4 September
Beta Leonis	2 September	25 September
Alpha Virginis	13 September	29 October
Alpha Boötis	5 November	22 October
Alpha Librae	10 October	21 November
Alpha Serpentis	17 November	21 November
Alpha Scorpii	6 November	16 December
Alpha Herculis	17 December	4 December
Alpha Ophiuchi	19 December	10 December
Sigma Sagittarii	11 December	23 January
Alpha Aquilae	17 January	11 January
Beta Capricorni	7 January	8 February
Epsilon Pegasi	7 February	13 February
Alpha Aquarii	6 February	3 March
Alpha Pegasi	27 February	9 March

Note: After Maunder, E. Heliacal risings and settings of stars. Observatory 1906, 29, 364–365.

https://doi.org/10.1515/9783111441245-023

III The constellations

The International Astronomical Union (IAU) has done a great service for the astronomy-education community by producing a set of attractive charts, one for each IAU-defined constellation. The project was managed by Lars Lindberg Christensen and developed by Gurvan Bazin. Raquel Yumi Shida was Head of the Web (development) Team. Administration was provided by Maria Rosaria D'Antonio with Database Management by Madeleine Smith-Spanier. The European Southern Observatory hosts the online version.

https://web.archive.org/web/20201216152749/https://www.iau.org/public/themes/constellations/

https://doi.org/10.1515/9783111441245-024

IV The Greek alphabets

α	Alpha
β	Beta
γ	Gamma
δ	Delta
ε	Epsilon
ζ	Zêta
η	Êta
θ	Thêta
ι	Iota
κ	Kappa
λ	Lambda
μ	Mu
ν	Nu
ξ	Xi
ο	Omikron
π	Pi
ρ	Rho
σ, ς	Sigma
τ	Tau
υ	Upsilon
φ	Phi
χ	Chi
ψ	Psi
ω	Omega

https://doi.org/10.1515/9783111441245-025

V Stars that rise and set at mid-northern latitudes

Latitude 30° N is assumed. The number following the star is its apparent magnitude. Stars are listed to magnitude 2.50. Epoch = 2000.

Alpha Canis Majoris	Sirius	−1.44
Alpha Carinae	Canopus	−0.62
Alpha Boötis	Arcturus	−0.05
Alpha Lyrae	Vega	0.03
Alpha Aurigae	Capella	0.08
Beta Orionis	Rigel	0.18
Alpha Canis Minoris	Procyon	0.40
Alpha Eridani	Achernar	0.45
Alpha Orionis	Betelgeuse	0.45
Alpha Aquilae	Altair	0.76
Alpha Tauri	Aldebaran	0.87
Alpha Virginis	Spica	0.98
Alpha Scorpii	Antares	1.06
Beta Geminorum	Pollux	1.16
Alpha Piscis Austrini	Fomalhaut	1.17
Alpha Cygni	Deneb	1.25
Alpha Leonis	Regulus	1.36
Epsilon Canis Majoris	Adhara	1.50
Alpha Geminorum	Castor	1.58
Gamma Crucis	Gacrux	1.59
Lambda Scorpii	Shaula	1.62
Gamma Orionis	Bellatrix	1.64
Beta Tauri	Elnath	1.65
Epsilon Orionis	Alnilam	1.69
Gamma Grucis	Alnair	1.73
Zêta Orionis	Alnitak	1.74
Gamma Velorum	Regor	1.75
Epsilon Ursae Majoris	Alioth	1.76
Epsilon Sagittarii	Kaus Australis	1.79
Alpha Persei	Mirfak	1.79
Delta Canis Majoris	Wezen	1.83
Êta Ursae Majoris	Alkaid	1.85
Thêta Scorpii	Sargas	1.86
Beta Aurigae	Menkalinan	1.90
Gamma Geminorum	Alhena	1.93
Delta Velorum	Koo She	1.93
Alpha Pavonis	Peacock	1.94
Beta Canis Majoris	Mirzam	1.98
Alpha Hydrae	Alphard	1.99
Gamma Leonis	Algieba	2.01
Alpha Arietis	Hamal	2.01
Beta Ceti	Diphda	2.04
Sigma Sagittarii	Nunki	2.05

https://doi.org/10.1515/9783111441245-026

(continued)

Thêta Centauri	Menkent	2.06
Beta Gruis	Al Dhanab	2.07
Alpha Andromedae	Alpheratz	2.07
Beta Andromedae	Mirach	2.07
Kappa Orionis	Saiph	2.07
Alpha Ophiuchi	Rasalhague	2.08
Beta Persei	Algol	2.09
Gamma Andromedae	Almach	2.10
Beta Leonis	Denebola	2.14
Gamma Centauri	Muhlifain	2.20
Iota Carinae	Aspidiske	2.21
Zêta Puppis	Naos	2.21
Alpha Coronae Borealis	Alphecca	2.22
Zêta Ursae Majoris	Mizar	2.23
Gamma Cygni	Sadr	2.23
Lambda Velorum	Suhail	2.23
Gamma Draconis	Eltanin	2.24
Alpha Cassiopeiae	Schedar	2.24
Delta Orionis	Mintaka	2.25
Beta Cassiopeiae	Caph	2.28
Epsilon Centauri	Nán Mén Yī	2.29
Delta Scorpii	Dschubba	2.29
Epsilon Scorpii	Wei	2.29
Alpha Lupi	Men	2.30
Êta Centauri		2.33
Beta Ursae Majoris	Merak	2.34
Epsilon Boötis	Izar	2.35
Epsilon Pegasi	Enif	2.38
Kappa Scorpii	Girtab	2.39
Alpha Phoenicis	Ankaa	2.40
Gamma Ursa Majoris	Phecda	2.41
Êta Ophiuchi	Sabik	2.43
Beta Pegasi	Scheat	2.44
Êta Canis Majoris	Aludra	2.45
Kappa Velorum	Markeb	2.47
Epsilon Cygni	Gienah	2.48
Alpha Pegasi	Markab	2.49

VI Northern circumpolar stars

Latitude 30° is assumed. The number following the star is its apparent magnitude.
Stars are listed to magnitude 3.10. Epoch = 2000.

Alpha Ursae Majoris	Dubhe	1.81
Alpha Ursae Minoris	Polaris	1.97
Beta Ursa Minoris	Kochab	2.07
Gamma Cassiopeiae	Cih	2.15
Alpha Cephei	Alderamin	2.45
Delta Cassiopeiae	Ruchbah	2.66
Êta Draconis	Aldibain	2.73
Gamma Ursae Majoris	Pherkad (Major)	3.00
Delta Draconis	Tais	3.07

https://doi.org/10.1515/9783111441245-027

VII Southern circumpolar stars

Latitude 30° S is assumed. The number following the star is its apparent magnitude.
Stars are listed to magnitude 3.10. Epoch = 2,00.

Alpha Centauri	Rigil Kentaurus	−0.27
Beta Centauri	Hadar	0.61
Alpha Crucis	Acrux	0.87
Beta Crucis	Mimosa	1.25
Beta Carinae	Miaplacius	1.67
Epsilon Carinae	Avior	1.86
Alpha Trianguli Australis	Atria	1.91
Alpha Muscae		2.69
Thêta Carinae		2.74
Beta Hydri		2.82
Beta Trianguli Australis		2.83
Alpha Hydri		2.86
Alpha Tucanae		2.87
Gamma Trianguli Australis		2.87
Upsilon Carinae		2.92
Beta Muscae		3.04

https://doi.org/10.1515/9783111441245-028

VIII Zodiac constellations used in astrology

The listing is in ecliptic coordinates, beginning with the historic First Point of Aries. These virtual constellations are construed so that each takes up an equal band of the ecliptic. They do not conform to the borders used by the IAU.

Aries	0–30°
Taurus	30–60°
Gemini	60–90°
Cancer	90–120°
Leo	120–150°
Virgo	150–180°
Libra	180–210°
Scorpio	210–240°
Sagittarius	240–270°
Capricorn	270–300°
Aquarius	300–330°
Pisces	330–360°

Note: After Clemency, Montelle. *Chasing Shadows: Mathematics, Astronomy, and the Early History of Eclipse Reckoning*. Baltimore: The Johns Hopkins University Press (2011).

https://doi.org/10.1515/9783111441245-029

IX Northern evening constellations by season

Epoch = 2000.

Constellations of the winter sky

Andromeda	Gemini
Aries	Horologium
Auriga	Lacerta
Caelum	Leo Minor
Camelopardalis	Lepus
Cancer	Lynx
Canis Major	Monoceros
Canis Minor	Orion
Carina	Pegasus
Cassiopeia	Perseus
Cepheus	Pisces
Cetus	Taurus
Columbia	Triangulum
Draco	Ursa Major
Eridanus	Ursa Minor
Fornax	

Constellations of the spring sky

Antilla	Gemini
Auriga	Hercules
Boötes	Hydra
Camelopardalis	Leo
Cancer	Leo Minor
Canes Venatici	Lynx
Canis Major	Monoceros
Canis Minor	Orion
Cassiopea	Perseus
Cepheus	Puppis
Coma Berenices	Pyxis
Corona Borealis	Sextans
Corvus	Ursa Major
Crater	Ursa Minor
Draco	Virgo

https://doi.org/10.1515/9783111441245-030

Constellations of the summer sky

Aquila	Lacerta
Boötes	Leo
Camelopardalis	Leo Minor
Canes Venatici	Libra
Capricornus	Lynx
Cassiopea	Lyra
Cepheus	Ophiuchus
Coma Berenices	Sagittarius
Corona Borealis	Scorpius
Corvus	Scutum
Cygnus	Serpens
Delphinus	Virgo
Draco	Vulpecula
Hercules	

Constellations of the fall sky

Andromeda	Hercules
Aquarius	Lacerta
Aquila	Lyra
Aries	Microscopium
Auriga	Perseus
Camelopardalis	Pisces
Capricornus	Pisces Australis
Cassiopeia	Sculptor
Cepheus	Scutum
Corona Borealis	Serpens
Cetus	Triangulum
Cygnus	Ursa Major
Delphinus	Ursa Minor
Draco	Vulpecula
Equuleus	

X Year, month, and day lengths

Year lengths

Eclipse year = 346 days, 14 h, 54 min
Lunar year = 354 days
Perfect year = 360 days
Sothic year = 365 days
Gregorian year = 365 days, 5 h, 48 min
Solar year = 365 days, 5 h, 48 min, 46 s
Julian year = 365 days, 6 h
Sidereal year = 365 days, 6 h, 9 min
Anomalistic year = 365 days, 6 h, 14 min

Month lengths

Nodical month = 27.21222 days
Sidereal month = 27.32166 days
Anomalistic 27.55455 days
Synodic month = 29.53059 days

Day lengths

Sidereal day = 23 h, 56 min, 4 s
Solar day = 24 h

https://doi.org/10.1515/9783111441245-031

XI Properties of naked-eye planets, satellites, and asteroids

The Earth is included for reference. Revolution periods are about the Sun, except for the Moon, which is about the Earth.

d = days

	Mercury	Venus	Earth	Moon	Mars	Vesta	Jupiter	Saturn	Uranus
Sidereal period of revolution	88.0^d	224.7^d	365.256^d	27.322^d	687.0^d	1,325.8^d	4,331^d	10,747^d	30,589^d
Synodic period of revolution	115.9^d	583.9^d	365.242^d	29.531^d	779.9^d	463.9^d	780^d	378^d	370^d
Orbital inclination to the ecliptic	7.0°	3.4°	0.0°	5.1°	1.8°	7.14°	1.3°	2.5°	0.8°
eccentricity	0.206	0.007	0.017	0.055	0.094	0.089	0.049	0.052	0.047

Note: Source: NASA

https://doi.org/10.1515/9783111441245-032

XII Apparent magnitudes of the planets[1]

This table assumes that the planet is not invisible due to conjunction. The maximum value for Saturn includes optimal ring tilt; the minimum value assumes edge-on rings, not contributing to the brightness of the planet.

		Arcus Visionis (for Heliacal Rise)[2]
Mercury	−1.9 to +3.8	13.0°
Venus	−4.7 to −3.5	5.7°
Mars	−2.9 to +1.8	14.5°
Jupiter	−2.9 to −1.6	9.3°
Saturn	−0.5 to +0.9°	13.0°
Uranus	+5.5 to naked-eye invisibility	
Neptune	Invisible to the naked eye	

1 After The naked eye planets in the night sky. Martin J. Powell, 2025. (Accessed March 9, 2025, at NakedEyePlanets.com.)

2 From Neugebauer, P. Tafeln zur berechnung der jährlichen auf- und untegänge der planeten. Astronomische Nachrichten 1938, 264, 6331, 313–322.

https://doi.org/10.1515/9783111441245-033

XIII   Selected annual meteor showers

Meteors should be visible for a few days before and after a shower's peak date.

	Peak date	Estimated hourly meteor count at peak	Meteoroid source (if known)
Quadrantids	3 January	30–60	
Lyrids	22 April	15–20	Comet Thatcher
Southern Delta Aquarids	28 July	15–20	
Perseids	12 August	80–100	Comet Swift-Tuttle
Orionids	22 October	20–25	Comet Halley
Leonids	17 November	15–20	Comet Temple-Tuttle
Geminids	14 December	50–100	Asteroid 3200 Phaethon

Source: NOAO.

https://doi.org/10.1515/9783111441245-034

XIV Selected naked-eye optical double stars

The order is in the increasing magnitude of the brightest star. Subscripts distinguish between the two stars. Each pair is separated by more than 5 arc minutes, but less than 30 arc minutes. Angular separation, in addition to magnitude, serves as an indicator of difficulty in visually splitting the two stars. Epoch = 2000.

m, apparent magnitude; a, angular separation between the two stars;
RA, right ascension; DEC, declination.

	m_1 and m_2	a	RA	DEC
Zêta Ursae Majoris	2.2 and 4.0	12′	$13^h\ 25^m$	+54 ° 51′
Iota Orionis	2.7 and 4.8	8′	$5^h\ 36^m$	−05° 54′
Mu Scorpii	3.1 and 3.6	6′	$16^h\ 52^m$	−38° 03′
Thêta Tauri	3.4 and 3.8	6′	$4^h\ 29^m$	+15° 58′
Zêta Scorpii	3.6 and 4.7	7′	$16^h\ 54^m$	−42° 21′
Alpha Capricorni	3.7 and 4.3	6′	$20^h\ 18^m$	−12° 32′
Delta Tauri	3.8 and 4.8	18′	$4^h\ 24^m$	+17° 34′
Omicron Cygni	3.9 and 4.8	6′	$20^h\ 13^m$	+46° 44′
Kappa Draconis	3.9 and 4.9	15′	$12^h\ 33^m$	+69° 47′
Omega Scorpii	4.0 and 4.3	15′	$16^h\ 6^m$	−20° 41′
Delta Lyrae	4.3 and 5.6	10′	$18^h\ 54^m$	+36° 54′
Pi Pegasi	4.3 and 5.6	10′	$22^h\ 10^m$	+33° 11′
Alpha Vulpeculae	4.6 and 5.9	7′	$19^h\ 28^m$	+24° 40′
Zêta Corvi	5.2 and 5.9	6′	$12^h\ 21^m$	−22° 18′
Nu Coronae Borealis	5.4 and 5.6	6′	$16^h\ 22^m$	+33° 48′

https://doi.org/10.1515/9783111441245-035

XV Naked-eye recurring novae

y, years; d, days; m, apparent magnitude

	Approximate period	Maximum brightness	Year last seen
T Coronae Borealis[1]	79–80^y	m = 2.0	1946 (some consider it to be overdue as of May 2024)
RS Ophiuchi	15–20^y	m = 4.3	2021
SS Leporis[2]	258^d	m = 4.8	
T Pyxidis (very difficult)	19^y	m = 6.3	1967

[1] or the Blaze Star
[2] Sort a cross between a recurrent nova and a variable star

https://doi.org/10.1515/9783111441245-036

XVI Selected naked-eye variable stars

These stars are in the order of increasing period. The change in brightness is at least half a magnitude. All are in the North Celestial Hemisphere. The longest period is 2 years (with the exception of Betelgeuse), and the shortest at least 2 days. Epoch = 2000.

E, eclipsing; P, pulsating; m, apparent magnitude; ~, about; d, days; y, years; RA, right ascension; DEC, declination

	Name	Type	m	Period	RA	DEC
Beta Persei	Algol	E	2.1 to 3.4	2.9 d	$03^h\,08^m$	+40° 57′
Lambda Tauri	Pectus	E	3.4 to 3.9	4.0 d	$04^h\,00^m$	+12° 29′
Delta Cephei		P	3.5 to 4.4	5.4 d	$22^h\,29^m$	+58° 25′
Êta Aquilae	Tiān Fú Sì	P	3.5 to 4.4	7.2 d	$19^h\,52^m$	+01° 00′
Zeta Geminorum		P	3.6 to 4.2	10.2 d	$07^h\,04^m$	+20° 34′
Beta Lyrae	Sheliak	P	3.3 to 4.4	13.0 d	$08^h\,50^m$	+33° 22′
R Lyrae		P	5.0 to 3.9	46 d	$18^h\,55^m$	+43° 54′
Alpha Herculis	Rasalgethi	P	2.7 to 4.0	~100 d	$17^h\,14^m$	+14° 23′
Êta Geminorum	Propus	P	3.9 to 3.2	~233 d	$06^h\,15^m$	+22° 30′
Gamma Cygni	Mira	P	3.3 to 14.2	407 d	$19^h\,51^m$	+32° 54′
W Boötis		P	4.7 to 5.4	450 d	$14^h\,43^m$	+26° 30′
Mu Cephei	Erakis	P	3.4 to 5.1	2 y	$21^h\,43^m$	+58° 47′
Alpha Orionis	Betelgeuse	P	0.3 to 0.9	~6 y	$05^h\,55^m$	+07° 24′

Note: After Isles, J. The top 12 naked-eye variable stars. Sky & Telescope 2010, 119, 3.

https://doi.org/10.1515/9783111441245-037

XVII Selected naked-eye open clusters

Epoch = 2000

RA, right ascension; DEC, declination.

Name	Constellation	RA	DEC
Ursa Major Cluster	Ursa Major	(central portion of the "Big Dipper" asterism)	
Alpha Persei Cluster	Perseus	$03^h 27^m$	+48° 48′
Baby Scorpion Cluster	Scorpius	$16^h 54^m$	−41° 50′
Butterfly Cluster	Scorpius	$17^h 41^m$	−32° 13′
Coma Cluster	Coma Berenices	$03^h 00^m$	+27° 59′
Double Cluster	Perseus	$02^h 20^m$	+57° 8′
Hyades	Taurus	$04^h 27^m$	+15° 52′
Jewel Box Cluster	Crux	$12^h 54^m$	−60° 22′
Little Beehive Cluster	Canis Major	$06^h 46^m$	−20° 51′
Omicron Velorum Cluster	Vela	$08^h 41^m$	−53° 02′
Pearl Cluster	Centaurus	$11^h 36^m$	−61° 37′
Pleiades	Taurus	$03^h 47^m$	+24° 7′
Praesepe	Cancer	$08^h 40^m$	+19° 59′
Ptolemy Cluster	Scorpius	$17^h 54^m$	−34° 48′
Southern Pleiades	Carina	$10^h 43^m$	−64° 24′
Wishing Well Cluster	Carina	$11^h 06^m$	−58° 40′

Note: With the use of the StarWalk application by Vito Technologies.

https://doi.org/10.1515/9783111441245-038

XVIII Selected naked-eye globular clusters

Epoch = 2000

RA, right ascension; DEC, declination.

	Constellation	RA	DEC
47 Tucanae	Tucana	$00^h 24^m$	−72° 05′
Caldwell 86	Ara	$17^h 41^m$	−53° 40′
Caldwell 93	Pavo	$19^h 11^m$	−59° 59′
M13 (Great Cluster)	Hercules	$16^h 42^m$	+36° 28′
M4	Scorpius	$16^h 24^m$	−26° 32′
M5	Serpens	$15^h 19^m$	+02° 05′
M22	Sagittarius	$18^h 36^m$	−23° 54′
Omega Centauri	Centaurus	$13^h 27^m$	−47° 29′

https://doi.org/10.1515/9783111441245-039

XIX Selected naked-eye nebulae

Nebulae marked with an asterisk are extremely challenging. Epoch = 2000.

RA, right ascension; DEC, declination.

Bright nebulae

	RA	DEC
California Nebula	$04^h\,03^m$	+36° 25′
Carina Nebula	$10^h\,45^m$	−59° 52′
Helix Nebula	$22^h\,23^m$	−20° 50′
Lagoon Nebula	$18^h\,05^m$	−24° 23′
North American Nebula*	$20^h\,59^m$	+44° 32′
Omega Nebula/Swan Nebula*	$18^h\,20^m$	−16° 11′
Orion Nebula	$05^h\,35^m$	−05° 23′
Tarantula Nebula	$05^h\,39^m$	−69° 06′

Dark nebulae

	RA	DEC
Coalsack	$12^h\,50^m$	−62° 30′
Great Rift	$20^h\,37^m$	+43° 32′
Pipe Nebula	$17^h\,27^m$	−26° 56′

https://doi.org/10.1515/9783111441245-040

XX Naked-eye galaxies

Galaxies marked with an asterisk are extremely challenging. Epoch = 2000.

RA, right ascension; DEC, declination.

	RA	DEC
Great Andromeda	$00^h\,43^m$	$+41°\,16'$
Large Magellanic Cloud	$05^h\,24^m$	$-69°\,45'$
Small Magellanic Cloud	$00^h\,55^m$	$-72°\,50'$
Triangulum Galaxy*	$01^h\,31^m$	$+30°\,40'$

https://doi.org/10.1515/9783111441245-041

Further reading

[1]   Barentine J. Uncharted constellations: Asterisms, single-source and rebrands. New York, NY, USA, Springer, 2016.

[2]   Berman, B. Secrets of the night sky: The most amazing things in the universe you can see with the naked eye. New York, NY, USA, William Morrow and Company, 1996.

[3]   Berry, R. Discover the stars: Starwatching using the naked eye, binoculars, or a telescope. Largo, MD, USA, Crown Books, 2010.

[4]   Bobrovnikoff, N. Astronomy before the telescope, I-III. Tucson, AZ, USA, Packart Publishing House, 1984–1990.

[5]   Brody, J. The enigma of sunspots: A story of discovery and scientific revolution. Edinburgh, UK, Floris Books, 2002.

[6]   Clark, D Stephenson, F. The historical supernovae. Oxford, UK, Pergamon Press, 1977.

[7]   Davidson, N. Sky phenomena: A guide to naked-eye observation of the stars. Spencertown, NY, USA, Lindisfarne Books, 2004.

[8]   De Meis, S. Heliacal phenomena: An astronomical introduction for humanists. Milan, Italy, Mimesis International, 2014.

[9]   Dehant, V. Mathews, P. Precession, nutation and wobble of the Earth: Illustrated edition. Cambridge, UK, Cambridge University Press, 2015.

[10] Evans, J. The history and practice of ancient astronomy. Oxford, UK, Oxford University Press, 1998.

[11] Fabian, S. Patterns in the sky: An introduction to ethnoastronomy. Prospect Heights, IL, USA, Waveland Press, 2001.

[12] Falck-Ytter, H. Aurora: The northern lights in mythology, history and science. Hudson, NY, USA, Bell Pond Books, 1985.

[13] Harrison, E. Darkness at night: A riddle of the Universe. Cambridge, MA, USA, Harvard University Press, 1987.

[14] Harvey, R. Night sky: Stargazing with the naked eye. London, UK, Amber Books, 2019.

[15] Henes, D. The Moon watcher's companion: Everything you ever wanted to know about the Moon, and more. New York, NY, Marlow and Company, 2002.

[16] Hetherington, B. A chronicle of pre-telescopic astronomy. Chichester, NY, USA, John Wiley and Sons, 1996.

[17] Hetherington, E Hetherington, N. Astronomy and culture. Santa Barbara, CA, USA, Greenwood Press, 2009.

[18] Hockey, T. The comet Hale-Bopp book. Shrewsbury, MA, USA, ATL Press, 1996.

[19] Judge, M. The dance of time: The origins of the calendar. New York, NY, USA, Arcade Publishing, 2004.

[20] Kaler, J. The ever-changing sky: A guide to the celestial sphere. Cambridge, UK, Cambridge University Press, 1996.

[21] Littman, M. The heavens on fire: The great Leonid meteor storms. Cambridge, UK, Cambridge University Press, 1998.

[22] McCully, J. Beyond the Moon: A conversational, common-sense guide to understanding the tides. London, UK, World Scientific, 2006.

[23] Reed, G. Dark sky legacy: Astronomy's impact on the history of culture. Buffalo, NY, USA, Prometheus books, 1989.

[24] Schaaf, F. Seeing the sky: 100 projects, activities, and explorations in astronomy. New York, NY, USA, John Wiley and Sons, 1990.

[25] Selin, H. Astronomy across cultures: The history of non-western astronomy. Dordrecht, Netherlands, Kluwer Academic Publishers, 2000.

https://doi.org/10.1515/9783111441245-042

[26] Upgren, A. Night has a thousand eyes: A naked-eye guide to the sky, its science, and lore. New York, NY, USA, Perseus Books, 1988.
[27] Westfall, J Sheehan, W. Celestial shadows: Eclipses, transits, and occultations. New York, NY, USA, Springer, 2015.
[28] Whitaker, E. Mapping and naming the Moon. Cambridge, UK, Cambridge University Press, 1999.
[29] Whitfield, P. The mapping of the heavens. San Francisco, CA, USA, Pomegranate Books, 1995.

Glossary

A

Acronical phase either acronical rise or acronical set

Acronical rise rising of a star, planet, constellations, or certain asterisms in the East as the Sun sets in the West

Acronical set setting of a star, planet, constellation, or certain asterisms in the West as the Sun rises in the East

Age of the Moon time since the last New Moon

Airglow extremely faint luminosity of the sky at night due to daytime chemistry in the Earth's upper atmosphere; alternately, phosphorescence caused by the Sun's ultraviolet light acting upon air

Air mass (in astronomy) parameterization of the amount of atmosphere through which you observe an astronomical object

Albedo (in physics) ratio of the amount of light reflected by a nonluminous surface and the amount of light shining upon that surface

Alignment (in astronomy) purposeful arrangement of two or more monuments, or a combination of monuments and a natural landscape feature, such that they will be co-linear with a celestial object

Almucantar parallel of altitude in the horizon-coordinate system

Altitude (in astronomy) minimum angle between the astronomical horizon and a celestial object

Analemma closed curve graphically representing the relationship between a location's equation of time and the Sun's declination

Angular momentum function of inertia and angular velocity, the latter being the speed and axial direction of rotation

Annual (in astronomy) having to do with the revolution of the Earth

Annular eclipse central solar eclipse in which the apparent disk of the Moon does not completely cover the apparent disk of the Sun

Anomalistic referring to the time it takes a solar-system body to travel from its perihelion (or aphelion) to its perihelion (or aphelion) again; alternately, referring to the time it takes a satellite to travel from its perigee (or apogee) to its perigee (or apogee) again

Ante meridiem time from midnight to noon

Anti-circumpolar celestial object never visible above your horizon

Anti-lunar lobe water in one of two tidal bulges made up of the oceanic deficit from the tidal troughs; alternately, half of the Earth's oblong-shaped hydrosphere, created by the tensional force of the Moon and Sun

Antipodean referring to the opposite hemisphere of the Earth; alternately, in the North, often meaning Australia and New Zealand

https://doi.org/10.1515/9783111441245-043

Anti-tail apparent spiked projection from a comet coma, opposite the tail, produced by the view of illuminated cometary material left in the comet's orbit and seen behind it

Anti-twilight arch colored band of refracted sunlight above the shadow of the Earth, seen in the East after sunset (or the West before sunrise)

Antumbra extension of the Moon's umbral cone past its point of convergence; alternately, the shadowing seen during an annular eclipse

Aphelion/aphelic point at which an object, orbiting the Sun, is furthest from the center of the Sun; alternately, the time at which an object, orbiting the Sun, is furthest from the center of the Sun

Apogee/apogean point in an Earth satellite's orbit at which it is farthest from the center of the Earth; alternately, the time at which an Earth satellite is farthest from the center of the Earth

Apparition (in astronomy) continuous interval during which a celestial object is visible in the sky

Appulse (in astronomy) two celestial objects temporarily at their closest to each other

Apside (in astronomy) one of two points farthest or nearest to a focus in an elliptical orbit

Archaeoastronomy/archaeoastronomer study of prehistoric peoples' practices regarding the celestial realm

Arcus visionis angular distance between a star (or planet) from the Sun when it is first (or last) visible above the horizon

Armillary (in astronomy) framework of rings meant to illustrate the medieval view of the Universe; alternately, a framework of rings used to make measurements upon the Celestial Sphere

Ascending node point in the orbit of the Moon at which it travels north through the plane of the ecliptic

Asterism any pattern of stars in the sky, regardless of constellation borders

Asteroid rocky and/or metallic body, smaller than a planet, orbiting the Sun

Astrology/astrological belief system involving prognostication based upon the positions of celestial objects

Astronomical dawn time at which the rising Sun reaches eighteen degrees below the horizon; alternately, the end of night's total darkness

Astronomical dusk time at which the setting Sun reaches eighteen degrees below the horizon; alternately, the beginning of night's total darkness

Astronomical horizon great circle ninety degrees from the zenith

Aurora australis colored light seen in the southern sky, caused by the interaction of solar radiation and molecules of the upper atmosphere

Aurora borealis colored light seen in the northern sky, caused by the interaction of solar radiation and molecules of the upper atmosphere

Averted vision looking slightly away from a faint object in order to expose a more light-sensitive portion of your cornea to it

Axis mundi rotational axis of the Earth

Azimuth/Azimuthal angle eastward from north of a celestial object in the sky

B

Backscatter (in physics) particles or waves reflected in the direction of their source

Backsight first of two objects in your line-of-sight directing your view toward a point in the sky

Baily's beads visual phenomenon seen when only the lowest points on the lunar limb allow photospheric sunlight to pass through, during a total (or nearly total) eclipse of the Sun

Bayer system method for identifying a star by a Greek letter followed by the genitive form of the constellation in which the star is found

Binary star two stars revolving about each other

Black drop effect sunlight is bent around a transiting body as the body appears to contact the limb of the Sun, thereby distorting its projected shape

Blue Moon (in calendrics) second Full Moon in a calendar month; alternately, an instance when the Moon appears blue due to atmospheric conditions

Bolide larger-sized meteoroid that explodes in the atmosphere

C

Cairn purposeful pile of stones made by humans

Callippic cycle time period in which the solar year and synodic month are nearly commensurate

Cardinality/cardinal degree to which an alignment or alignments correspond to the cardinal directions (north, south, east, and west)

Celestial referring to the astronomical sky

Celestial meridian meridian passing through due north and south, as well as through the zenith

Celestial sphere all radial directions away from the center of the Earth

Centerline (of a total solar eclipse) set of points corresponding to the locations of maximum eclipse duration at each longitude along an eclipse path

Central eclipse center of the Moon's apparent disk intersects the center of the Sun's apparent disk as seen from the Earth; alternately, the Moon intersects the axis of the Earth's shadow cone as seen from the Earth

Chromosphere/chromospheric thin layer of plasma just above the solar photosphere; alternately, the red, hot, but also tenuous, portion of the Sun first seen after second contact during a total solar eclipse

Circumpolar (in astronomy) celestial object continually visible above your horizon

Civil calendar (in astronomy) not celestially based; alternately, legal or used in day-to-day affairs

Civil dawn moment the rising Sun reaches altitude −6 degrees

Civil dusk moment the setting Sun reaches altitude 0 degrees

Civil month (in astronomy) not celestially based; alternately, legal or used in day-to-day affairs

Civil twilight period during which the Sun is at altitudes 0 degrees to −6 degrees; alternately, the time between sunset (or sunrise) and nautical dusk (or dawn)

Civil year (in astronomy) not celestially based; alternately, legal or used in day-to-day affairs

Clair obscure high contrast due to the juxtaposition of light and shadow

Color blindness with more nuance called Color Vision Deficiency, it is the impaired ability to see certain or all colors

Colure meridian that intersects a solstice or equinox

Coma (in astronomy) halo of sublimated gas and dust liberated from a comet nucleus, produced by the heat experienced by the comet's approach to the Sun

Comet Solar-system body, smaller than a planet, made principally of ice and gas and further characterized by an eccentric orbit

Comet tail gas molecules and dust grains blown away from a comet coma by a wind of electrically charged particles flowing outward from the Sun

Compass rose circle divided by four, eight, sixteen, or thirty-two equally spaced direction indicators

Conjunction (in astronomy) two astronomical objects close to each other in the sky; alternately, two astronomical objects, at least one of which appears to move on the Celestial Sphere, sharing the same right ascension

Constellation two-dimensional pattern of (usually) proximate stars recognized as unique by an entire culture; alternately, a region on the Celestial Sphere that includes one of 88 exclusive groupings of stars

Contact (in an eclipse) In a solar eclipse, the limb of the Moon appears to intersect a limb of the Sun; alternately, in a total or partial lunar eclipse, the limb of the Moon appears to intersect the edge of the Moon's umbra

Coriolis effect (in physics) apparent deflection of a linear-moving object when viewed from within a rotating reference frame, such as the surface of the Earth

Cornea/corneal transparent, outermost layer of the eye, which protects the anatomy within

Corona/coronal broad, outermost layer of the Sun; alternately, the white, very hot, but also very tenuous, irregular portion of the Sun seen during a total solar eclipse

Cosmology/cosmologist study of the physical Universe

Crescent Moon lunar phase that you see when the illuminated hemisphere of the Moon is mostly pointed away from you; alternately, the lunar phase that you see when the Moon's terminator curves in the direction of the lunar limb

Cross-quarter (day) referring to dates equally distant in time from each sequential pair of Solstices and Equinoxes; alternately, referring to the location of the Sun on the Celestial Sphere when it is equally distant in ecliptic longitude from the points corresponding to each sequential pair of Solstices and Equinoxes

Culmination (in astronomy) diurnal time (or location) when a celestial object is at its highest altitude in your sky; alternately, for a star, its location at the time it transits the celestial meridian

Cusp (in astronomy) location on the apparent disk of the Moon where the terminator intersects the limb

Cylindrical projection mapping projection of a sphere onto a concentric cylinder

D

Danjon limit minimum angle between the Sun and Moon at which the Moon is visible; alternately, the youngest or oldest Moon visible

Danjon scale Five consecutive whole numbers representing the darkness of a total lunar eclipse (0 = darkest, 4 = little difference from the Full Moon)

Dark adaptation ability of the eye/brain to adjust to low levels of illumination, so as to making seeing in various ambient brightnesses easier

Declination angular distance on the Celestial Sphere, north or south of the Celestial Equator, measured in positive degrees (North Celestial Hemisphere) and negative degrees (South Celestial Hemisphere)

Deep-sky (object) extended object in the sky physically beyond the Solar System, e.g., nebulae, star clusters, and galaxies

Defect of illumination percentage of the Moon's disk that is unilluminated; alternately, the maximum fraction of the Moon's diameter that is unilluminated

Descending node point in the orbit of the Moon or a planet at which it is traveling south through the plane of the ecliptic

Diamond ring effect visual phenomenon that occurs when only a single low point on the lunar limb allows photospheric sunlight to pass through, during a total (or nearly total) eclipse of the Sun

Dichotomy (in astronomy) inferior planet's orbital position with respect to the Earth such that, viewed in your sky, the planet's brightness is (partly) due to solar illumination over exactly half of its disk

Dies caniculares (former) time of year beginning with the heliacal rise of Sirius; alternately, the warmest part of the year

Dispersion (in physics) as different wavelengths (colors) of light are refracted though slightly different angles, the appearance of different wavelengths (colors) produced by a single light source appears to come from slightly different directions

Diurnal (in astronomy) having to do with the rotation of the Earth

Diurnal libration geometrical phenomenon allowing you to see slightly more than fifty percent of the Moon, due to the fact rotation produces slightly different line-of-sights depending upon time of day

Dock-pathed description of a constellation, asterism, or extended object for which your observation of its rise or set is late because you must await nightfall or is interrupted by daybreak

E

Earthshine light from the dayside of the Earth reflected by the nightside of the Moon

Eclipse limit angular distance from a node, measured in degrees of ecliptical longitude, within which an eclipse is possible

Eclipse magnitude any time during a solar eclipse, the ratio of the apparent diameter of the Moon to the apparent diameter of the Sun

Eclipse obscuration any time during a solar eclipse, the fraction of the area belonging to the Sun that is in eclipse

Eclipse path that portion of the Earth's surface experiencing a given total eclipse

Eclipse season two periods during a given year during which some kind of eclipse is possible

Eclipse year time between successive syzygies of the Sun with the same lunar node

Eclipsing binary two stars orbiting about each other such that their plane of revolution intersects your line of sight; alternately, what appears to be a single star varying due to the inability to resolve two stars in which one periodically adds to and/or subtracts from their combined brightness

Ecliptic/ecliptical the annual path of the Sun on the Celestial Sphere; alternately, the plane in which the Earth orbits the Sun

Ecliptical coordinates 1) ecliptical latitude, measured in degrees, with zero on the ecliptic and increasing northward, plus 2) ecliptical longitude, measured in degrees, with $0° = 360°$ at the First Point of Aries and increasing eastward (counterclockwise as viewed from the North)

Elongation (in astronomy) angular distance between the Sun and a planet on the Celestial Sphere

Embolismic month name given to a month inserted into the calendar through intercalation

Embolismic year one that includes an embolismic month

Emersion (in astronomy) time during which a body emerges from behind another, or the other passes from in front of it, and after which an occultation is ended

Endogenic process (in geology) natural surface morphology created by the energy of a body's internal heat or self-gravity

Epact number of days since New Moon at the beginning of the civil year

Ephemeris (usually) tabular record of the position of celestial objects on the Celestial Sphere at different times

Epoch (in astronomy) date to which a time-varying parameter is referenced

Equation of time difference between solar time and the civil time in use

Equator/equatorial set of points equidistant from the poles of a rotating body

Equinox/equinoctial one of two locations on the Celestial Sphere where the ecliptic intersects the Celestial Equator; alternately, the time at which the Sun crosses the Celestial Equator

Ethnoastronomy/ethnoastronomer a branch of cultural anthropology studying peoples' beliefs and practices regarding the celestial realm

Exeligmos interval over which two similar eclipses are visible from the same longitude on the Earth; alternately, three Saros cycles

Exogenic process (in geology) natural surface morphology created by an energy source from without, e.g., solar heating or asteroid/comet impact

Extinction (in astronomy) diminution of a celestial object due to the passage of its light through the atmosphere

F

Farside hemisphere of the Moon facing away from you

First point of Aries location on the Celestial Sphere occupied by the Sun on the (northern) Vernal Equinox

First Quarter Moon lunar phase that you see when half the illuminated hemisphere of the Moon is pointed away from you; alternately, the lunar phase that you see when the Moon's terminator passes through the midpoint of its apparent disk

Foresight second of two objects in your line-of-sight directing your view toward a point in the sky

Foucault pendulum one in which the plane of its oscillation appears (is allowed) to rotate in the reference frame of the Earth's surface

Full month (in astronomy) longest month (as a count of days) in a luni-solar calendar

Full Moon lunar phase that you see when the entire illuminated hemisphere of the Moon is pointed toward you; alternately, daylit lunar phase that you see when the Moon's terminator coincides with the limb

G

Gamma during a solar eclipse, the distance (in Earth radii) between the intersection of the Moon's shadow-cone axis with the Earth and the point on the Earth's surface orthogonal to the Sun/Earth line

Gegenschein oval shaped, white sky glow in the East just after twilight, and in the West just before twilight, caused by sunlight reflecting off of interplanetary dust

Geocentricity/geocentric Earth-centered

Gibbous Moon lunar phase that you see when the illuminated hemisphere of the Moon is mostly pointed toward you; alternately, lunar phase that you see when the Moon's terminator curves outward in the direction of the dark lunar limb

Gleam only a vertex of the Sun is above the horizon

Globe (in geography) sphere onto which another (usually larger) sphere is mapped; alternately, a model of a physical, spherical object

Globular (star) cluster densely spaced, spherical grouping of tens of thousands to millions of stars that are physically bound to each other

Gnomon part of a sundial raised above the horizontal dial itself

Grand conjunction massing of the three, naked-eye, superior planets

Grattan-Guinness cycle period after which two eclipses occur on the same date

Great comet comet characterized by exceptional brightness, longevity in the sky, and breadth upon the Celestial Sphere

Great year (in astronomy) length of time it takes the Earth's rotational axis to precess once; alternately, the period taken by any point on the ecliptic to make one complete circle upon the celestial sphere due to the Precession of the Equinoxes

Green flash (in astronomy) optical phenomenon seen shortly before the disappearance of the solar vertex at sunset (shortly after the appearance of the solar vertex at sunrise), when atmospheric refraction preferentially directs the green component of sunlight into your line of sight

Gregorian calendar version of the Julian luni-solar calendar 1) in which the leap-year rule causes the calendar year to remain close to the Earth's period of revolution for a longer period of time, and 2) that introduced a correction to make the Vernal Equinox again correspond well with its traditional date and that of the Easter holiday

H

Heliacal phase heliacal rise or set

Heliacal rise when a star, planet, constellation, or certain asterisms rise just before the Sun; alternately, the time at which a star, planet, constellation, or certain asterisms first emerge(s) from the glare of the morning Sun

Heliacal set when a star, planet, constellation, or certain asterisms set(s) just after the Sun; alternately, the time at which a star, planet, constellation, or certain asterisms make it(s) last appearance in the sky, before becoming lost in the glare of the evening Sun

Heliocentricity/heliocentric Sun centered

Highlands (in astronomy) high albedo, heavily cratered, large-scale surface features on the Moon

Hollow month month of shorter duration (as a count of days) than a full month in a luni-solar calendar

Horizon (in astronomy) intersection of sky and ground

Horology/Horologist (in technology) referring to time; alternately, the craft of making mechanical timekeeping devices

Horoscope attempt to forecast your future based upon the positions of the Sun, Moon, and planets on (or near) the zodiac

Hour angle smallest angle between a right ascension and your central meridian (usually expressed in units of positive or negative hours)

Hybrid eclipse central solar eclipse that appears total at some locations along its path and annular at other locations along its path

I

Immersion (in astronomy) time during which a body passes behind another, or the other body passes in front of it, and subsequently becomes occulted

Impact crater (in astronomy) circular depression formed on the surface of a body by the collision of a small solar-system object

Inclination (in astronomy) angle between the plane of a solar-system body's orbit and the plane of the ecliptic, or the plane of a satellite's orbit to the celestial equator of it's primary

Inex period between saroses

Inferior conjunction that which occurs when an inferior planet is between you (or close to being so) and the Sun

Inferior planet planet the orbital radius of which is smaller than that of the Earth

Insolation solar energy that reaches a given surface

Intercalation/Intercalated (in astronomy) insertion of an embolismic day or month into a luni-solar calendar

Intercardinal point one of four directions midway between the cardinal directions; alternately, northeast, southeast, northwest, and southwest

Interplanetary dust tiny particles (grains) pervading the Solar System

Iris (in anatomy) colored part of the eye, the muscles in which adjust the size of the pupil

J

Julian calendar Roman luni-calendar in which month lengths (in days) are fixed according to a leap-year rule and no embolismic month is intercalated

Julian date that which is equal to the continuous day count since 24 November 4714 BC/BCE

K

L

Libration in latitude that caused by the obliquity of the Moon with respect to its orbital plane about the Earth

Light pollution sky glow caused by inefficient, unnecessary, and/or trespassing artificial lighting

Limb (in astronomy) edge of the apparent disk displayed by a sphere seen at a distance; alternately, a great circle upon a (distant) sphere, in a plane perpendicular to your line of sight

Limb darkening (in solar astronomy) central part of Sun's disk appears brighter (and redder) than the edge

Limiting (stellar) brightness that of the faintest star you can see under given sky conditions

Line of apsides that of which the major axis of an elliptical orbit is a subset; alternately, that passing through perihelion and aphelion, or perigee and apogee

Line of nodes that passing through the nodes of an orbit, the plane of which is inclined to the plane of the orbit

Luminary disk-shaped objects in the sky that can be seen with the naked-eye; alternately, the Sun and the Moon

Luni-solar calendar one in which an attempt is made to keep the number of months (measured in days) in the year nearly commensurate (or, on average, nearly commensurate) with the length of the year; alternately, one in which an attempt is made to integrate both the apparent revolution period of the Sun and revolution period of the Moon into a single system to denote lengths of time greater than that of the day

Luni-tidal interval interval between the Moon's Celestial Meridian transit and your next high tide

M

Mansion (in astronomy) set of asterisms designed such that the Moon passes through one each day of the sidereal month; alternately, the lunar equivalent of the zodiac

Mare (in astronomy) low albedo, lightly cratered, large-scale surface features on the Moon; alternately, lunar surface features produced by the largest impacts

Massing three or more planets near each other in the sky

Meteor light in the sky produced by a high-speed meteoroid entering the Earth's atmosphere and incinerating in whole or part due to atmospheric friction

Meteor shower increase in the average number of meteors seen per unit interval of time; alternately, a pattern of meteors that appears to radiate from a single point on the Celestial Sphere

Meteorite remains of a meteoroid that survives passage through the Earth's atmosphere and reaches the ground

Meteoroid a rocky and/or metallic body, smaller than an asteroid, in a semi-random orbit about the Sun; alternately, the body that produces a meteor when it intercepts the Earth

Meteorological season one of four starting on the first day of March, June, September, or December, regardless of the length of the month

Metonic cycle period after which the phase of the Moon repeats on the same day of the year

Moon illusion a psychological effect that causes the Moon to appear larger very near the horizon than near the zenith

N

Nadir point on the Celestial Sphere ninety degrees below the horizon; alternately, the point on the Celestial Sphere directly beneath the observer and 180 degrees from the zenith

Nautical dawn moment the rising Sun reaches altitude −12 degrees

Nautical dusk moment the setting Sun reaches −6 degrees

Nautical twilight period during which the Sun is at altitudes −6 degrees to −12 degrees; alternately, the time between nautical dusk (or nautical dawn) and astronomical dusk (or civil dawn)

Neap tide (in oceanography) low-amplitude tide occurring when the Moon is at quadrature; alternately, a low-amplitude tide occurring seven days after and before a spring tide

Nearside hemisphere of the Moon facing toward you

Nebula/nebular large cloud of either luminous or nonluminous gas and/or dust in space

New moon lunar phase that you see when the entire unilluminated hemisphere of the Moon is pointed toward you; alternately, the dark lunar phase that you see when the Moon's terminator coincides with the limb

Night path visible in the sky from rise until set

Node/nodal (in astronomy) one of two intersections of an astronomical body's orbital plane with the ecliptic plane

Nodical cycle regression of the lunar nodes through 360° every 18.6 years; alternately, the time between standstill maximum and the next standstill maximum (or standstill minimum and the next standstill minimum)

Nodical month mean period between successive passages of the Moon through its ascending node (or its descending node)

North star star of reasonable brightness that lies so close to the North Celestial Pole that it changes altitude and azimuth very little due to diurnal motion; alternately, a star which can be used as an approximate visual marker for the location of the Celestial Pole

Nova transient outburst of light from a star due to an explosion in its outer layer; alternately, a sudden (usually) aperiodic increase in a star's brightness

Nucleus (in planetary astronomy) solid core of a comet

Nyctalopia medical condition characterized by the inability to see or see well in low light levels; alternately, night blindness

O

Oblate (in geometry) spheroid flattened at the poles

Obliquity (in astronomy) angle between a body's rotational axis and a perpendicular to its orbital plane

Occultation (in astronomy) celestial object passes in front of another so that the more distance object is lost from view

Olber's paradox question: If the Universe is infinitely large and infinitely old (or effectively so), why does the sky not appear mostly dark, inasmuch as every line of sight should intersect a star

Open (star) cluster tens to thousands of young stars "born" from the same nebula and therefore yet in physical proximity to each other

Opposition (in astronomy) geometry when two bodies are on opposite sides of the Celestial Sphere; alternately, the time at which two bodies are on opposite sides of the Celestial Sphere

Optical double two stars not in physical proximity but coincidentally in nearly the same line-of-sight

Orientation (in astronomy) arrangement of two or more monuments or a combination of monuments and natural landscape features, such that they are co-linear with a celestial object

Orthogonal forming a right angle (90°)

P

Parallax apparent change in an object's position due to observing it from two different locations

Pareidolia seeing images in random visual data

Passage grave tomb accessed by long-but-narrow tunnel

Penumbra/penumbral transition zone between complete shadow and complete illumination

Perfect calendar 360 days per year

Perigee/perigean point in an Earth satellite's orbit at which it is closest to the center of the Earth; alternately, the time at which it is reaches this point

Perihelion/perihelic point at which an object, orbiting the Sun, is closest to the center of the Sun; alternately, the time at which it reaches this point

Petroglyph image created by removing part of a rock surface

Photopic vision produced by cellular photoreceptors in the eye that have evolved for use in high-light environments

Photoreceptor/photoreceptive (in the eye) retinal rods or cones (cells) that convert light into neural impulses

Photosphere/photospheric outside of total eclipse, the visible layer of the Sun

Physical libration (in astronomy) apparent oscillation of the Moon's disk about its axis, due to differences in the lunar synchronous-rotation speed and the varying lunar speed in its orbit about the Earth

Pingré cycle time after which two eclipses occur on the same date and day of the week

Planetarium (in astronomy) physical model of the planetary system

Planetary system composed of the eight major planets

Pole/polar (in astronomy) one of two points located where the surface of body and its axis of rotation intersect

Position angle apparent angular relationship of two celestial objects with respect to north; alternately, apparent angular relationship between a celestial object's axis with respect to north

Post meridiem time from noon to midnight

Precession/precessional (of the Equinoxes) wobble of the axis mundi, exhibiting a property common to all rotating objects

Prime vertical vertical circle passing through due east and west

Prograde (in astronomy) counter-clockwise movement on the Celestial Sphere, reckoned as if looking down upon the Earth's north pole

Proleptic applied retroactively to the past

Prominence (in astronomy) temporary structure of dense, magnetically confined, glowing plasma extending from the solar photosphere into the corona

Proper motion angular change in position of a star, with respect to other stars, on the Celestial Sphere

Pupil (in anatomy) aperture of the eyeball, the size of which is regulated by the expansion and contraction of the iris

Purkinje effect description of the transition from high-illumination vision to low-illumination vision, in which red sensitivity lags blue sensitivity; alternately, description of the transition from low-illumination vision to high-illumination vision, in which red sensitivity lags blue sensitivity

Q

Quadrant (in astronomy) instrument consisting of a graduated quarter circle with a sight along one edge and (often) a plumb bob hanging from the right-angle corner as a reference

Quadrature (in astronomy) two celestial bodies separated by a right angle on the Celestial Sphere; alternately, a solar-system body with an elongation of ninety degrees

Quarter day Vernal Equinox, Summer Solstice, Autumnal Equinox, and Winter Solstice collectively

Quinduodecile (in astronomy) point or time at which two celestial bodies are 150 degrees from each other

R

Ray (in astronomy) high-albedo, superficial, filament radiating from an impact crater on the Moon

Refraction/refractive (in physics) change in direction of a light wave as it passes from one medium to another at an angle (with respect to the orthogonal at the interface)

Resolution (in astronomy) ability to visually distinguish two objects

Retardation (in astronomy) difference in time between two moonrises (or moonsets) on successive days

Retina/retinal back wall of the eyeball, lined inside with photoreceptive cells

Retrograde (in astronomy) clockwise movement on the Celestial Sphere, reckoned as if looking down upon the Earth's North Pole

Right ascension east-to-west angular distance on the Celestial Sphere, starting at the First Point of Aries, measured in (one of twenty-four) hours of arc

S

Saros particular pattern of eclipses that both starts and ends

Saros cycle repeating time period over which certain intervals related to eclipses are commensurate

Satellite (in planetary astronomy) solar-system body orbiting a larger mass solar-system body other than the Sun

Scotopic vision produced by cellular photoreceptors in the eye that have evolved for use in low-light environments

Seasonal lag difference between the date of maximum or minimum insolation and the date of climatologically averaged high or low temperature

Seeing (in astronomy) degradation of view of an astronomical object due to scattering of light, caused by differing refractive properties belonging to individual atmospheric cells passing between the observer and object

Selenography/selenographer study and mapping of the lunar surface

Semanex period after which two eclipses occur on the same day of the week

Sextile (in astronomy) point or time at which two celestial bodies are sixty degrees from each other

Shadow bands alternating dark and light ripples seen on the ground shortly before second contact and/ or shortly after third contact during a total eclipse of the Sun

Sidereal measurement or motion that uses the Celestial Sphere as a reference point

Solar anomaly difference between the time determined by the position of the Sun in the sky and the civil time read from a clock

Solar system Sun and all bodies in orbit about it; alternately, the Sun and all objects gravitationally bound to it

Solstice/solstitial one of two locations on the Celestial Sphere where the apparent path of the Sun reaches a north-south inflection point; alternately, the time at which the Sun reaches a north-south inflection point

Spring tide (in oceanography) high-amplitude tide occurring when the Moon is at syzygy; alternately, a high-amplitude tide occurring seven days after and before a neap tide

Standstill (in astronomy) point on the horizon at which the Moon rises (and sets) farthest north, or rises (and sets) farthest south, during a given month

Stationary point (in astronomy) location on the Celestial Sphere at which an astronomical object changes from prograde to retrograde motion (or retrograde motion to prograde motion)

Stellar magnitude measurement of a star's brightness on a logarithmic scale

Stellar phase heliacal rise (or set) or acronical rise (or set)

Sub-lunar lobe water in one of two tidal bulges made up of the oceanic deficit from the tidal troughs; alternately, half of the Earth's oblong-shaped hydrosphere, created by the tensional force of the Moon and Sun

Sundial horological instrument that uses the shadow cast by a gnomon upon a dial

Sunspot (in astronomy) dark-in-contrast patch, appearing temporarily on the solar photosphere, associated with a local increase in intensity of the Sun's magnetic field; alternately, the source region on the Sun for solar prominences and flares

Sunspot cycle periodic change in the average number of sunspots visible on the solar disk; alternately, the approximately eleven-year interval between episodes of increased solar activity (prominences, flares)

Superior conjunction that which occurs when an inferior planet is on the other side of the Sun (or close to being so) from you

Superior planet planet the orbital radius of which is larger than that of the Earth

Supernova one-time stellar explosion that quickly makes the star look terrifically brighter or a non-naked-eye star to suddenly become a naked-eye star

Synchronous rotation that demonstrated by an astronomical body with a rotation period equal to its revolution period; alternately, astronomical body that always keeps the same face toward the object about which it revolves

Synodic (in astronomy) measurement or motion that uses the Sun as a reference point

Syzygy alignment, or near alignment, of three celestial bodies; alternately, alignment, or near alignment, of the Earth, Moon, and Sun

T

Terminator apparent curve on the apparent disk of an object shining due to reflected light that separates the illuminated part of the body from the self-shadowed part; alternately, on a rotating solar-system body, the demarcation between that portion experiencing daytime and that portion experiencing nighttime

Tetrad four consecutive total eclipses of the Moon

Third Quarter Moon lunar phase that you see when half the illuminated hemisphere of the Moon is pointed away from you; alternately, the lunar phase that you see when the Moon's terminator recedes past the midpoint of its apparent disk

Tidal day period over which exactly two high (or low) tides take place

Tide regular rise and fall of the sea level

Time zone one of twenty-four regions, semi-longitudinally depended, in which it is agreed to use a common civil time, with each one-hour apart

Totality period between second and third contact during a total eclipse of the Sun

Transit (in astronomy) body of smaller apparent size passes between the Earth and another body of larger apparent size

Transparency (in astronomy) degree to which the atmosphere attenuates (absorbs) light; alternately, the clarity or opacity of the atmosphere

Trine point or time at which two celestial bodies are 120 degrees from each other

Tritos cycle repeating time period over which certain intervals related to eclipses are commensurate (superseded in use by the Saros cycle)

Tropical year interval between the repetition of a given equinox or solstice

True libration real wobble of the Moon; alternately, an actual oscillation of the lunar mid-disk

U

Umbra/umbral inner, darkest zone of a shadow surrounded by the penumbra

Uranography/uranographer celestial cartography

V

Variable star one that changes in brightness (usually) on a periodic basis; alternately, (normally) a star that expands and contracts

Vertex (in astronomy) highest point in altitude of an extended celestial object[1]

Vertical circle great circle on the Celestial Sphere passing through your zenith

W

Wane/waning (in astronomy) to become less

Wax/waxing (in astronomy) to become more

X

Y

Year star In China, an imaginary "star" with a sidereal period equal to that of Jupiter, but traveling around the Celestial Sphere in the retrograde direction

Z

Zenith/zenithal point on the Celestial Sphere at ninety degrees altitude; alternately, the point directly above the observer and 180 degrees from the nadir

---

1 There are other definitions of vertex used in astronomy.

Zenith angle that between a point in the sky and the zenith

Zodiac/zodiacal band around the Celestial Sphere that incorporates nine degrees on either side of the ecliptic; alternately, twelve constellations situated around the Celestial Sphere, covering the ecliptic, imbued with significance in (especially) natal astrology

Zodiacal light triangularly shaped, white sky glow in the West just after twilight, and in the East just before twilight, caused by sunlight scattering interplanetary dust

Index

Table of contents (Brief)